Behavioral Neurobiology
an integrative approach

Günther K.H. Zupanc
International University Bremen

Foreword by Theodore H. Bullock
University of California, San Diego

OXFORD
UNIVERSITY PRESS

OXFORD

UNIVERSITY PRESS

Great Clarendon Street, Oxford OX2 6DP

Oxford University Press is a department of the University of Oxford.
It furthers the University's objective of excellence in research, scholarship,
and education by publishing worldwide in

Oxford New York

Auckland Bangkok Buenos Aires Cape Town Chennai
Dar es Salaam Delhi Hong Kong Istanbul Karachi Kolkata
Kuala Lumpur Madrid Melbourne Mexico City Mumbai Nairobi
São Paulo Shanghai Taipei Tokyo Toronto

Oxford is a registered trade mark of Oxford University Press
in the UK and in certain other countries

Published in the United States
by Oxford University Press Inc., New York

British Library Cataloguing in Publication Data

Data available

Library of Congress Cataloging in Publication Data

Data available

ISBN 0-19-870056-3

1 3 5 7 9 10 8 6 4 2

Typeset by Newgen Imaging Systems (P) Ltd., Chennai, India
Printed in Great Britain by
Antony Rowe, Chippenham, Wiltshire

Books are to be returned on or before
the last date below.

Be

7-DAY
LOAN

2 5 OCT 2004

1 6 1 1 0 4

LIBREX —

To my wife Marianne
and our children Frederick, Christina, and Daniel

About the author

Günther K.H. Zupanc is Full Professor of Neurobiology at the International University Bremen, Germany. After graduating in both biology (1985) and physics (1987) from the University of Regensburg, Germany, he did his graduate studies in Neurosciences in the laboratory of Walter Heiligenberg at the University of California, San Diego, where he received his Ph.D. in 1990. Subsequently, he worked as a Research Biologist at the Scripps Institution of Oceanography in La Jolla, California (1990–1992), headed a Junior Research Group at the Max Planck Institute for Developmental Biology in Tübingen, Germany (1992–1997), and was on the faculty of the University of Manchester, England (1997–2002). Zupanc was also Visiting Professor at the University of Ottawa, Canada; Visiting Scientist at the Salk Institute for Biological Studies, San Diego, the University of Chicago, and the Max Planck Institute for Behavioral Physiology, Seewiesen; and adjunct faculty member of the University of Tübingen. He has taught lecture and laboratory courses in Behavioral Neurobiology at both the undergraduate and the graduate level to hundreds of students in the USA and in Europe. He has received numerous awards for his research and his contributions to the public understanding of science. His research focuses on the exploration of cellular mechanisms underlying behavioral and neuronal plasticity in teleost fish. In addition to a large number of articles, other book publications include: Fish and Their Behavior (1982); *Praktische Verhaltensbiologie* (Editor; 1988); *Fische im Biologieunterricht* (1990); Adult Neurogenesis: A Comparative Approach (Editor; 2002).

Foreword

How profoundly human it is; how deeply characteristic of our species, ethologically speaking—to wonder at, to study and investigate behavior of our own and other species—toward understanding, in the sense of accounting for, the actions, the appetites, the drives, alternative modes, and sensory guidance that we observe. How rewarding we find it to think of hypotheses, test and discard them, and achieve a degree of understanding—at one or more levels—and report it to our colleagues!

In all its angles and aspects this human urge to embrace, comprehend, and explain what we and other species do—our behavior—is now recognized as a field of endeavor called neuroethology, rivaling those other distinctively human traits such as cooking, dancing in innovative ways, recounting the past and imagining the future, making each other laugh and making music!

It strikes me as some kind of a pinnacle—to bring together between book covers what's been learned and how it was done so that we can exercise another very human trait—wonder at it!

One reason for this somewhat obtuse opening is that it underlines a strongly felt view of Günther Zupanc, that the raison d'etre of such a book need not depend on its potential relevance to practical human concerns, such as medicine or the psychology of human aggressivity, but, in today's world, upon simple curiosity and knowing for its own sake. This latest summing up follows worthily the path pioneered by Jörg-Peter Ewert (*Neuro-Ethologie: Einführung in die neurophysiologischen Grundlagen des Verhaltens*; Springer-Verlag, Berlin, 1976), Jeff Camhi (Neuroethology: *Nerve Cells and the Natural Behavior of Animals*; Sinauer, Sunderland/Massachusetts, 1984), and most recently Tom Carew (*Behavioral Neurobiology*: The Cellular Organization of Natural Behavior; Sinauer, Sunderland/Massachusetts, 2000), each in its own style. Exploiting the freedom of a foreword writer, I would like to underline some of the features of special importance in Günther Zupanc's treatment—without implying that they are neglected in previous books!

First and foremost, it is a feature of the subject, that it deals with biodiversity in the extreme. The consequence of evolution is an accumulation of diversity, particularly in behavior—more conspicuously and significantly, and more inviting to analysis and explanation, than form, color or pattern. The usual distillate of biology—that "life is genes propagating genes"—is a serious misrepresentation because evolution has created diversity in what animals do between generations; life consists of diverse states and actions—of course including keeping alive and reproducing. But the big picture is missed if we don't hold up for scrutiny the different ways of doing these things in beetles that burrow, butterflies that migrate, corals that luminesce and lions that sleep away most of the day.

Zupanc has recognized the importance of organizing the wealth of detail in a reader-friendly way—with Leitmotifs in special categories and sidebars. He has also recognized the importance of descriptive natural history preceding and leading into reductionist analysis—neuroethology begins with adequate ethology.

Particularly important in neuroethology, and well represented in this treatment, is comparison—comparison of sensory stimuli that trigger, of pathways and central structures involved, of background state dependence and alternative tricks for cancelling self-generated signals. Comparison is the essence and applies to different taxa, ontogenetic stages and readiness states. Nature appears rarely to use a single mechanism for all animals that exhibit similar behavior—nor does she use a large number of alternate mechanisms, but several is the norm. The limitation in our knowledge is a limitation of research endeavor perhaps because there is less glamor in looking at the same behavior in other taxa than in looking at hitherto unstudied behavior.

Another feature that stands out in the present treatment though frequently underplayed in the primary literature is that every case is identified with a broad or basic issue. This is complemented by showing how some species is particularly favorable for the given study and closer analysis.

A distinctive feature of this book is its concern with the history of each strand of the fabric of neuroethology, the drama and the dependence on serendipity and mindset. The results of this historical research are distributed throughout the text, often in sidebars or boxes. May this book illumine the science, influence the way investigators proceed, and draw new ones into the field.

La Jolla   *Theodore Holmes Bullock*
June 2003

Preface

About the book

This book is based on two courses that I designed and taught upon joining the faculty of the School of Biological Sciences of the University of Manchester in 1997. One of these Behavioral Neurobiology courses was targeted at a beginner's level, the other at an advanced level. Over the five years that followed, they were taken by several hundreds of students coming from a wide range of degree programs. The students constantly rated both courses among the very best within the School. It was this positive interaction, as well as the students' request to make available a text covering these courses, that encouraged me to write this book. These students also provided me with invaluable feedback on how to stimulate interest in the subject, without compromising the quality of the science taught.

Like the Manchester courses, this book is primarily designed for undergraduate and graduate students, but postdoctoral scientists and instructors of such courses are also likely to benefit from it. I have assumed that the students have some basic knowledge in biology, physics, and mathematics. However, courses in animal behavior or neurobiology are not a prerequisite. Concepts and approaches from these disciplines that are important for understanding Behavioral Neurobiology are introduced in Chapter 3.

Any scientific discovery can be fully understood only within its historical context. I have, therefore, also included a chapter (Chapter 2) on the historical development of Behavioral Neurobiology, and added to several chapters a description of the work and life of those who have pioneered this development. I am grateful to the Royal Society in the UK for the award of an History of Science grant that enabled me to collect the information necessary to write Chapter 2.

The approach used in the book has been to focus on a few selected systems that, in my opinion, best illuminate key principles. These examples are then discussed in depth. Unavoidably, such an approach leads to the exclusion of a number of other excellent studies. I apologize to those whose work has been neglected.

Learning tools

Each of the eleven chapters is organized in a similar way, and contains a number of valuable learning tools:

- An **introductory section** outlines the major topic covered in the chapter.
- The **margin notes** highlight important points and define general biological, chemical, or physical terms used in the main text.

- Key concepts, such as the structure and function of central pattern generators, are explained in depth in several **text boxes**.
- Each chapter is supplemented by a **summary** and a list of **key reviews** suggested for additional reading.
- For those who wish to explore further the field covered in the individual chapters, a list of **additional review articles** and of **important original research papers** is provided at the end of the book.
- The **questions** at the end of each chapter may be used by the students to check their knowledge, or by the instructors to set exams.

Companion web site

This book is accompanied by a companion web site which includes the following:

- Answers to the questions presented at the end of each chapter.
- Many illustrations from the book in downloadable form.
- A library of web links to web pages of relevance to the content of this book.

The companion web site is freely available at **www.oup.com/uk/booksites/biosciences**

Acknowledgments

The foundation to write this book was laid during my own graduate education. Four teachers have been particularly influential. The late Walter Heiligenberg, in whose laboratory at the Scripps Institution of Oceanography of the University of California, San Diego (UCSD) I had the privilege to work both as a Ph.D. student and as a postdoctoral fellow. Ted Bullock, also of UCSD, has been a source of inspiration since my graduate student days. He made many suggestions on the book manuscript, and he also kindly contributed the Foreword. Larry Swanson, then at the Salk Institute for Biological Sciences in La Jolla, California, now at the University of Southern California at Los Angeles, taught me how to use neuro-anatomy as a tool to analyze the structural basis of behavioral control mechanisms. Equally important, he also showed me that even the teaching of a difficult subject can be a source of joy for both instructor and student. Len Maler of the University of Ottawa, Canada, taught me, through many joint projects, how to integrate behavioral, anatomical, and physiological data to gain an appreciation of how the central nervous system generates behavior.

In the course of writing this book, I greatly benefitted from the advise of the following colleagues: Jonathan R. Banks (University of Manchester, UK); W. Jon P. Barnes (University of Glasgow, UK); Robert C. Beason (University of Louisiana, Monroe, USA); Franz Huber (Starnberg, Germany); Masakazu Konishi (California Institute of Technology, Pasadena, California, USA); my friend Jürg Lamprecht (Max-Planck-Institut für Verhaltensphysiologie, Seewiesen, Germany), who, sadly, saw the beginning, but not the end of this book; Ken Lohmann (University of North Carolina, Chapel Hill, USA); Anne Lyons (Oxford, UK); Eve Marder (Brandeis University, Walthan, Massachusetts, USA); Joachim Ostwald (Eberhard Karls

Universität Tübingen, Germany); Alan Roberts (University of Bristol, UK); David R. Skingsley (Staffordshire University, Stoke-on-Trent, UK); Wim A. van de Grind (Universiteit Utrecht, The Netherlands); and Neil V. Watson (Simon Fraser University, Burnaby, British Columbia, Canada).

My friend Cecilia Ubilla (UCSD) spent numerous hours on the manuscript to optimize my style of writing. Her comments from the perspective of a naive reader, on the one hand, and an experienced professional writer and teacher, on the other, were instrumental in polishing up the text.

Another factor that proved to be indispensable in writing this book was the excellent collaboration with the staff of Oxford University Press. Esther Browning proposed this project and got me started. Jon Crowe, as Editor, combined, in a particularly pleasant way, two seemingly diametrical features, patience and perseverance, to make me complete the manuscript.

Finally, my special thanks are due to my wife Marianne, who read through the manuscript many times and alerted me whenever clarity seemed to be compromised. Without her support, this book would not have been possible.

Bremen *Günther K.H. Zupanc*
September 2003

Contents

Foreword   vii
Preface   ix

1 Introduction   1

Neuroethology: the synthesis of neurobiology and ethology   2
Choosing the right level of simplicity   3
Quantifying behavior: a prerequisite for neuroethological research   5
Finding the right model system   6
Summary   9
Recommended reading   9
Questions   9

2 The study of animal behavior: a brief history   11

Introduction   11
The roots of the study of animal behavior   12
    Aristotle and the Middle Ages   12
    Mind–body dualism   13
The new era in the study of animal behavior   14
    Evolutionary theory and comparative approaches   14
    Experimental and objective approaches   14
    The rise of ethology   16
    Mechanistic schools   18
Relating neuronal activity to behavior: the establishment of neuroethology   24
    The beginnings   24
    The breakthroughs   29
    The future   30

Summary   32
Recommended reading   33
Questions   33

3   The tools and concepts of behavioral neurobiology   35

Introduction   35
Neurobiology: basic concepts and experimental approaches   35
   Cell theory versus reticular theory   35
   Neurons and nervous systems   36
   Synapses   37
   Resting potential   39
   Generation of action potentials   43
   Conduction of action potentials   44
   Synaptic potentials and synaptic transmitters   46
   Sensory systems   49
   Elucidation of neuronal connections   51
   Localization of tissue constituents   53
   Immunohistochemistry   54
   In situ hybridization   56
   Selective elimination of cells   58
Ethology: basic concepts and experimental approaches   59
   Sign stimuli and releasing mechanisms   59
   Supernormal stimuli   60
   Law of heterogenous summation   62
   Gestalt principle   64
   Importance of condition of recipient   65
   Motivational effect of releaser   67
   Communication   67
Summary   75
Recommended reading   77
Questions   78

4   Spatial orientation and sensory guidance   79

Introduction   79
Classification of orienting movements   80
Orienting behavior without a nervous system   81
   Cellular mechanisms of taxis behavior in paramecians   83
   Phobotaxis   84
   Galvanotaxis   85
Geotaxis in vertebrates   88
   Effective physiological stimulus   90

Physiological properties of hair cells   91
Transduction mechanism   93
Echolocation in bats   94
The beginnings of echolocation research   95
Classification of bat ultrasound   96
FM signals: distance estimation   98
CF signals: Doppler shift analysis   100
Adaptations of the auditory system   103
Counter sonics: the prey's adaptations   105
Summary   107
Recommended reading   109
Questions   109

5   Neuronal control of motor output: swimming in toad tadpoles   111

Introduction   111
The behavior   112
A physiological approach to study swimming behavior   113
The spinal circuitry controlling swimming   115
Operation of the swimming circuitry   117
Coordination of oscillator activity along the spinal cord   118
Summary   119
Recommended reading   120
Questions   120

6   Neuronal processing of sensory information   121

Introduction   121
Recognition of prey and predators in the toad   122
The model system   122
The natural behavior   123
Dummy experiments   124
In search of feature detectors   127
The visual system   128
Recording experiments   129
Stimulation experiments   131
Connections with other brain regions   131
Directional localization of sound in the barn owl   132
The behavior   132
Experimental approach   134
Accuracy of orientation response   136

Physical parameters of sound involved in head orientation   137
The cochlear nucleus: parallel processing of time and intensity information   142
The laminar nucleus: computation of interaural time differences   144
The posterior lateral nucleus: computation of interaural intensity differences   146
The lateral shell: convergence of timing and intensity information   148
External nucleus: formation of a map of auditory space   148
The final step in sensory processing: formation of an auditory–visual map   149
Summary   150
Recommended reading   152
Questions   153

7 Sensorimotor integration: the jamming avoidance response of the weakly electric fish, *Eigenmannia*   155

Introduction   155
The system and its components   157
Electric organs and electroreceptors   157
Electrolocation   159
The jamming avoidance response   161
Behavioral experiments   161
Determination of the sign of the frequency difference without internal reference   161
Behavioral rules governing the jamming avoidance response   164
Neuronal implementation   169
Electrosensory processing I: electroreceptors   169
Electrosensory processing II: electrosensory lateral line lobe   170
Electrosensory processing III: torus semicircularis   173
Electrosensory processing IV: nucleus electrosensorius   174
Motor control   174
Reflections on the evolution of the jamming avoidance response   176
Summary   177
Recommended reading   179
Questions   179

8 Neuromodulation: the accommodation of motivational changes in behavior   181

Introduction   181
Neuronal plasticity as the basis of motivational changes in behavior   183
Structural reorganization   183
Dendritic plasticity: seasonal changes in chirping behavior of weakly electric knifefish   184
Seasonal variation in dendritic morphology of motoneurons in white-footed mice   188
Structural reorganization mediated by glial cells   191
When to use 'structural reorganization'?   192

Biochemical switching	192
Modulation of the stomatogastric ganglion	192
Modulation of crayfish aggressive behavior	195
Modulation of the modulators	196
What makes modulators suitable for neuromodulation?	196
When to use 'biochemical switching'?	197
Summary	197
Recommended reading	198
Questions	199

9   Large-scale navigation: migration and homing   201

Introduction	201
Modes of migration	202
Genetic control of migratory behavior	205
Homing	207
Approaches to study animal migration and homing	208
Mechanisms of long-distance orientation in birds	209
Sun compass	210
Star compass	212
Magnetic compass and maps	215
Olfactory navigation in homing pigeons	224
Homing in salmon	225
Life cycle of salmon	225
Precision of homing	228
Transplantation experiments	228
The olfactory imprinting hypothesis	228
Laboratory experiments	231
Imprinting to artificial substances	231
Ultrasonic tracking	233
Hormonal regulation of olfactory imprinting	234
The model	236
Possible neuronal mechanisms of olfactory imprinting	236
Open sea navigation: sun-compass orientation	237
Open sea navigation: magnetoreception	237
Orientation in sea turtles	239
Life cycle of sea turtles	239
Orientation on the beach	240
Magnetic orientation	241
Orientation on land	242
Waves as an orientation cue	242
Summary	245
Recommended reading	247
Questions	248

10   Communication: the neuroethology of cricket song   251

Introduction   251
Biophysics of cricket songs   253
Mechanism of sound production   253
Neural control of sound production   255
Behavioral analysis of auditory communication   256
Perception of auditory signals   259
Song recognition by auditory interneurons   262
Temperature coupling   265
    The problem   265
    The songs   266
    Effect of temperature on calling song   266
    Effect of temperature on calling song recognition   268
Genetic coupling in cricket song communication   269
Summary   273
Recommended reading   275
Questions   275

11   Cellular mechanisms of learning and memory   277

Introduction   277
Explicit and implicit memory systems   278
The cell biology of an implicit memory system: sensitization in *Aplysia*   280
    Why *Aplysia*?   282
    The gill-withdrawal reflex   282
    Neural circuit of the gill-withdrawal reflex   283
    Molecular biology of short-term sensitization   285
    Molecular biology of long-term sensitization   286
The cell biology of an explicit memory system: the hippocampus of mammals   289
    Nonhuman models to study explicit memory   290
    The structure of the hippocampus   291
    Place cells in the hippocampus   292
    Long-term potentiation   293
    Molecular biology of LTP in the mammalian hippocampus   294
New neurons for new memories   296
    Adult neurogenesis and memory formation in the hippocampus of rodents   296
    Adult neurogenesis and spatial learning in the avian hippocampus   300
Summary   301
Recommended reading   303
Questions   304

References   305

Figure and table acknowledgments   315

Index   331

Introduction 1

▦ Neuroethology: the synthesis of neurobiology and ethology
▦ Choosing the right level of simplicity
▦ Quantifying behavior: a prerequisite for neuroethological research
▦ Finding the right model system
▦ Summary
▦ Recommended reading
▦ Questions

This book introduces the reader to the fascinating field of **neuroethology**. As other disciplines studying animal behavior, a major task of neuro-ethology is to understand the causal factors that lead to the production of behavior. There are two principal approaches to achieve this goal. One approach aims at a 'software' explanation of behavior. As shown in Fig. 1.1, the animal is treated as a black box, which, in response to a biologically relevant stimulus, generates a behavior. Such an approach is used by all behavioral sciences, including ethology, one of the founder disciplines of neuroethology. The second approach, which is employed by neuroethology, aims at understanding how the central nervous system translates the stimulus into behavioral activity. In other words, neuro-ethology seeks a 'hardware' explanation of behavior by elucidating the structure and the function of the black box.

> Neuroethology attempts to understand how the central nervous system controls the natural behavior of animals.

Figure 1.1 The black-box approach. Scientific disciplines restricting their research to the behavioral level treat the animal as a black box that, upon stimulation with a biologically relevant stimulus, produces a behavioral pattern. Such disciplines, thus, provide 'software' explanations of behavior. In contrast, neuroethology attempts to give 'hardware' explanations by exploring the structure and function of the black box, in relation to the production of behavior. (Courtesy: G. K. H. Zupanc.)

Neuroethology: the synthesis of neurobiology and ethology

Neuroethology has its roots in both **neurobiology** and **ethology**. The synthesis of these two disciplines, which created a new area of study, was and continues to be challenging. This is mainly due to the rather diametric approaches employed by the two founding disciplines. Neurobiologists have traditionally worked on anesthetized animals, isolated parts of tissue, or even single cells. They are primarily interested in the structure and function of such particular cells or tissues. The species is often chosen based on technical considerations, such as the presence of large nerve cells, and the ease by which the animal preparation can be obtained. Ethologists, on the other hand, employ a **whole animal approach,** with the animal kept under conditions as natural as possible. Preferably, at least part of their observations should be conducted in the field. If this is not possible, then the animal is transferred to, or bred in the laboratory, where it is kept under semi-natural conditions to minimize the occurrence of unnatural behavior.

Despite the obvious differences between neurobiology and ethology, the success of neuroethology is based on the incorporation of a blend of neurobiological and ethological approaches into its own scientific armory. Particularly, its focus on 'natural' and biologically relevant behavioral patterns makes neuroethology distinct from other disciplines studying the neural basis of behavior. As part of the overall strategy, this should include investigations of the animal in its natural habitat. The researcher can then simulate the field conditions in the laboratory and apply more natural stimuli in the experiment than would be possible if studying the behavior in the laboratory only.

In recent years, such **field studies** have been eased by many technological developments, such as the availability of battery-powered laptop computer, which allow the researcher to characterize the animal's natural behavior with an unprecedented degree of precision. With the enormous advances made in the miniaturization of instruments, it is not unthinkable anymore that neuroethologists will, at one point in the future, be able to obtain physiological recordings from animals living a relatively normal life in their natural habitat!

On the other hand, neuroethologists investigate rather simple behaviors. This is sometimes to the disappointment of ethologists, who are typically interested in more complex behaviors. However, the intrinsic conceptual and technical difficulties make such a self-applied restriction not only unavoidable, but also desirable.

In the latter respect, one could compare the situation of today's neuroethology with that of physics in the seventeenth century. The initial

Neuroethological research combines both neurobiological and ethological approaches.

restriction to simple models to analyze the motion of objects (neglecting, for example, air resistance when examining falling objects) led to the discovery and establishment of many fundamental principles, such as the Newton's laws of motion. An attempt in the early days of mechanics to analyze more complex systems, although closer to reality, would almost certainly have failed and tremendously delayed the further development of the physical sciences.

Choosing the right level of simplicity

Progress in neuroethology is crucially dependent upon choosing the right level of simplicity. Thus, although the ultimate goal of neuroethology is to understand the neural mechanisms underlying behavior, it would at present not be sensible to examine the entire behavior of an animal.

In any animal, the behavior consists of many individual elements. It involves not only what is generally associated with behavior, movements of the body in particular, but may also include specific body postures, production of sound, color changes, electric discharges, and even glandular activity, for example the secretion of pheromones. The entire behavioral repertoire of an animal is called the **ethogram**. As immediately evident, this entire set of behavior is too complex to be analyzed by the neuroethologist, or even by the ethologist. Therefore, the behavior of an animal has to be split into its individual components, referred to as the individual behaviors or behavioral patterns. As a first step, the total behavioral repertoire is often divided into major groups representing functional categories, for example 'sleep,' 'feeding,' 'courtship,' or 'aggression.' Yet, these categories are still too large to be quantified and analyzed in a meaningful way. This makes a further subdivision necessary.

Ethogram: the entire behavioral repertoire of an animal species.

> *Example: Cichlids are a family of more than two thousand teleost fishes. They are well known for their highly developed aggressive, courtship, and parental care behavior. As illustrated in Fig. 1.2, the behavior subsumed under the term 'aggression' actually consists of a number of individual behavioral patterns, including chasing, butting, frontal display, lateral display, and mouth wrestling.*

Obviously, by dividing the behavior of an animal further and further down, a hierarchical arrangement results. Niko Tinbergen, one of the founders of ethology (see Chapter 2), called the different levels within this hierarchical system **levels of integration**. The lateral display of cichlids, for example, involves alignment of the fish beside its opponent, typically

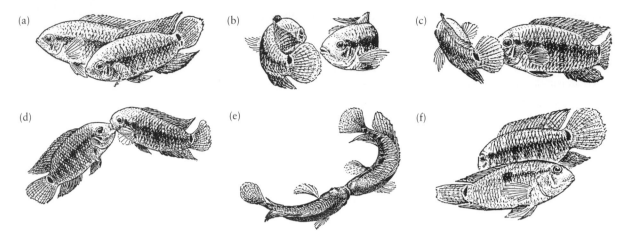

Figure 1.2 Aggressive behavior of the blue acara (*Aequidens pulcher*), a teleost fish of the cichlid family. (a) Lateral display. The fish align beside each other, spread the dorsal, anal, and pelvic fins, and intensify the coloration of their bodies. These threats are accompanied by light tail beats. (b) Circling. While circling each other, the fish have the ventral part of the mouth lowered. (c) Tail beating. One fish beats its tail against the head of the opponent. (d) Mouth grasping. The fish grasp each other at the lower or upper mandible. (e) Mouth pulling. Each fish tries to pull the opponent, after having grasped each other's mouth. (f) Defeat. The fish in the front gives up. It folds its fins, adopts a pale coloration, and swims off. (After **Wicker, W.** (1968).)

Dividing the behavior of an animal into components of decreasing complexity results in a hierarchical arrangement with different levels of integration.

within a few centimeters, and a rapid, powerful sideways thrust of the tail. During this display, the dorsal, anal, and pelvic fins are erect and the opercula extended. The erection of the fins, on the other hand, is defined by the action of the individual fin rays, whose movement is the result of the contraction of muscle fibers controlled by neural motor units. Figure 1.3 illustrates this stepwise subdivision of behavior into smaller and smaller elements.

While it hardly makes sense to undertake a study aimed at elucidating the neural basis of 'aggression' in a cichlid fish, one would likely succeed, with the techniques available, to identify the structures within the central nervous system that control the movements of the dorsal fin. Operation at this lower level of integration not only reduces the number of neuronal structures involved in the control of a behavioral pattern, but also provides the investigator with behavioral elements typically much better defined than those encountered at a higher level of integration. This makes it markedly easier to quantify behavioral patterns—a prerequisite for many types of analysis at both the behavioral and the neurobiological level.

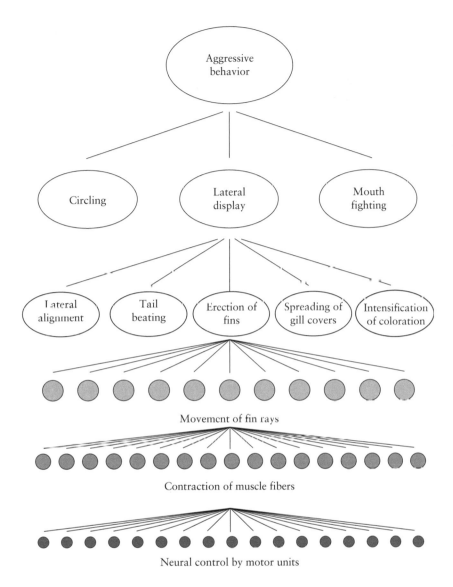

Figure 1.3 Splitting the lateral display, an aggressive behavioral pattern of cichlid fish, into individual elements. This leads to a hierarchical arrangement in which various levels of integration are distinguished. (Courtesy: G. K. H. Zupanc.)

Quantifying behavior: a prerequisite for neuroethological research

In general, behavioral patterns can be quantified using their rate of occurrence ('how often is the dorsal fin erected?'), their duration ('how long is the dorsal fin kept in an erect position?'), and/or their intensity ('is the dorsal fin erected maximally, with the fin rays almost perpendicular to the dorsal edge of the fish's body, or do the fin rays adopt positions intermediate between the maximal and the minimal angle?'). A similar attempt to quantify the lateral display would be very difficult, particularly because

A behavioral pattern can be
quantified using its rate of
occurrence, duration, and/or
intensity.

of the complexity of the different actions involved in the execution of this behavior. Since not all individual actions are necessarily executed simultaneously during lateral display, a main problem would be to identify the endpoints of this behavior. Lack of such information makes it virtually impossible to determine the parameters, 'rate of occurrence' and 'duration'. Also, measurement of the intensity would be difficult: is the behavior more intense when a larger number of individual actions are displayed, or when the degree of execution of individual patterns is maximized?

Finding the right model system

The above considerations underline the importance of choosing the right **model system**. Such systems are *not* primarily studied to provide insights into the neural mechanisms underlying the behavior of the respective species. Rather, their characterization enables the neuroethologist to extract principles applicable to many, if not all, animals. This is possible, because there are, probably in any case, only a finite number of solutions to a given behavioral problem. Keeping the body oriented, for example, requires the analysis of geophysical invariants, but the number of options available is limited to a very few, such as the measurement of the direction of the incident light or of the animal's angle relative to gravity (see Chapter 4).

Many solutions to such problems were invented very early in evolution, so that frequently the neural implementation of these solutions are **homologous** in different species. Such homologous developments are a major reason why many fundamental cellular mechanisms underlying learning and memory are very similar among animals, including both vertebrates and invertebrates. On the other hand, these universals of life make it possible to establish principles of learning and memory processes by studying the rather simple neural network of the sea slug *Aplysia*, although ultimately most researchers would like to understand more complex systems, including those of humans (see Chapter 11).

Ideally, when choosing a model system, the behavior under scrutiny should be simple and robust, readily accessible and ethologically relevant. The animal displaying this behavioral pattern should be inexpensive, suitable for examination in the laboratory, easy to maintain, and possible to breed. These requirements are far from trivial to meet, and the right choice of a suitable model system always demands profound knowledge in both animal biology and husbandry.

Example: Two orders of teleost fish produce, by means of a specialized organ, electric discharges of low voltage. As will be demonstrated in detail in Chapter 7, these so-called electric organ discharges are quite

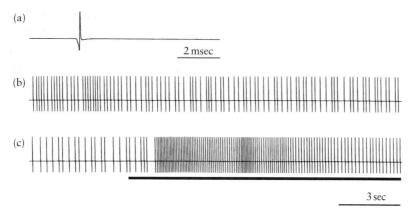

Figure 1.4 Electric organ discharge of the elephant nose, *Gnathonemus petersii*. (a) Each discharge results in a brief electric pulse, which is highly constant in terms of duration, amplitude, and waveform. (b) Resting discharge pattern at slower time scale. As the individual pulses are highly constant, this behavior can readily be quantified by counting the number of pulses (here represented as vertical bars) produced per second and by analyzing the pattern of discharge. In the example shown, approximately five pulses per second are generated, and the intervals between the individual pulses are somewhat irregular. (c) After stimulating an isolated fish with the discharges of a second elephant nose (indicated by the horizontal bar below the trace), the discharge pattern changes significantly. In the example shown, the fish discharges at almost three times the resting rate, and the pattern of pulse production becomes more regular. The result of this experiment underlines the ethological relevance of the electric organ discharge. (Courtesy: G. K. H. Zupanc, and J. R. Banks.)

simple, in terms of their biophysical properties, highly robust, and they can readily be monitored by placing recording electrodes near the fish. Their rate of occurrence, duration, and intensity can easily be measured, thus allowing the researcher to quantify this behavior. This is illustrated by Fig. 1.4, which shows the discharge pattern of the elephant nose (Gnathonemus petersii). This mormyriform fish produces very brief electric pulses, which are, even over hours and days, highly stable in terms of their physical appearance (Fig. 1.4a). However, the fish are able to modulate the discharge pattern, for example by altering the pulse repetition rate, or by changing the mode of regularity (Fig. 1.4b,c). These modulations are, for example, used to encode information in the context of intraspecific communication, such as during aggressive encounters (Fig. 1.5a–f). In addition, the fish employ their discharges for object detection. The electric behavior, therefore, meets the above requirement of ethological relevance. Moreover, many weakly electric fishes can be kept in aquaria under semi-natural conditions; several species have even been successfully bred in the laboratory. Taken together, these properties make them ideal subjects for neuroethological research.

Figure 1.5 The elephant nose shown in various behavioral situations. (a) At rest, the fish likes to stay in caves or, in this case, in a clay pipe. In such a situation, the fish typically emits only a few pulses, separated by rather irregular intervals. (b) If a second fish is introduced into the tank, both fish stay initially at a short distance from each other. (c) They then frequently adopt a head-to-tail stance alongside each other, with their 'chin' appendage (sometimes incorrectly called a 'nose') projecting rigidly forward. (d) If the intruder fails to move off, the territory-holder attacks and rams the intruder. (e) Finally, the intruder is beaten by the territory-holder. The defeated fish turns light brown. Now, both fish have curled in their chin appendages. The social interactions shown in (b)–(e) are accompanied by specific patterns of electric organ discharges which clearly differ from the resting discharge. (f) Elephant nose fish, like other electric fish, can be stimulated by mimics of the discharges of a neighboring fish played back via an electric fish model. This model consists of plexiglass (perspex) rods with built-in electrodes. (Courtesy: G. K. H. Zupanc.)

This and other examples of good neuroethological model systems are discussed in detail in the following chapters of this book. Their exploration over the last decades has greatly advanced our understanding of how the brain controls behavior. Moreover, and equally important, this research has also deepened our appreciation for the biology of the whole animal.

Summary

▪ Neuroethology attempts to understand the neural mechanisms governing animal behavior by employing approaches derived from ethology and neurobiology.

▪ A crucial requirement for neuroethological research is the choice of suitable model systems. These systems are characterized by behavioral patterns which, although simple, robust, and readily accessible, are also ethologically relevant.

Recommended reading

Martin, P., and Bateson, P. (1993). *Measuring behaviour: an introductory guide.* Second Edition. Cambridge University Press, Cambridge.
An excellent guide to the principles and methods of quantitative studies of behavior, with emphasis on techniques of observation, recording, and analysis.

Tinbergen, N. (1951). *The study of instinct.* Oxford University Press, London.
A classic. Even half-a-century after publication of its first edition, still a source of inspiration.

Questions

1.1 You intend to apply for a research grant to develop a novel neuroethological model system. What animal species and which behavior would you propose to examine? Justify your choice using the above criteria.

1.2 Songbirds produce, particularly during the breeding season, characteristic songs which subserve a variety of behavioral functions. How would you split the songs into individual behavioral patterns? How would you quantify these behaviors?

1.3 Any model system is a simplification of reality. Using one of the model systems presented in this book, discuss aspects which, in your opinion, were neglected for the sake of simplicity. How would have inclusion of one or several of these aspects impeded neuroethological analysis? On the other hand, how could consideration of such aspects provide more realistic explanations of the neural mechanisms underlying behavior?

The study of animal behavior: a brief history

2

■ Introduction

■ The roots of the study of animal behavior

■ The new ear in the study of animal behavior

■ Relating neuronal activity to behavior: the establishment of neuroethology

■ Summary

■ Recommended reading

■ Questions

Introduction

The interest of man in animal behavior is certainly as old as mankind itself. Especially before the rise of civilization, knowledge about the behavior of animals was often a matter of survival. However, in spite of this natural curiosity, it took surprisingly long to establish a scientific discipline that applied the rigor of empirical research to the study of animal behavior. One of the reasons for this rather slow and cumbersome development was that, until modern times, such studies tended to be compromised by **anthropocentric approaches**—animal behavior was interpreted from a human point of view, and only too often research results were used to 'prove' the superiority of human behavior. Neuroethology, whose conception was possible only after the advent of both well-defined ethological concepts and sophisticated neurobiological techniques, originated even later and was firmly established only toward the end of the twentieth century.

True understanding of present developments is possible only by knowing the past. In this chapter, we will, therefore, take a closer look at the

Figure 2.1 Marble head of Aristotle in the *Kunsthistorisches Museum* in Vienna (Austria). (Courtesy: Kunsthistorisches Museum, Vienna, Austria/Bridgeman Art Library.)

historical development of both the study of animal behavior and neuroethology, including some of their major figures.

The roots of the study of animal behavior

Aristotle and the Middle Ages

First attempts to produce a systematic account of animal behavior can be dated back to the Ancient Greeks, especially to **Aristotle** (384–322 BC). Figure 2.1 shows a marble head of him. Aristotle can be considered the father of **natural history** and the founder of **zoology**. His work on animals covered not only anatomical, physiological, and developmental aspects, but psychological as well. His detailed descriptions indicate that he spent an enormous amount of time carefully watching many species of animals. Also, the collection of his observations and the number of species categorized are immense; and the scientific quality of his work, measured by the standards of his time, can be considered very high.

The approach Aristotle employed is characterized by the following adjectives:

- **inductive**: general conclusions are drawn from observations of particular instances
- **comparative**: particular phenomena are analyzed in various animal species and compared with each other
- **anecdotal**: general, even far-reaching, conclusions are based on single, isolated observations; no attempt is made to replicate observations or to verify the significance of data using statistical approaches
- **vitalistic**: all features of living forms are ascribed to an imminent vital principle called *psyche* ('soul') in Greek or *vis vitalis* ('force of life') in Latin. This rather mysterious force driving behavior cannot be studied by scientific methods
- **teleological**: behavior, like morphological properties of the organism, is explained as the expression of an all-pervading design of a *telos* (Greek: purpose, goal). Obviously, such an approach cannot explain features of biological phenomena that are not optimized to serve a certain purpose. However, as we know today, behavior, similar to morphological structures, also reflects its evolutionary history, and not just its purpose. As a striking example of imperfection in design, we will discuss, in Chapter 7, the neural implementation of the jamming avoidance response of some weakly electric fish
- **anthropomorphic**: analysis of behavior is performed from an *anthropos* (Greek: human) point of view, rather than from an

objective one. This becomes evident, for example, when certain character traits of human beings, such as jealousy, courage, and nobility, are attributed to animals.

While the former two approaches are also characteristic of modern ethology, the latter four are nowadays regarded as rather unscientific. However, as we will see, they have dominated the study of animal behavior for more than 2000 years.

The period following Aristotle, for almost 2000 years, witnessed not only a halt in the initial progress made in the study of animals and their behavior, but even a decline in this area, as well as in sciences in general. During the 'dark' Middle Ages, experimentation and scientific thinking were largely suppressed. Also, mainly influenced by the rising Christian theology, a dichotomy was drawn between the 'rational soul' of man and the 'sensitive soul' of beasts. The authority of the writings of Aristotle on natural history was mainly used to find evidence of the superiority of mankind, rather than viewing his texts as a stimulus for further observations.

A remarkable exception to this development was **Albertus Magnus** (ca 1206–1280), a German Dominican, bishop, scientist, philosopher, and theologian. Although devoted to the reconciliation of reason and faith through the fusion of Aristotelianism and Christianity, he was also—in contrast to most of his contemporaries—critical of the work of Aristotle, and emphasized the importance of independent observations and of experimentation. As he put it, 'the aim of natural science is not simply to accept the statements of others, but to investigate the causes that are at work in nature.' Based on this principle, he made, during numerous excursions, many biological observations of high originality, especially in the field of zoology.

Mind–body dualism

Although Albertus Magnus was considered an extraordinary genius by his contemporaries and by posterity, his work could not prevent the adherence to Aristotle as a source of authority by most scholars in the following centuries. This attitude was only gradually overcome in the sixteenth and seventeenth centuries. These, as well as the following centuries, were marked by an abundance of new discoveries, especially in the physical sciences. In the biological sciences, emphasis was on classification, morphology, physiology, and embryology. Despite this scientific awakening, the impact on the study of animal behavior as a scientific discipline was rather minor. However, triggered by the enormous success in applying the laws of physics to explain not only physical phenomena but also biological processes—such as the circulation of blood as done by

Aristotle's approach: although inductive and comparative on the one side, is also anecdotal, vitalistic, teleological, and anthropomorphic on the other side.

Mind–body dualism: mind is completely separate from matter.

William Harvey (1578–1657)—a revolt began against the traditional vitalistic interpretation of the organism. The chief exponent of this movement was **Descartes** (1596–1650) with his attempt to explain bodily processes purely on **mechanistic** grounds. Animals were viewed as natural machines. Mind, which included conscious and psychic functions, was assumed to be completely separate from matter, a concept referred to as **mind–body dualism**.

Darwin: principles of evolution apply not only to morphological characteristics, but also to behavior.

The new era in the study of animal behavior

Evolutionary theory and comparative approaches

Despite its mechanistic outlook, Descartes' doctrine did not cause conflict with the orthodox theology of his times, as it did not challenge man's primacy in terms of his mental and psychic capabilities. It took another 200 years to seriously question the distinct position of man among living organisms. This was done by the **evolutionary theory** of **Charles Darwin** (1809–1882) (Fig. 2.2). His ideas were published in various books, including *On the Origin of Species by Means of Natural Selection* (1858), *Variation of Animals and Plants under Domestication* (1868), *Descent of Man* (1871), and *The Expression of the Emotions in Man and Animals* (1872). As a central dogma, the evolutionary theory proposes a continuity of both morphological and behavioral characteristics within the living world, including man. These publications triggered an enormous interest in behavioral observations from a **comparative** perspective, a theme that would become central to ethology. Moreover, the initially still anecdotal, and sometimes even anthropomorphic, approach was gradually replaced by objective and systematic methods. Important figures of this development were **Douglas Spalding** and **Conway Lloyd Morgan**.

Experimental and objective approaches

Figure 2.2 Charles Darwin in 1881. (Courtesy: Bettmann CORBIS.)

Spalding, who was presumably born about 1840 in London to working-class parents and died in 1875 from tuberculosis, had only a few years of scientific activity. His work received wide recognition for a short time after his death, but was then mostly forgotten for more than half a century, until it was rediscovered in the 1950s. Spalding combined his original behavioral observations with carefully designed **experiments**, thus proceeding far beyond the anecdotal approach that dominated the study of animal behavior in the nineteenth, and even part of the twentieth, century. His major contributions were on the **development of behavior**. By using a self-constructed incubator, he hatched chicken eggs to study the influence of visual and acoustic experience on the maturation of behavior. Spalding is also credited with the first experimental study of the **following-response** and the **critical period** of birds, which laid the foundation to the **imprinting** concept of ethology. Further, he worked on the nature of instinct and anticipated, to a certain degree, the **releaser** concept. The latter has been seminal to both ethology and neuroethology.

Among those who were greatly inspired by Spalding's work, especially by his experimental approach, was Conway Lloyd Morgan (1852–1936) (Fig. 2.3) of the University of Bristol in England. His contributions had a major impact on the further development not only of ethology, but also of psychology. In particular, he stressed the need for operational definitions and for replication of experiments—requirements that are obvious today, but not at Morgan's time when anecdotal and subjective approaches were still widely used. This encompasses what has become known as **Morgan's Canon**, published in 1894. It is summarized in his own words as follows: 'In no case may we interpret an action as the outcome of the exercise of a higher psychical faculty, if it can be interpreted as the outcome of the exercise of one which stands lower in the psychological scale.' In other words, a behavioral pattern of an animal should not be interpreted in terms of 'higher' intellectual capabilities, if a 'simpler' explanation is possible as well. The major significance of this requirement has been to help avoid biased and anthropomorphic interpretation of animal behavior. On the other hand, strict adherence to this (unproved) 'law' could easily lead to misinterpretation of behavior. The goal of any study of animal behavior has to be to explain the complex mechanisms of behavior correctly—even if this may not be the simplest interpretation.

Figure 2.3 Conway Lloyd Morgan. (Courtesy: University of Bristol Special Collections.)

> *Example: In conditioning experiments, pigeons were trained to discriminate between two-dimensional patterns rotated at different angles in the plane of presentation. After a learning phase, they performed similarly well as do humans in such experiments. These results have been used as evidence for the existence of intelligence in pigeons. An alternative explanation, following Morgan's Canon, is that pigeons, as air-borne animals, are frequently confronted with the problem of viewing ground-based objects at different angles. They have, therefore, developed sensory mechanisms enabling them to readily identify objects, even if these objects, for example, the projection of trees or houses, are viewed under different rotational angles.*

In the twentieth century, especially within the first 50 or 60 years, the study of animal behavior took two, quite separate routes. One of these developments culminated in the establishment of **ethology** as a biological discipline, in Europe. The other led to the foundation of several psychological schools, such as **comparative psychology** and **behaviorism**, in North America. All three disciplines are characterized by the attempt to use an objective approach. The main difference between them lies on the conditions under which they observe and analyze behavior: ethology

Douglas Spalding and Conway Lloyd Morgan introduced experimental and objective approaches to the study of animal behavior.

stresses the importance of observations of animals under **natural** (or **semi-natural**) **conditions**, whereas the more psychologically oriented schools prefer to analyze behavior under the stringent conditions of the **laboratory**.

The rise of ethology

In the first decades of the twentieth century, the study of animal behavior grew particularly rapidly in the USA. Main contributors to this development were **Charles Otis Whitman** (1842–1910), **Wallace Craig** (1876–1954), **William Morton Wheeler** (1865–1937), and **Karl Spencer Lashley** (1890–1958).

Whitman was a prominent biologist of his times. He held a chair position in the Department of Zoology at the University of Chicago and acted as the first director of the Marine Laboratory at Woods Hole. Although in his lifetime he published just two papers in the field of animal behavior, he exerted a major influence on the development of ethology. Already in 1898, he wrote that '... instincts and organs are to be studied from the common viewpoint of phyletic descent.' The results of his detailed research on the behavior of pigeons were published posthumously.

Among Whitman's students were Craig and Wheeler. Craig continued the studies of his mentor on the behavior of pigeons. He was the first to distinguish between **appetitive behavior** and **consummatory action**. According to this classic ethological concept, changes in the animal's internal state (e.g. in water balance caused by dehydration) are sensed by the brain and result in a build-up of 'drive.' This build-up shows itself externally as a state of agitation called appetitive behavior, which involves search for a suitable external stimulus (e.g. water). When this stimulus is encountered, it triggers a consummatory action (e.g. drinking). This leads to a reduction, and finally to a cease, in the corresponding appetitive behavior.

Wheeler was the first to use the term 'ethology' in the English literature, in a paper published in *Science* in 1902. He became widely recognized for his detailed descriptions of the social life of insects. Wheeler also developed a classification scheme of insect societies. Among his achievements related to animal behavior was the discovery of **trophallaxis**. This term describes the donation of salivary secretions by larvae of social wasps to their adult, winged sisters. The secreted substances are important means for recognition and communication within colonies. As such, his discovery prepared the grounds for investigation of **pheromones**.

Lashley (Fig. 2.4) worked in his early career with the later founder of behaviorism, John B. Watson (see below) on the homing of birds in the field and on the development of monkeys. He became both a renowned primatologist and a distinguished neurophysiologist. Among his achievements is the attempt to reveal the neural basis of learning in rats and to analyze the function of the cortex in the brain. He also made detailed observations of the behavior and the social relationships in primates.

Figure 2.4 Karl Spencer Lashley. (Courtesy: National Academy of Sciences of the USA.)

In Europe, it was especially the work of **Oskar Heinroth** (1871–1945) that inspired many. Heinroth (Fig. 2.5), a distinguished ornithologist and director of the Berlin Zoo, proposed that **behavioral patterns can be used to analyze systematic relationships between species**—just as morphological patterns can be used for taxonomic classification. In an influential paper published in 1911, Heinroth furthermore demonstrated that young goslings follow the first moving object they see after hatching. The young geese subsequently follow this object with preference to any other object. Normally, this object is the parent goose. Goslings reared by a human foster parent, thus, prefer this person to any conspecific. Heinroth called this phenomenon **imprinting**. In the same paper, he was also the first to use the term 'ethology' in the modern sense, namely, as the 'study of natural behavior.'

Another influential scholar of this pre-ethological time was **Jakob von Uexküll** (1864–1944). Based on his wealth as a 'Baltic Baron,' he largely funded his studies himself, which enabled him to maintain an exceptional degree of independence. His non-conformism was also expressed in many of his ideas, including the proposal that the subjective world in which an animal lives is quite different from the objective physical environment. He called this subjective world the *Umwelt*. Although the term, literally translated from German, means 'environment,' we will, in agreement with other authors, use it to indicate an animal's subjective world, and use the term 'environment' to denote the objective physical world in which the organism lives. This concept has had an enormous impact not only on sensory physiology, but also on the rising ethology.

Example: The visual spectrum of honeybees is, compared to that of humans, shifted into the UV range. Bees make use of this sensory ability when searching for food, as some flowers have markings at their petals, the so-called **honey guides** *(Fig. 2.6), visible in the UV range of light only, which guide the bee to the flower's source of nectar. On the other hand, honey bees are blind in the red range of light. Their Umwelt is, thus, quite different from that of humans, although both humans and bees live in the same physical environment.*

The final establishment of ethology as a new and independent discipline was achieved by **Konrad Lorenz** (see Box 2.1) and **Niko Tinbergen** (see Box 2.2) between the 1930s and the 1950s. They are, therefore, considered the founders of ethology. Their major accomplishment was to provide a greater conceptual framework for their own observations and experiments, as well as for those of others, including Heinroth, Whitman, Craig, and Morgan. Many concepts of this early theoretical framework centered on the nature of **innate behavior**. A milestone in this development was the publication of *The Study of Instinct* by Niko Tinbergen in 1951, which comprised the first textbook of ethology.

Figure 2.5 Oskar Heinroth. (Courtesy: AKG, London.)

(a)  (b)

Figure 2.6 Honey guides. (a) The drawing was made after a photograph taken on regular film. The flowers appear white, without any distinct pattern. (b) The same flowers taken through a photographic system sensitive to ultraviolet light. Now a striking color pattern, the so-called honey guides, are revealed. These two records approximate the visual pattern perceived by a human and a honey bee, respectively, and, thus, demonstrate the differences in the *Umwelt* between the two species. (After **McFarland, D.** (1993).)

Mechanistic schools

Ethology, as the study of the natural behavior of animals, was founded by Konrad Lorenz and Niko Tinbergen, who provided a conceptual framework for this new biological discipline.

At the same time that ethology gradually grew, several more mechanistically oriented schools emerged. Their members were often affiliated with psychology rather than biology departments. All these different schools shared with ethology the objective approach toward the study of behavior. In contrast to ethology, their emphasis was, typically, on laboratory-based experiments, often conducted on only one or a few 'model systems' (frequently rats or pigeons).

One of the first of these mechanistic schools was established by **Jacques Loeb** (1859–1924) with his **theory of tropism**. Loeb, a Prussian, first studied philosophy in Berlin, but disillusioned with his professors, whom he regarded as 'wordmongers,' he turned to biology and received an M.D. degree from the University of Strasbourg in 1884. In 1891, he emigrated to the USA, where he taught at Bryn Mawr College, the University of Chicago, the University of California, and, from 1910 until his death, at the Rockefeller Institute for Medical Research (now Rockefeller University). He became well known for experiments in which he demonstrated that echinoderm larvae could be chemically stimulated to develop in the absence of fertilization. This and other studies made him a leading proponent of a mechanistic conception of biology. After studying the movements of plants, he proposed to also explain animal

BOX 2.1 Konrad Lorenz

Konrad Lorenz with graylag geese imprinted to follow him, rather than their mother. (Courtesy: Nina Leen/Timepix.)

Konrad Zacharias Lorenz is, together with Niko Tinbergen, regarded as the founder of ethology. He was born in Vienna (Austria) in 1903. Like his father, a famous orthopedic surgeon, Lorenz studied medicine, first at Columbia University in New York, then at the University of Vienna. It was during his time at the Institute of Anatomy in Vienna, working with the well-known anatomist Ferdinand Hochstetter, that he became exposed to the idea of revealing evolutionary descent by comparing homologous morphological structures of various animals. Using the underlying principle, Lorenz soon introduced a similar approach to the study of animal behavior to elucidate evolutionary traits by comparing homologous behavioral patterns.

Despite his training in medicine, recognized by the award of an M.D. degree in 1928, his greater passion was for biology, watching birds and fish in particular. During these years, he conducted most of his studies in his parents' home—actually more a kind of a castle built on a huge property in Altenberg, a small village near Vienna. His investigations resulted in the award of a Ph.D. in zoology in 1933. It was around that time that Konrad Lorenz met the German ornithologist **Oskar Heinroth** and the Dutch naturalist **Niko Tinbergen**, the latter being considered the co-founder of ethology. Influenced by these two men, Lorenz published in the 1930s a series of pioneering papers, especially on avian behavior. These publications laid the foundations for the definition of several key concepts in ethology.

In 1940, Konrad Lorenz was appointed to the Chair of Comparative Psychology at the University of Königsberg, an imminent position that goes back to the famous German philosopher Immanuel Kant as its first holder. However, one year later, Lorenz was drafted into the army medical service. In 1944, Lorenz was taken prisoner by the Russian army and released only in 1948. During his imprisonment, he wrote the manuscript to a book that, when published 25 years later under the title *Behind the Mirror: A Search for a Natural History of Human Knowledge*, became a bestseller.

Lorenz's major contribution after the Second World War was the establishment of the Max Planck Institute for Behavioral Physiology in Seewiesen near Munich, of which he was director from 1961 to 1973. In addition to his enormous influence on the development of ethology, an achievement for which he received—together with Niko Tinbergen and Karl von Frisch—the Nobel prize in 1973, he published a large number of bestsellers. They include popular books, such as *King Solomon's Ring* and *Man Meets Dog*, as well as *On Aggression*, a highly controversial attempt to explain human aggression as the result of a spontaneously active drive.

Konrad Lorenz died in Altenberg in 1989.

BOX 2.2   Niko Tinbergen

Niko Tinbergen. (Courtesy: The Nobel Foundation.)

Among the 'fathers' of ethology, Konrad Lorenz has often been characterized as the theorist who, based on observations of animals kept under semi-natural conditions in captivity, formulated major ethological concepts. In contrast, Niko Tinbergen was primarily the field biologist who carefully collected an impressive amount of data on the **natural behavior of animals in the wild**, and who demonstrated a remarkable ability to devise simple, yet highly informative, **field experiments**.

Nikolaas ('Niko') **Tinbergen** was born in 1907 in The Hague, in the Netherlands. While still in school, he developed a keen interest in bird watching and animal photography. However, when studying biology at the University of Leiden, he did not impress his professors—often he missed classes because he went bird watching or played hockey instead. The latter he did so well that he even became a member of the Dutch national hockey team.

In his doctoral thesis, Tinbergen studied the behavior of bee-hunting digger wasps. This work formed the basis for subsequent important publications on wasps' homing and hunting behavior, as well as their ability to learn landmarks. Following the completion of his graduate studies, Tinbergen went, with his wife Elisabeth whom he had just married, on an unusual honeymoon—a Dutch expedition to the Scoresby Sound region of East Greenland. This one-year trip resulted in the publication of monographs on the snow bunting and the red-necked phalarope.

Tinbergen spent the period between 1933 and 1942 at the University of Leiden, where he carried out a number of influential studies on the behavior of various animals, including digger wasps, sticklebacks, and herring gulls. In 1938, Konrad Lorenz invited Tinbergen to work with him at his home in Altenberg, where they subsequently produced a classic piece of research on the **egg-retrieval response of nesting geese**.

During the years from 1942 to 1945, Niko Tinbergen was imprisoned in a German internment camp, after he and other faculty members of the University of Leiden had protested against the dismissal of Jewish professors. Despite all the difficulties, he wrote, during those years, several children's books and prepared the draft of *Social Behaviour in Animals*, which was published in 1953.

After the war, he returned to Leiden where he became a full professor at the University in 1947. In 1953, he accepted a lectureship at the Zoology Department of the University of Oxford. This move tremendously accelerated the establishment—in Great Britain—of the new field of ethology, until then dominated by continental scientists. Elected to a Chair in Animal Behavior in 1966, Tinbergen stayed in Oxford until the end of his life in 1988. During this period at Oxford, he broadened his research interests to include studies on human behavior. His ethological investigations on **childhood autism**, conducted in collaboration with his wife, have generated significant interest from psychologists.

Tinbergen's pioneering work, which includes the publication of the first textbook of ethology (*The Study of Instinct*) in 1951, was recognized by the award of the Nobel prize for Medicine and Physiology, together with Konrad Lorenz and Karl von Frisch, in 1973.

behavior in terms of tropisms—involuntary orienting movements. It was particularly his experimental approach that impressed many scientists of his time. This was largely the case because people viewed Loeb's theory as a counter movement to the previously dominating vitalistic, anthropomorphic, and anecdotal approaches. On the other hand, his attempt to universally explain animal and human behavior in terms of tropisms turned out to be insufficient—as we know today, behavior is too variable and the underlying physiological mechanisms too complex for a single, rather simple, universal theory to account for all of them.

While the theory of tropism did not survive its major proponent, another mechanistic theory, the **reflex theory** of **Ivan Petrovitch Pavlov**, did. Pavlov (1849–1936) (Fig. 2.7) was a Russian physiologist who was awarded the Nobel Prize for Medicine in 1904 for his work on the physiology of digestion. His influence was especially pronounced in the USA after the publication of the English translation of his book *Conditioned Reflexes*. The discovery that made Pavlov famous was the phenomenon of the **conditioned reflex**. The type of set-up used by Pavlov to demonstrate this behavior is shown in Fig. 2.8. When he presented a dog with a piece of food (**unconditioned stimulus**), the dog salivated. Then, upon repeated occasions, a bell was sounded (**conditioned stimulus**) just before the food was presented. This stimulation regime led to a pairing between the conditioned stimulus and the unconditioned stimulus. After a certain number of such paired presentations, the bell

Figure 2.7 Ivan Petrovich Pavlov. (Courtesy: The Nobel Foundation.)

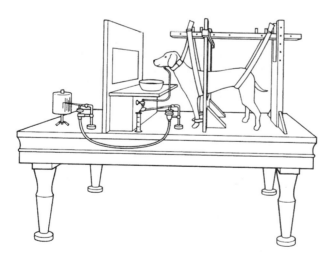

Figure 2.8 Set-up used by Ivan Pavlov for the conditioning of dogs. A hungry dog is restrained in a harness. During the training phase, delivery of food (unconditioned stimulus) is preceded by the sound of a bell (conditioned stimulus). After a while, the dog starts to salivate upon presentation of the conditioned stimulus alone. The salivating response is recorded by inserting a tube into the salivary duct and collecting the saliva. (After **McFarland, D.** (1993).)

Figure 2.9 John B. Watson in 1932. (Courtesy: Nickolas Muray/Getty Images.)

alone was sufficient to elicit salivation—even in the absence of food. This process of association between the unconditioned stimulus and the (previously neutral) conditioned stimulus is referred to as **classical conditioning**.

Based on the conditioned reflex as a simple form of learning, the American school of **behaviorism** attempted to explain behavior, and also more complex forms of learning, primarily as a series of reflex chains. Behaviorism was established by **John B. Watson** (1878–1958) (Fig. 2.9), especially through the publication of his paper *Psychology as the Behaviorist Views It* (1913) and his book *Psychology from the Standpoint of a Behaviorist* (1919). His emphasis on rigorous laboratory experimentation and the rejection of any subjective approach in the study of behavior had a major impact upon psychology in the first half of the twentieth century. Watson's earlier work included investigations which, in many respects, can be characterized as 'pre-ethological.' He was probably the first who demonstrated communication via chemical signals in rats. Also, over three summers, he conducted field studies on the behavior of noddy and sooty terns. Most significantly, in an article published in *Harper's Magazine* in 1912, he clearly accepted the existence of innate components of behavior in animals. This notion is in contrast to the standpoint Watson adopted later in his life when he attempted to account for behavior (including human behavior) primarily in terms of learning and experience.

The second important figure of behaviorism was **Burrhus Frederic Skinner** (1904–1990) (Fig. 2.10). The phenomenon he explored in great

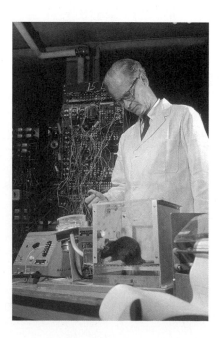

Figure 2.10 Burrhus Frederic Skinner in 1964. (Courtesy: Nina Leen/Timepix.)

Figure 2.11 Rat in a Skinner box. Upon pressing a lever, a pellet of food is delivered into the cup. (After **Manning, A.** (1972).)

detail was **operant conditioning**, also called **instrumental conditioning**. This type of conditioning is characterized by learning of a novel response through **trial and error**. In contrast to classical conditioning, in which animals associate with reward a novel stimulus, instrumental conditioning leads to certain actions whose performance is rewarded.

To analyze operant conditional behavior, Skinner used an apparatus that has become known as the **Skinner box**. An example of such a device is shown in Fig. 2.11. It consists of a box in which the animal (e.g. a rat or a pigeon) can manipulate an object. A Skinner box constructed for rats often has a lever that can be depressed by the rat. Upon depression, a door may open to release a food pellet for the rat. After the rat has pressed the lever the first time by accident, it will learn very quickly to perform this behavior in order to become rewarded with food. The experimental manipulation leading to a change in the rat's probability to depress the lever is called **reinforcement**. This term corresponds to what traditionally has been referred to as 'reward'.

The use of such **operational** nomenclature, as well as definition of behavior in operational terms, is a logical consequence of Skinner's view of the valid scientific approach to the study of behavior. It is only the objectively measurable changes in behavior that can be analyzed, but not subjective sensations, such as the feeling associated with the receipt of reward. One of the primary goals of behaviorism is, therefore, to identify reinforcement contingencies that result in changes in behavior. In contrast to Watson, Skinner did not aim at revealing neuronal or hormonal substrates under-lying such behavioral processes. In his opinion, experimental analysis of all behavioral operations possible would be sufficient to finally lead to a complete understanding of behavior.

It was behaviorism's emphasis on the importance of learning processes that caused, especially in the 1960s and 1970s, considerable tension between this school of thought and ethology. At an extreme, some behaviorists claimed that any form of behavior is learned, thus denying the existence of genetic determinants, as they were preferentially studied by ethologists. These radical behaviorists believed any animal or human to be born as a *tabula rasa* (a Latin term for 'blank writing-tablet')—an idea that traces back to the empiricism of the British philosopher John Locke (1632–1704). According to this notion, it is experience, that is, learning processes, that determines what is 'on the tablet.' This rather dogmatic view has led to an intense, sometimes even polemic, dispute over the relative importance of genes ('nature') and environment ('nurture'), known as the **nature versus nurture controversy**. This discussion had an enormous effect not only in biology and psychology, but also in the social sciences and in education.

Despite this controversy, the stringent empirical approach employed by behaviorism had, especially in North America, a positive impact upon the further development of ethology. While, during its classic phase, ethological research was often restricted to descriptive investigations, and, sometimes, even major conclusions were drawn based upon judgmental and not very quantitative observations, later studies made extensive use of experimental approaches and statistical methods. Also, the initial emphasis on innate behavior gradually lost its importance, and many ethologists included learned behavior in their studies. What continues to distinguish ethology from other disciplines is its attempt to study biologically relevant forms of animal behavior under conditions as natural as possible, and to include comparative approaches into its investigations.

Mechanistic schools share with ethology the objective approach; however, in contrast to ethology, their emphasis is on laboratory-based experiments.

Relating neuronal activity to behavior: the establishment of neuroethology

The beginnings

As early as 1951, Niko Tinbergen wrote in *The Study of Instinct* that it is the job of the ethologist to carry the analysis of behavior down to the level studied by the the physiologist. Despite his call, initial progress in this area of research, which is now known as **neuroethology**, was rather slow. One reason for this is that the problems encountered in relating the activity of individual neurons, or assemblies of neurons, to a biologically relevant behavior of the whole animal are tremendously complex.

Starting shortly after the beginning of the twentieth century, rapid progress was made in the elucidation of the physiological properties of neurons. This led to the establishment of strong neurophysiological schools, particularly in Great Britain and the USA. Prominent leaders in this development were Charles Sherrington, Alan Hodgkin, Andrew Huxley, Ernst and Berta Scharrer, John Eccles, Roger Sperry, and Bernard Katz. Their work generated an enormous body of information, for example, on the physicochemical factors that define the resting potential of neurons, on the production and propagation of action potentials, and on the structure and function of synapses. However, the work of these neurophysiologists was largely independent of ethological questions. One of the very few exceptions was the German **Erich von Holst** (Box 2.3), who, toward the end of his life, jointly with Konrad Lorenz headed the Max Planck Institute for Behavioral Physiology in Seewiesen. Among his studies, especially the investigations on endogenous control of rhythmic movements and his brain stimulation experiments (see below) were of seminal importance for the development of neuroethology.

BOX 2.3   Erich von Holst

Konrad Lorenz (left) and Erich von Holst (right) at the Max Planck Institute for Behavioral Physiology in Seewiesen. (Courtesy: Gerhard Gronefeld.)

Among the great scientists who have made outstanding contributions to the development of neuroethology, **Erich von Holst** is probably the most ingenious.

His life is characterized by fundamental and highly original discoveries, enormous experimental skills, and absolute dedication to whatever he did. It was also he who pioneered physiological experiments conducted on whole animals, rather than on isolated organs.

Born in Riga in 1908, he studied in Kiel, Vienna, and Berlin. In his Ph.D. thesis, published in 1932, he analyzed physiological mechanisms controlling the movements of earthworms. According to the predominant view of his times, the peristaltic waves traveling in rostro-caudal direction across the earthworm's body are caused by **reflex chains**: mechanical deformation of one segment induces mechanical deformation of the following segment, and so on. If this theory were correct, the elimination of one or several segments, while leaving the ventral nerve cord intact, should lead to cessation of the contraction waves generated by the longitudinal muscles. However, von Holst's experiments demonstrated the opposite: the waves of contraction are transmitted through the ventral nerve cord even across gaps created by the experimenter in the body's segments. This suggests that the **pattern of movement is controlled by endogenous rhythms in the central nervous system**, rather than by reflex chains. He, thus, disproved the then very popular hypothesis that all behavior can be explained as a results of the action of reflex chains.

In 1946, Erich von Holst was appointed to a Zoology Chair at the University of Heidelberg, in 1948 he became one of the directors of the Max Planck Institute for Marine Biology in Wilhelmshaven, and from 1957 on he headed the newly established Max Planck Institute for Behavioral Physiology in Seewiesen near Munich. The idea for the foundation of this center for behavioral physiological and ethological research dates back to 1936 when von Holst met Lorenz for the first time. This meeting was of historical importance. It was especially the proposal of a **central pattern generator** (although von Holst did not use this rather modern term) that intrigued Lorenz, as he saw in this finding the physiological correlate of spontaneously occurring behavior—an idea vehemently rejected at that time by the school of behaviorism.

Among other research projects that made von Holst famous were his studies on **relative coordination**, that is, the influence of different groups of neurons on the temporal pattern of rhythmic activity; the **biophysics of avian flight**; and the **physiology of the vertebrate labyrinth**. Later in his life, he added to these experimental investigations more and more theoretical studies. One of the results of this work was the formulation of the **reafference principle**, a model to explain how animals and humans can distinguish moving retinal images caused by movements of objects from those generated by movements of the body. Together with his associate Horst Mittelstaedt, he proposed that the brain produces a 'copy' of the efferent command information controlling the body movement, and this efference copy is compared with the incoming (afferent) sensory information mediated by retinal stimulation. In the last years of his life, Erich von Holst turned, once again, to a new research theme. Using **electrical brain stimulation** techniques, he studied mechanisms of neural control of behavior patterns in chicken.

It is also characteristic of von Holst that, in addition to his scientific work, he spent many years of his life on a completely different subject. He was not only an excellent viola player, but he also designed and built himself revolutionary new violas. They were asymmetric instead of symmetric, a feature that combines excellent sound characteristics with the comfort of the rather small size of the instrument's body.

Suffering from heart disease since childhood, von Holst tried to achieve in the time available as much as he could. Yet, many projects remained unfinished when he died in 1962, at the age of only 54.

On the other hand, the physiological properties of sensory organs or neurons involved in the sensory processing were examined largely by separate schools. Most of the centers involved in this type of research were located in the USA and Germany. Many of their researchers, such as David Hubel and Torsten Wiesel in the USA, studied the basic properties of such sensory neurons in one or a very few particularly favorable model systems, such as the responses of neurons to arbitrary, simple stimuli, and did not pursue the relation of these properties to the natural stimuli. Yet, at the same time, schools with a strong zoological foundation emerged. Their research was driven by the desire to link the physiological properties of sensory systems, in a variety of taxonomic groups, to behavioral function. Important figures in this movement were **Karl von Frisch** (see Box 2.4) in Germany, who established conditioning paradigms as powerful tools to explore the sensory capabilities of the whole animal; **Hansjochem Autrum** (1907–2003), also in Germany, who studied the properties of sensory organs, particularly in arthropods, at the physiological level; and **Ted Bullock** (see Box 2.5) in the USA, who, more than

BOX 2.4   Karl von Frisch

Karl von Frisch. (Courtesy: Bettmann/CORBIS.)

Only a very few scientists have achieved what **Karl von Frisch** did: to become accepted as a leading figure by the scientific community, while reaching a degree of popularity usually reserved to artists and writers. The work on which both von Frisch's scientific success and his popularity is based upon centers around the dance language of honey bees, although he has made pioneering discoveries also in many other areas of sensory and behavioral physiology.

Karl von Frisch was born in Vienna, Austria, in 1886. His family, to which renowned professors and medical doctors belonged, provided for him an intellectually stimulating environment and encouraged his leaning toward research and scholarship. This interest was further reinforced by the extended periods he spent in the family's summer home in Brunnwinkl on Lake Wolfgang, where he collected specimens for his 'little zoo' and made his first scientific observations. Later in his life, he frequently retreated there to perform his experiments in the peaceful and harmonious surrounding of the lake. And it was also Brunnwinkl which provided him refuge during the two world wars.

After completion of secondary school and yielding to his father's request, he enrolled in 1905 at the medical school of the University of Vienna. However, disenchanted with medicine, he transferred after two years to the Zoological Institute of the University of Munich, Germany, to study under the famous Richard von Hertwig. At one of the frequent excursions to the Dolomites, von Frisch was assigned to study the behavior of solitary bees. This marked the beginning of a lifetime interest.

In 1910, von Frisch obtained his Ph.D. degree with a thesis on color adaptation and light perception of minnows. After various academic positions, he was appointed Professor of Zoology and Director of the Institute of Zoology at the University of Munich in 1925. He remained affiliated with this internationally renowned institution until his retirement in 1958, except for a short interruption after the Second World War.

Among his epoch-making contributions to biology were the demonstration of **color vision in bees** and of **hearing in fish**, the discovery of *Schreckstoff* (alarm substance) **in minnows,** and a detailed analysis of the **bee language,** including the round and waggle dances. Many of his discoveries disproved up-to-then widely accepted dogmas, such as the belief, at the beginning of the twentieth century, that fish and all invertebrates are color-blind. It was also Karl von Frisch who introduced **conditioning paradigms** to sensory physiology. These paradigms have proven to be extremely powerful tools which use the natural behavior of animals to examine their sensory capabilities.

It is a characteristic of Karl von Frisch to have been not only an ingenious scholar, but also an excellent science communicator. His books are among the very best ever written in biology and include *The Dancing Bees: An Account of the Life and Senses of the Honey Bee* and *Man and the Living World.*

In 1973, at the beginning of his 88th year, von Frisch was, together with Konrad Lorenz and Niko Tinbergen, awarded the Nobel Prize for Physiology and Medicine. He died in Munich in 1982 at the age of 95.

BOX 2.5   Theodore Holmes Bullock

Theodore H. Bullock. (Courtesy: Scripps Institution of Oceanography.)

One of the fathers of neuroethology, whose efforts were instrumental in the establishment of this scientific discipline in North America, is **Theodore ('Ted') Holmes Bullock**. Bullock was born to American Presbyterian missionaries in 1915 in Nanking, China, where he grew up until the age of 13. For his undergraduate education, he attended Pasadena Junior College and the University of California at Berkeley, majoring in zoology. His Ph.D. thesis, also conducted at Berkeley, was on the anatomy and physiology of acorn worms. After four years at Yale University, with summers at the Marine Biology Laboratory at Woods Hole on Cape Cod, he assumed an Assistant Professorship at the University of Missouri School of Medicine at Columbia, followed by faculty positions at the University of California at Los Angeles and the University of California at San Diego, where he still serves as Professor Emeritus in Neurosciences.

Among the features that distinguish Ted Bullock are his exceptionally diverse research interests, covering not only numerous research themes, but also a huge number of taxonomic groups. This has provided the basis for a comprehensive analysis of **brain evolution**. One of the themes to which Bullock has returned, time and again, is the **importance of DC and low-frequency electric fields for intercellular communication** in the brain. This phenomenon is still largely unexplored in nonhuman species, partially because many of the corresponding concepts and techniques, such as electroencephalography, have primarily been developed for application to the human brain. However, Bullock's research suggests that the effects of such fields arising from individual cells, or synchronized populations of cells, may be significant in modulating those mediated by conventional forms of intercellular communication, which employ all-or-none electrical impulses or chemical transmitter substances and neuromodulators. Other pioneering scientific contributions of Bullock include the discovery of two new sensory modalities: the **facial pit of pit vipers as an infrared receptor** which enables the snakes to detect temperature changes caused by prey animals whose temperature differs from that of the background surfaces; and **electroreceptors in weakly electric fish** and others which respond even to extremely weak electric signals with high specificity (see Chapter 7).

It is also characteristic of Ted Bullock that, throughout his life, he has been a true cosmopolitan. He never ceased to encourage international collaborations and the exchange of ideas across nations, and he himself has visited many laboratories and participated in a large number of international expeditions. Furthermore and rather unusual, he has stimulated others to enter new, promising research areas, without necessarily pursuing himself further work in this field. This is exemplified by the pioneering work of Ulla Grüsser-Cornehls and Otto-Joachim Grüsser on **complex recognition units** in the frog's optic tectum (see Chapter 6), which was initiated during a visit of the two German researchers to the laboratory of Ted Bullock in California.

In addition to his academic work, Ted Bullock has served as president of several societies, including the American Society of Zoologists, the Society for Neuroscience, and the International Society for Neuroethology. His book *Structure and Function in the Nervous Systems of Invertebrates*, co-authored by Adrian Horridge, is widely regarded as the most comprehensive and authoritative review of this topic ever written.

anyone else, championed the comparative approach, and whose research on a large number of different taxonomic groups led to the discovery of pit organs in pit vipers and of electroreceptors in weakly electric fish and other electrosensory animals. Each of these three prominent figures established a large school from which many distinguished neuroethologists originated.

However, due to technical limitations, the work of both the neurophysiologically oriented schools and the sensory physiology groups was restricted to the analysis of a few elements within the entire neural chain involved in the sensory processing and the generation of the motor output of a given behavior. Thus, it did not lead to an integrative understanding of the neural mechanisms underlaying a specific behavior.

Another reason for the initially slow development of neuroethology was that several, in themselves quite promising attempts to relate the action of neuronal assemblies to the behavior of the whole animal remained, for a considerable amount of time, rather isolated and were only reluctantly used by others. This was, for example, the case with the **focal brain stimulation technique**. This approach was developed and championed by **Walter Rudolf Hess** (1881–1973) of the University of Zurich, Switzerland, who was awarded the Nobel Prize in Physiology and Medicine in 1949. Using the focal brain stimulation technique, he examined in great detail how regions within the diencephalon control vegetative functions and various behaviors of cats.

The power of this approach for neuroethological research was demonstrated in the late 1950s and the beginning of the 1960s by Erich von Holst through an extensive series of experiments. He showed that stimulation of certain brain areas in chickens can evoke specific behavioral patterns. The exact type of the behavior elicited, and the degree of expression, depend upon both the intensity of the current applied and the presence or absence of an external stimulus. Stimulation of the hypothalamus in an alert chicken, for example, triggers attack of a stuffed weasel. The intensity of this attack behavior depends upon the current applied—without electrical stimulation the stuffed weasel evokes no special response.

The 'fathers' of neuroethology: Karl von Frisch, Hansjochem Autrum, Erich von Holst, and Ted Bullock.

The breakthroughs

Significant advances towards the establishment of neuroethology as a scientific discipline were made only in the 1970s and 1980s. This success was due to both the advent of new neurobiological methods and to the focus on simple and robust forms of behavior. The new methods allowed researchers to routinely trace neural pathways by employing a variety of *in vivo* and *in vitro* techniques; to characterize individual neurons through immunolabeling and *in situ* hybridization; to perform intracellular recordings and combine them with intracellular labeling techniques; and to apply selective agonists and antagonists of transmitters to characterize the physiological and pharmacological properties of neurons. The list of behaviors investigated by neuroethologists was rather short and included, in particular, rhythmic motor patterns. Even movements of internal organs, such as those of intestines, were intensively studied. Although the latter types of behavior are hardly considered worth studying by the majority of ethologists, their analysis led to the formulation of many useful concepts, such as those of identified neurons, central pattern generators, and modulators.

A breakthrough in the establishment of neuroethology was possible by focusing on simple and robust forms of behavior, and by applying modern neurobiological methods to explore the entire chain of sensory and neural mechanisms underlying these behaviors.

The major breakthroughs achieved making use of these new approaches included the work of **Jörg-Peter Ewert** on neural correlates of prey recognition in toads (see Chapter 6), of **Walter Heiligenberg** on the jamming avoidance response in weakly electric fish (see Chapter 7), of **Mark Konishi** on the sound localization in barn owls (see Chapter 6), of **Eve Marder** on the modulation of the motor pattern of the stomatogastric ganglion in decapod crustaceans (see Chapter 8), and of **Franz Huber** on auditory communication in crickets (see Chapter 10).

The formal establishment of neuroethology as a new discipline was completed by the appearance of its **first textbook** entitled *Neuroethology* by Jörg-Peter Ewert (first published in German in 1976 and in English in 1980) and by the foundation of the **Society for Neuroethology**, which held its first congress in Tokyo in 1987.

The future

In 1975, Edward O. Wilson set forth in his epoch-making book *Sociobiology: The New Synthesis* his vision of the future of behavioral biology. His prediction was that 'both [ethology and comparative psychology] are destined to be cannibalized by neurophysiology and sensory physiology from one end and sociobiology and behavioral ecology from the other.'

What is the status of this development a quarter of a century later? And, where will neuroethology head in the next 25 years?

As predicted by Wilson, neuroethology has, indeed, increasingly incorporated **neurophysiological** and **sensory physiological** techniques and concepts to explore the neural basis of behavior, frequently down to the level of single neurons or even single types of channels. As the data generated through such experiments become more and more complex, combining the information to produce testable models will be essential. In this development, **computational neuroscience**, a flourishing theoretical discipline, will play an important role.

In addition, **neuroendocrinology** has also significantly gained in terms of its impact upon neuroethology. Its influence is likely to increase even further due to the enormous advances made in the molecular characterization of hormones, peptidergic releasing factors, and their receptors. This has triggered the generation of novel pharmacological receptor agonists and antagonists that, yet largely unexplored, provide exciting tools to investigate the involvement of specific hormonal and peptidergic factors, as well as their receptors and receptor subtypes, in behavioral processes. The numerous investigations in this area have also made the seemingly clear-cut boundary between endocrine factors and transmitters to become more graded and continuous (see Chapter 8).

The largest gain over the next decades is to be expected from **molecular genetics**. In some favorable cases, it has now become possible to pinpoint the execution of certain behaviors of an animal down to the expression of single genes, or even single-point mutations of a gene.

Example: Naturally occurring strains of the nematode Caenorhabditis elegans are distinguished by their difference in feeding behavior. While worms of one group are solitary feeders, move slowly on a bacterial lawn, and disperse across it, individuals of the other group feed socially, move rapidly on bacteria, and aggregate together. As has been shown by Mario de Bono and Cori Bargmann of the University of California, San Francisco, this behavioral difference is due to two naturally occurring alleles of a single gene that differ by a single nucleotide. This gene, called npr-1, encodes a G protein-coupled receptor resembling the receptors of the neuropeptide Y family; it is referred to as NPR-1. One of the resulting two isoforms of this receptor, NPR-1 215F, contains the amino acid phenylalanine (designated 'F' in the one-letter code of amino acids) at position 215; this isoform is found exclusively in social strains. In the second isoform, called NPR-1 215V, the phenylalanine at position 215 is replaced by a valine (designated 'V' in the one-letter code of amino acids); this isoform is found exclusively in solitary strains. Position 215 is in the third intracellular loop of the putative receptor. This region is important for G-protein coupling in many seven-transmembrane receptors. Mutations in this loop are likely to result in changes in the presumptive neuropeptide pathway, which appear to generate the natural variation in behavior. The hypothesis that the difference in behavior is caused by this mutation in the NPR-1 protein has received strong support by transgenic experiments. They demonstrated that an NPR-1 215V transgene can induce solitary feeding behavior in a wild type social strain.

Does the increasing importance of molecular approaches and the focus of neuroethology on a few selected model systems and (seemingly) simple behaviors make behavioral research redundant? Certainly not! Neuroethological studies continuously generate an abundance of specific behavioral questions that wait to be answered through ethological approaches. On the other hand, results of such behavioral investigations, based on observations of and experiments on natural behavior often stimulate further research also at the neurobiological level. For these studies, molecular biology and neurobiology provide powerful tools through which behavior can be studied.

Like many areas of biology, neuroethology is at the verge of a new era. The current rapid development in computational neuroscience, neuro-endocrinology, and molecular genetics will provide exciting opportunities also for neuroethology. However, the success of neuroethology will not be determined just by the application of the corresponding techniques. Rather, its strength lies on integrating the information acquired through the different approaches for a true understanding of how the brain controls natural behavior in the whole animal. The ability to perform this **integration** will distinguish neuroethology from other disciplines.

> A distinctive feature of neuroethology is the attempt to gain an integrative understanding of how the brain controls behavior in the whole animal.

Summary

■ The study of animal behavior has its roots in natural history, which can be traced back to Aristotle. However, until the end of the nineteenth century, behavioral research was dominated by anecdotal approaches, as well as vitalistic, teleological, and anthropomorphic interpretation. Investigations were mainly concerned with finding evidence for the superiority of mankind, thus drawing a sharp dichotomy between the 'rational soul' of man and the 'sensitive soul' of beasts.

■ A more objective approach toward the study of behavior emerged only at the time of the announcement of Charles Darwin's evolutionary theory. His publications proposed a continuity of both morphological and behavioral characteristics within the living world, including man. This triggered an enormous interest in behavioral observations from a comparative perspective, thus marking the beginning of ethological research.

■ In ethology, the comparative observations were typically conducted under natural or semi-natural conditions. This was supplemented by the application of experimental approaches, as well as by the use of operational definitions and objective interpretations. Final establishment of ethology as an independent discipline within biology was achieved by Konrad Lorenz and Niko Tinbergen in the period between the 1930s and the 1950s. Their merit was to having placed the observational and experimental data accumulated at that time within a conceptual framework.

■ Parallel to the emergence of ethology, several mechanistically oriented schools developed. The theory of tropism of Jacques Loeb attempted to explain behavior in terms of tropisms—involuntary orienting movements. The reflex theory of Ivan Petrovitch Pavlov had a pronounced effect especially on the development of behaviorism. One direction within behaviorism led by John B. Watson proposed to explain behavior in physiological terms based on reflexes as the elementary behavioral unit. A second direction within behaviorism established by Burrhus F. Skinner stressed the importance of experimental manipulations and operational definitions to understand behavior. Main emphasis of this research direction was on learned behavior. The denial of innate components of behavior by some of the proponents of behaviorism resulted, for a considerable amount of time, in a sharp dispute with ethology, known as the 'nature versus nurture controversy'.

■ Neuroethology emerged only at the end of the twentieth century when scientists started to apply modern neurobiological concepts and techniques to the elucidation of neural mechanisms underlying simple forms of behavior in animals. Important contributions to this development originated from both neurophysiology and sensory physiology. Eminent figures in the early stages of this development were Karl von Frisch, Hansjochem Autrum, Erich von Holst, and Ted Bullock.

Recommended reading

Klopfer, P. H. (1974). *An introduction to animal behavior: ethology's first century. Second edition*. Prentice-Hall, Englewood Cliffs/New Jersey.
An introduction to ethology written from an historical perspective, containing a wealth of information on the development of scientific disciplines studying animal behavior.

Thorpe, W. H. (1979). *The origins and rise of ethology: the science of the natural behaviour of animals*. Heinemann Educational Books, London.
A vivid historical and personal account of the study of animal behavior, written by a British ethologist who contributed to this history himself.

Questions

2.1 Discuss the major historical reasons for the very slow progress made in applying an objective approach to the study of animal behavior. At what point was this difficulty finally overcome? Name at least three scientists who have made major contributions toward the establishment of the study of animal behavior as an empirical scientific discipline.

2.2 Two major scientific disciplines that had an enormous impact on the study of animal and human behavior in the twentieth century are behaviorism and ethology. What do these disciplines have in common, in what respect do they differ, and what approaches do they use? Name at least two prominent figures in each of the two disciplines and describe briefly their major contributions.

2.3 Sketch the historical development of neuroethology. Include in your essay a brief summary of the work of two prominent scientists who have greatly influenced the establishment of neuroethology as an independent discipline.

The tools and concepts of behavioral neurobiology

3

■ Introduction
■ Neurobiology: basic concepts and experimental approaches
■ Ethology: basic concepts and experimental approaches
■ Summary
■ Recommended reading
■ Questions

Introduction

The last chapter has demonstrated that behavioral neurobiology has its roots both in the behavioral sciences and in neurobiology. As a consequence, its own approaches are often derived from these two disciplines. The aim of this chapter is, therefore, to discuss some of the key concepts and methodologies of ethological and neurobiological research, with special emphasis on those tools that are of immediate relevance to behavioral neurobiology.

Neurobiology: basic concepts and experimental approaches

Cell theory versus reticular theory

While nowadays there is no doubt that the brain is composed of individual cells, it took a rather long time to firmly establish the **cell theory**. In the nineteenth century, a common hypothesis was that the brain is made up of nerve cells forming long thin processes that are fused together to form a continuous network—just as arteries and veins are linked by capillaries.

Cell theory: the nervous system is composed of individual cells.

Reticular theory: the nervous system forms a continuous network ('reticulum') of fused processes.

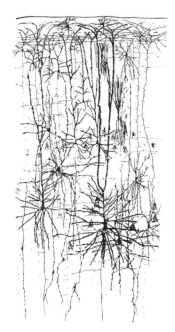

Figure 3.1 The principal neuronal types in the different layers of the motor cortex of a newborn human infant. The composite figure is based on drawings of individual neurons stained with the Golgi method. (After **Marin-Padilla, M.** (1987).)

Application of the Golgi method to histological sections of the nervous system provided strong support for the cell theory.

1 µm (micrometer) = 10^{-6} m.

This has become known as the **reticular theory** of nervous organization. By contrast, the opposing cell theory proposed the existence of entirely separate entities ('cells'). According to the latter hypothesis, the branches of the nerve cells terminate in **free nerve endings**.

The cell theory received an enormous boost after Camillo Golgi (1843–1926), an Italian physician, who invented a new histological staining technique, the **Golgi method,** in 1873. Its essential feature is a silver impregnation of all the components of a nerve cell, thus permitting visualization of a black neuron against a transparent background of unstained tissue. Even more important is the capricious staining behavior of this method: in a section, only a tiny fraction of the cells present are stained by the Golgi method (Fig. 3.1). The reasons for this staining selectivity are unknown. However, what may at first sight be considered a disadvantage turned out to be the major strength of the Golgi method. Due to the enormous structural complexity of the nervous system, a more complete staining of the tissue would make a study of individual cells utterly impossible.

It was the ingenious Spanish neurohistologist Ramón y Cajal (1852–1934) who realized the potential of the Golgi method to provide strong evidence in support of the cell theory. He further improved the staining technique and applied it to many parts of the nervous system. His reconstructions clearly showed individual cells. However, although the existence of 'discontinuities' in the nervous system was also strongly suggested by physiological findings, the cell theory received its final confirmation only in the 1950s when electron microscopy revealed the structure of such gaps at synapses.

Neurons and nervous systems

Neurons constitute the elementary units of nervous systems. Based on their size, shape, and the branching pattern of their processes, a large, and sometimes confusing variety of different neuronal types is defined. It is not the purpose of this section to consider this terminology further; rather, we will try to extract some of the salient features typical of neurons.

As in other cells, the **cell body** or **soma** (plural: **somata**) is the region of the cell that surrounds and includes the nucleus. The cytoplasmic region of the cell body that surrounds but excludes the nucleus is called the **perikaryon**. Depending on the animal and the type of neuron, the size of the soma ranges between approximately 5 µm and more than 1000 µm in diameter. The **nucleus,** as in other eukaryotic cells, contains the DNA and its associated proteins. During development, the nucleus provides DNA for mitotic replication. This function ceases, as fully differentiated neurons do not mitotically divide any more. A second important function of the nucleus, which is not restricted to developmental stages, is to transcribe DNA into RNA in order to synthesize proteins.

In most neurons, the cell body gives rise to several extensions called **dendrites**. They vary greatly in number, length, and branching pattern according to the type of neuron. Together with the cell body, dendrites constitute the major domain of the neuron that receives synaptic input. As a consequence, the arborization of the dendrites largely defines the receptive field of the neuron. An important dendritic specialization of certain neurons is the presence of **dendritic spines**, which are projections of the membrane from the surface of the dendrites. Typically, one spine is contacted by one chemical synapse. On some types of neurons, several tens of thousands of dendritic spines are present.

The second type of process found in neurons is the **axon**. Unlike dendrites, the axon may travel over considerable distances (up to several meters in some large animals). This process arises from a cone-shaped extension (called **axon hillock**) either of the soma or of a major dendrite. Axons, together with the cell body, constitute the major conducting unit of the neuron. The **projection** of a neuron is a term to describe the route of an axon from the area of origin to its target region.

> *Example: In fishes, amphibians, reptiles, and birds, ganglion cells of the vertebrate retina send axons to the **optic tectum** (the homologue of the **superior colliculus** in mammals). The ganglion cells are, therefore, said to project to the optic tectum. This axonal pathway is also referred to as a **retinotectal projection**. In an equivalent expression, retinal ganglion cells are said to **innervate** the optic tectum.*

In general, neurons that convey information from sensory organs in the periphery to the central nervous system are referred to as **afferent neurons**. Neurons involved in the control of motor action performed by muscles and glands are called **efferent neurons**. **Interneurons** are nerve cells whose cell bodies, dendrites, and axons are confined to the central nervous system. They may project over long distances. The term **local interneuron** indicates that the distance over which the axon travels is rather short—often just a few hundreds of micrometers.

Synapses

Generally, an axon has a number of branches called **collaterals**. Axonal enlargements (**varicosities**) specialized to transmit information to other cells are called **end feet, terminals,** or **boutons**. The entire point of contact between the two cells is known as the **chemical synapse** (Fig. 3.2). The terminal branches of an axon may form synapses with hundreds of other

The major parts of the neuron: soma, dendrites, and axon.

The projection of a neuron defines the route of its axon from the site of origin to the target region.

Figure 3.2 Illustration of a neuron with its various points of contact. Both the cell body and the dendrites receive synaptic input from excitatory synapses (white triangles) and inhibitory synapses (black triangles). The corresponding somatic and dendritic regions serve as postsynaptic elements. At its terminal region, the axon of the neuron forms synaptic contact with the cell bodies and dendrites of three other cells. There, the neuron is considered presynaptic, whereas the three other cells are postsynaptic. Also, note the insulation of the axon by myelin sheaths. They are interrupted at the nodes of Ranvier. For diagrammatic purposes, the relative dimensions of the axon and the synapses are considerably distorted. In reality, the axon is very thin and often extremely long. Moreover, the terminal branches of the axon may form synapses with as many as a thousand other neurons. (After **Kandel, E. R., and Schwartz, J. H.** (1985).)

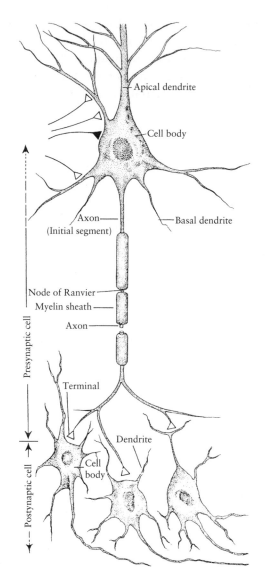

neurons. Each synapse consists of the following three major components:

- **presynaptic bouton,** which contains **transmitter substance** packaged in a large number of **synaptic vesicles;**
- the **postsynaptic element** of the recipient cells, which incorporates specific receptor elements for binding of the transmitter molecules;
- a space 20–30 nm wide between the two cells (the **synaptic cleft** or **synaptic gap**).

In the course of synaptic transmission, the synaptic vesicles fuse with the presynaptic membrane, and release their content into the synaptic cleft. The neurotransmitter molecules diffuse across the synaptic cleft to the postsynaptic neuron, where they bind to the receptor molecules

1 nm (nanometer) = 10^{-9} m.

embedded in the plasma membrane of the postsynaptic cell. The binding of the neurotransmitter molecule increases the permeability of the post-synaptic membrane to specific ions.

The structure of the chemical synapse makes this type of junction **polarized**: communication is primarily **unidirectional**; it occurs from the presynaptic element to the postsynaptic element.

A second type of specialized junction between two nerve cells is called an **electrical** or **electrotonic synapse**. The structure of the electrical synapse is identical to that of the **gap junction** found in many non-neuronal types of tissue. Both electrical synapse and gap junction are characterized by a close apposition of the plasma membranes of the two cells, leaving a gap of only 2–4 nm. The two cells are connected by tiny tubular structures that enable small molecules to pass directly from one neuron to the other. Thus, these areas of contact form a low-resistance site of communication between the two neurons—current flow arriving at the presynaptic membrane spreads directly into the postsynaptic neuron. In principle, this signal transfer can occur in either direction. It is, thus, **bidirectional**. Moreover, and again in contrast to the chemical synapse, there is no delay in the transmission of the signal across the electrical synapse, since no encoding of an electrical signal into a chemical signal, and back into an electrical one, is necessary.

The higher speed of signal transmission makes the electrical synapses well suited for the control of rapid movements, for example, those occurring in defensive behavior. On the other hand, chemical synapses, although slower, are more **plastic**. Their activity may be modulated by various factors, including the history of the previous activity and special substances called neuromodulators.

Axons may be surrounded by an insulating **myelin sheath** (Fig. 3.2). These sheaths originate from non-neuronal cells, rather than from the neuron itself. These non-neuronal cells belong to the class of **glia**. In the peripheral nervous system, it is the **Schwann cells** that provide the myelin. In the central nervous system, this function is performed by the **oligodendrocytes**. The myelin sheaths are interrupted by regions called **nodes of Ranvier**. A major function of the myelin sheath is to facilitate rapid impulse conduction in the axon. In vertebrates most axons are myelinated, whereas in invertebrates the majority of axons are unmyelinated. Loss of myelin has devastating effects on brain function, as evident in patients suffering from multiple sclerosis, a disease involving degeneration of myelin.

Resting potential

Nerve cells are specialized in conducting electric signals along the membranes of their axons. Recently, methods have been developed to measure membrane potentials by means of **voltage-sensitive fluorescent**

Synapses are specialized contact zones in the nervous system where one neuron communicates with another.

Myelin sheaths, originating from non-neuronal cells, provide insulation of axons and facilitate rapid impulse transmission.

dyes. The latter approach is particularly useful when studying electric events in organelles and cells that are too small to allow the use of microelectrodes (see below), for example in the case of dendritic spines, or when mapping variations in membrane potentials across excitable cells. One class of these probes operate based on changes in their electronic structure, and, thus, their fluorescent properties in response to alterations in the surrounding electric field. Probes of a second class exhibit potential-dependent changes in their transmembrane distribution, which are accompanied by changes in their fluorescent properties.

While the potentiometric optical probes are rather new developments, traditionally the generation and conduction of electric signals has been studied by intracellular and extracellular recordings using glass electrodes. Most of our knowledge about the physiology of excitable cells is based on this approach. **Intracellular recordings** were first made around the 1950s. To record from the inside of neurons, special glass micropipettes, called **microelectrodes**, are used (Fig. 3.3). These electrodes are made from glass capillaries heated in the middle and quickly pulled apart, so that fine capillary tubes result. The tip of these tubes is open and has a diameter of less than 1 μm. This allows the investigator to impale the neuron's cell body without destroying it and to place the tip of the microelectrode inside the cell. The capillary is filled with a salt solution, potassium chloride for example, to conduct electricity. At the wide end, a wire connected to both the salt solution and an amplifier is inserted. A second extracellular electrode is placed in the fluid that baths the tissue or the cell from which the recordings are made. By convention, the intracellular electrode is connected via the amplifier to the positive input terminal of an oscilloscope. The extracellular electrode is connected to the negative input terminal.

Intracellular recording: the tip of a microelectrode is placed inside the cell.

If both electrodes are placed outside a cell, no potential difference is recorded. The moment the intracellular electrode impales a cell, a potential difference of approximately -60 to -80 mV, with the inside being negative, does occur. This indicates that a neuron is polarized at rest. The resulting potential difference between the inside and the outside is called the **resting (membrane) potential**.

The membrane potential is caused by a **differential distribution of ions across the plasma membrane**, particularly of K^+, Na^+, and Cl^- ions. The unequal distribution is actively maintained by ionic pumps and exchangers. This results in K^+ to be at a much higher concentration inside the cell than outside the cell. Na^+ and Cl^- are more concentrated outside the cell. Due to these unequal concentrations, the ions tend to diffuse down their concentration gradients through specialized channels, called **ionic channels** or **ionophores**, which bridge the plasma membrane.

In the case of K^+, this movement causes the inside of the cell to become more negative, since the membrane is impermeable to the large anions

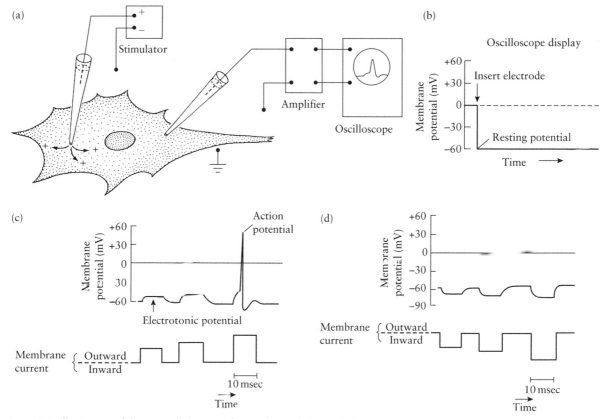

Figure 3.3 Illustration of the intracellular recording and stimulation technique.
(a) Arrangement of the stimulation and recording electrodes. A glass microelectrode, with a very fine tip, is placed inside the cell. A second electrode is positioned outside the cell. After amplification, the potential measured by the two electrodes across the cell membrane is fed into an oscilloscope. For stimulation, a second pair of electrodes is used. One of these electrodes is inserted into the cell, while the other is left outside in the surrounding fluid. These two electrodes are connected to a stimulator that can pass current into or out of the cell. (b) No current is applied. Initially, both recording electrodes are left outside the cell, thus, no potential difference between the two electrodes is measured. As soon as the microelectrode impales the cell, a potential difference of about -60 mV, reflecting the resting membrane potential, is recorded. (c) Injection of positive charge into the cell results in a decrease of the membrane potential. This is also referred to as depolarization. The degree of this depolarization varies proportionally with the amount of positive current passed into the cell. However, as soon as a threshold of approximately 15 mV of depolarization is reached, a graded potential is no longer recorded. Rather, a new, active type of electrical response is generated. This type of signal is called action potential. (d) Reversal of the direction of current flow by withdrawing positive charge from the inside of the cell leads to an increase in the potential difference measured across the cell membrane, thereby hyperpolarizing the cell.
(After **Camhi, J. M.** (1984) and **Kandel, E. R., and Schwartz, J. H.** (1985).)

Figure 3.4 The equilibrium potential as the result of a concentration gradient and voltage gradient across the membrane. In neurons, K$^+$ ions are actively concentrated inside the cell. Due to a diffusion process, they flow down the concentration gradient from the inside to the outside of the cell. However, the negative potential inside the cell counteracts this diffusion by attracting the positively charges K$^+$ ions. At the equilibrium potential, these two forces balance one another. In the squid giant axon, this potential is −76 mV. (After **Zigmond, M. J., Bloom, F. E., Landis, S. C., Roberts, J. L., and Squire, L. R.** (eds.) (1999).)

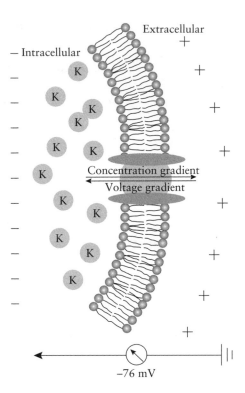

inside the cell, preventing them from following the K$^+$ ions across the membrane. This leads to an increase in negative force inside the cell, so that at one point the electrostatic attraction of the positively charged K$^+$ ions by the negative membrane potential inside the cell will balance the thermodynamic forces causing the K$^+$ ions to move down the concentration gradient. At this point, there is no net flow of K$^+$ ions. The membrane potential at which this equilibrium is reached is referred to as the **equilibrium potential**. This situation is summarized in Fig. 3.4.

Among other factors, the equilibrium potential depends on the relative concentrations of the respective ion inside and outside the cell, and can be calculated using the Nernst equation (named after the German physico-chemist Walter Nernst). For a monovalent, positively charged ion at a temperature of 20 °C, the equation can be simplified into the following form:

$$E_{ion} = 58.2 \log\frac{[\text{ion}]_{outside}}{[\text{ion}]_{inside}}$$

where E_{ion} denotes the equilibrium potential, $[\text{ion}]_{outside}$ the concentration of the respective ion outside the cell, and $[\text{ion}]_{inside}$ the concentration of the ion inside the cell.

Exercise: *Calculate the equilibrium potential of the K$^+$ ions for a classic physiological preparation, the squid giant axon. The concentrations of K$^+$ are 400 mM inside and 20 mM outside the axon.*

Solution: Using the above simplified Nernst equation, an equilibrium potential of −76 mV, with the inside negative, results.

Similar consideration apply to other ions. In particular, Na^+ and Cl^- are important due to their relatively high concentrations. Therefore, the membrane potential is somewhat in-between the equilibrium potentials of these three ions. In case of the squid giant axon, typically a membrane potential of approximately −65 mV is measured.

Generation of action potentials

If, during intracellular recordings, positive current pulses are passed through the membrane of the cell with the aid of an additional intracellular (**stimulation**) **electrode**, a reduction in the potential difference across the membrane is observed. This reduction, called **depolarization**, brings the membrane potential closer to zero. An increase in membrane potential, on the other hand, is referred to as **hyperpolarization**.

For small current pulses, the membrane potential V increases roughly linearly with the increase in current flow I in accordance with Ohm's law:

$$V = R \cdot I,$$

whereby R denotes the membrane resistance. Since within a certain range any increase in current flow is truly reflected by a proportional increase in membrane potential, the resulting voltage differences are termed **graded potentials**.

If, however, the depolarization reaches a critical **threshold** level, the potential across the membrane does not continue to increase linearly with current. Rather, a completely new type of signal called **action potential** (often also referred to as **impulse** or **spike**) is generated. The action potential is characterized by a rapid increase in voltage, crossing the zero mark, and becoming transiently positive (**overshooting potential**). The membrane potential, then, rapidly returns toward the resting value, often accompanied by a brief undershooting.

Action potentials are distinguished in that they are not graded. Rather, they have a constant amplitude and are of constant duration. As long as a depolarizing stimulus causes the membrane potential to exceed the threshold voltage, an action potential is produced. If, however, the threshold value is not reached, no action potential at all is generated (**all-or-nothing property**).

New action potentials cannot be generated before the previous one is completed. There is also a brief **refractory period** following each action potential. As a consequence, the rate of impulses produced has an upper limit of about 1000 spikes/sec.

1 mV (millivolt) = 10^{-3} V (volt).

An action potential is a transient depolarization of the membrane potential generated in an all-or-none fashion.

In the neuron, action potentials are initiated by a rapid influx of the ions into the cell through activation of Na^+ channels. The reversal of the depolarizing potential is achieved by inactivation of the Na^+ channels and by efflux of K^+ ions through activation of K^+ channels. The inactivation of the Na^+ channels is the cellular cause of the refractory period. As soon as the inactivation of these channels is removed, the portion of the cell with these channels is ready for the next action potential.

Conduction of action potentials

The conduction of action potentials along an axon can readily be demonstrated by **extracellular recordings**. Two wires are hooked, a small distance apart from each other, on the outside of the axon. These wires are connected via an amplifier to an oscilloscope. Then, the following can be observed (Fig. 3.5):

- Before the action potential has arrived at the electrodes, no potential difference between the two wires is observed.
- As the negative charge on the outside of the cell passes the first electrode, this electrode becomes negative, in relation to the second electrode.
- As the negative charge passes on, the second electrode also becomes more and more negative, thus diminishing the potential difference between the two electrodes. This becomes visible as a reversal of polarity.
- As the negative charge propagates further, the negative charge recorded by the second electrode exceeds more and more the negative charge encountered at the first electrode. At this point, the first electrode 'sees' a positive charge with reference to the second electrode.
- Further propagation of the action potential beyond that point gradually reduces this potential difference.
- When, finally, the negative charge has completely passed the two electrodes, the potential difference reaches zero again.

Extracellular recording: two electrodes are placed outside the cell.

The overall result of this recording is displayed on the oscilloscope as the typical waveform associated with an action potential, although the actual shape depends on several factors, including the exact placement of the electrodes. On a slower time scale, the entire width of an action potential is reduced to a thin stick visible as a rapid deflection from the base line. This slow display mode allows the experimenter to visualize simultaneously many spikes and to analyze both their rate of production and their inter-spike interval pattern.

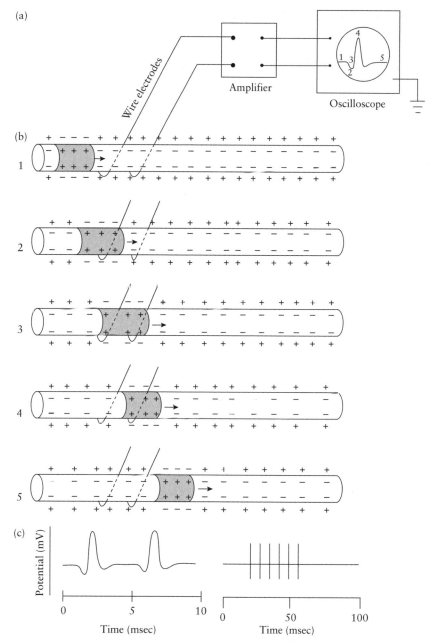

Figure 3.5 Principle of the extracellular recording technique and the conduction of an action potential. (a) Two wire electrodes are hooked on the outside of an axon. The signal obtained is amplified and fed into an oscilloscope. (b) The spreading of an action potential alongside the axon (its location is indicated by the gray-shaded portion) is shown at five consecutive moments. The resulting fives phases of the action potential are indicated by the corresponding numbers in the oscilloscope trace. (c) Oscilloscope traces of action potentials at faster (left) and slower (right) time scales. (Courtesy: G. K. H. Zupanc.)

Synaptic potentials and synaptic transmitters

The concept of the chemical synapse is intimately linked to that of neurotransmitters. Based on the proposal of Santiago Ramón y Cajal that the nervous system is composed of individual nerve cells (see above), the English neurophysiologist Charles Sherrington introduced, in 1897, the term **synapse** to describe the structure mediating transmission of nerve impulses from one cell to the next. However, it still remained unknown how information was transmitted across the synaptic cleft between two cells. Observations especially by a group of English physiologists headed by John Langley at the beginning of the 1920s suggested the involvement of chemical substances in this process. This was confirmed when the Austrian pharmacologist Otto Loewi published the results of his classic experiment in 1921. He demonstrated that a non-stimulated frog heart can be stimulated by applying ventricular fluid from a stimulated heart onto it. Since then, a large number of chemical substances, called **transmitters,** have been identified and characterized.

Transmitters are released from a specialized zone, often situated in the terminal region of the axon of the presynaptic neuron, and subsequently bind to specific receptors at the postsynaptic neuron. The main effect of the binding of the transmitter substance to the postsynaptic receptor is the generation of a **postsynaptic potential** (Fig. 3.6). If this potential is depolarizing, and thus brings the membrane potential closer to the threshold for generation of action potentials, it is called an **excitatory postsynaptic potential** (EPSP). If, on the other hand, the binding of the transmitter substance keeps the membrane potential away from reaching the threshold for spike generation, it is termed **inhibitory postsynaptic potential** (IPSP). As a final consequence, the effect of EPSPs is to increase the probability of generation of action potentials in the postsynaptic cell, whereas IPSPs decrease the probability of production of action potentials in the postsynaptic cell.

Excitation and inhibition are caused by ion flow through channel proteins incorporated into the postsynaptic membrane. These ionophores form the receptor complex, together with the transmitter-binding component. In general, binding of excitatory transmitter leads to opening of channels that are relatively nonspecific for the cations Na^+, Ca^{2+}, and K^+. The Na^+ and Ca^{2+} ions have inward concentration gradients across the membrane, letting these ions flow into the cell. The concentration gradient for K^+ is in the opposite direction, causing K^+ ions to move out of the cell. Since the permeability for Ca^{2+} is low compared with that of the other two ions, the EPSP is largely due to the simultaneous flows, in roughly equal proportions, of the Na^+ ions into the cell and the K^+ ions out of the cell. The action of these two ions generates a combined equilibrium potential of the EPSP that lies roughly midway between the equilibrium

Binding of transmitter to postsynaptic receptors leads to the generation of excitatory postsynaptic potentials (EPSPs) or inhibitory postsynaptic potentials (IPSPs).

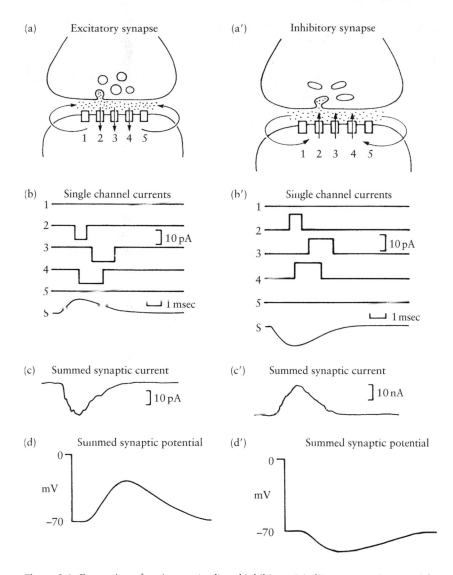

Figure 3.6 Generation of excitatory (a–d) and inhibitory (a′–d′) postsynaptic potentials (EPSPs and IPSPs, respectively). In each of the two cases, transmitter is released into the synaptic cleft by fusion of the presynaptic vesicles with the presynaptic membrane (a, a′). The transmitter diffuses across the synaptic gap and binds to some postsynaptic receptors (here: the receptors labeled 2–4). This opens conductance channels, which lead, at the excitatory synapse, to inward flow of positive ions, and to an outward flow across neighboring parts of the postsynaptic membrane. Recordings of the activity of the individual channels by means of the so-called patch-clamp technique reveals opening of the channels 2–4, as indicated by the step-wise increase of negative (b) and positive (b′) current. These currents of the individual channels sum up to produce a summed synaptic current (c, c′). Finally, the summed synaptic currents generate the summed EPSP and IPSP, respectively (d, d′). (After **Shepherd, G. M.** (1988).)

potential for Na^+ and K^+, namely at approximately 0 mV. Thus, a depolarization of the membrane potential occurs.

Inhibitory transmitters, by contrast, activate either Cl^- or K^+ channels. Since Cl^- ions are at much higher concentration on the outside of the cell than on the inside, opening of Cl^- channels leads to movement of Cl^- into the cell and, thus, to a net increase of the negative charge on the inside of the membrane. As a result, the membrane becomes hyperpolarized, producing an IPSP. A similar effect is achieved by an outward flow of K^+ ions.

Some transmitters can cause either EPSPs or IPSPs, depending on the type of ionophore opened in response to binding of the transmitter molecule. Further, generation of EPSPs is possible not only by opening ion channels, but also by closing K^+ channels that are open at rest. In contrast to channel opening, the closing of receptor channels is achieved through cyclic AMP or other second messenger systems. The time course of this action is much slower (typically in the order of seconds or minutes) than that of EPSP generation, due to the opening of channels (typically in the order of milliseconds). Such slow synaptic actions play an important role in modulation of neuronal activity.

Commonly, synapses are grouped into various types according to the transmitter released by the presynaptic element. They are further divided into subtypes based on the effect of various **agonists** that mimic the action of the transmitter molecule or of various **antagonists** that inhibit the action of the transmitter molecule. Application of agonists and antagonists has proven an extremely powerful tool to study details of the processes involved in synaptic transmission.

> *Example: Synapses using the transmitter **acetylcholine** are called* **cholinergic**. *They are subdivided into two types,* **nicotinic and muscarinic**, *named after the agonists nicotine and muscarine, respectively, that simulate the action of acetylcholine, but bind to different types of cholinergic receptors.*

According to their chemical nature, various types of transmitters are distinguished:

- **Acetylcholine** is the major transmitter at the neuromuscular junction. It is also used by some neurons in various regions of the central nervous system.

- **Amino acid transmitters** include substances like **glutamate, glycine,** and **γ-aminobutyric acid (GABA)**. Glutamate is one of the major excitatory transmitters in the central nervous system. The receptors of the glutamatergic synapses fall into two main categories. The first category is distinguished by binding of the agonist *N*-methyl-*D*-aspartate (**NMDA receptors**), the second by binding of quisqualate and kianate (**non-NMDA receptors**). Iontophoresis of glutamate or one of these agonists into brain regions with glutamatergic synapses causes depolarization of neurons. GABA and glycine are common inhibitory transmitters in the brain.

- **Biogenic amines** are molecules with an amine group. They form three subgroups: **catecholamines** (**dopamine; norepinephrine**, also referred to as **noradrenaline;** and **epinephrine**, also known as **adrenaline**); **serotonin**, also called **5-hydroxytryptamine**; and **histamine**. Dopamine and serotonin often act as **modulators**. Thus, they exert a modulatory action on other neurotransmitters, rather than having a direct excitatory or inhibitory effect of their own.

- **Neuropeptides** have only been recognized in the last few decades to be involved in synaptic transmission. Originally, their distribution was thought to be restricted to the hypothalamus. Since the 1970s and 1980s, however, an increasing number of neuropeptides have been discovered, mainly by immunohistochemistry and *in situ* hybridization (see below), in many extrahypothalamic areas of the vertebrate brain and in the nervous system of numerous invertebrates. Currently, about a hundred neuropeptides have been isolated and characterized. Like dopamine and serotonin, they commonly act as neuromodulators. In Chapter 8, we will discuss some of the reasons that make them so well suited to exercise such a function.

Common transmitters/modulators: acetylcholine, GABA, glycine, dopamine, norepinephrine, epinephrine, serotonin, histamine, and a large number of neuropeptides.

Sensory systems

Sensory organs act as interfaces between an animal's environment and its central nervous system. They convert a specific form of energy from the external or internal environment into neuronal activity. This process is called **transduction**. The type of energy able to evoke such a neuronal response differs according to the **sensory modality**. In the auditory system of mammals, for example, hair cells of the cochlea transduce mechanical energy. In the visual system, rods and cones of the retina transduce light energy.

The cells within sensory systems that perform this task are specialized neurons called **receptor cells**. These cells transmit information to ganglion cells that send axons to the central nervous system. In some sensory systems, the axons of the ganglion cells are specialized and form the receptive structure. In other systems, the auditory system for example, the signals produced by the receptor cell are transmitted to a process of the ganglion cell. In the visual system, an interneuron is interposed between the retinal receptor cells and the ganglion cells. Since one receptor cell, typically, sends, information to more than one ganglion cell (principle of **divergence**), and one ganglion cell receives information from more than one receptor cell (principle of **convergence**), a considerable amount of computation is performed by the time a signal reaches the ganglion cell.

Non-neuronal cells within the sensory organ form accessory structures. Lens and cornea in vertebrate eyes, for example, are the product of such non-neuronal cells. They are important for the generation of an image on the retinal cells.

Sensory organs may not only respond to energy differing according to the particular sensory modality involved; they also mediate different **qualities** of perception within this modality. Therefore, different subtypes of receptors exist that transduce energy associated with these different qualities.

Example: Many animals possess the ability to distinguish light of different wavelengths (which corresponds to the subjective perception of different colors). This sensory ability is based on the existence of different subtypes of receptor cells (in human, three types of cones) that are maximally sensitive to light of different wavelength.

In all sensory systems examined thus far, the result of the transduction process is a change in the conductance of membrane channels, which leads to a change in the membrane voltage of the receptor cell. This voltage is referred to as **receptor potential** or **generator potential**. Like the EPSPs and IPSPs at chemical synapses, it is graded.

The individual steps of the transduction process are especially well studied in the visual system. Absorption of photons (the elementary physical units of light) by receptor cells triggers a cascade of biochemical processes. At the end of this cascade, a massive closure of Na^+ channels occurs, causing a decrease in the resting potential (typically, -20 to -40 mV, as measured in complete darkness), and thus, hyperpolarization.

In other sensory systems, the transduction of energy may lead to an opening of Na^+ channels, and, therefore, to depolarization. And in still other systems, the response can be even biphasic, that is, one type of stimulation causes a hyperpolarizing response, whereas a different type of stimulation elicits a depolarizing response from a receptor cell.

In any sensory system, the process of transduction involves considerable amplification of the energy associated with the stimulus.

J = joule; A = ampere; V = volt; 1 pA (picoampere) = 10^{-12} A.

Example: The energy of one photon of 500 nm wavelength, which is equivalent to $4 \cdot 10^{-19}$ J, is sufficient to excite an isolated rod cell. The receptor response consists of changes in current of roughly 1 pA and in potential of roughly 1 mV, and lasts for several seconds. Since the energy is defined as the product of current, voltage, and time (1 joule = 1 ampere·volt·second), this change in energy is equivalent to roughly $1 \cdot 10^{-14}$ J. Thus, amplification at least by factor 10 000 occurred!

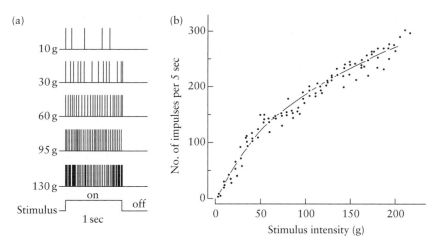

Figure 3.7 Pressure receptor in the foot of a cat. Pressure of various strengths is applied for 1 sec. Impulse activity is recorded from afferent fibers. (a) Examples of recordings. The stimulus intensity is indicated on the left. (b) The number of impulses produced in response to a pressure stimulus increases with increasing stimulus intensity. (After **Penzlin, H. (1980).**)

Upon transmission to second-order neurons, the graded receptor potential is translated into a series of impulses (Fig. 3.7), a mechanism commonly referred to as **frequency coding**. Thus, the amplitude of the receptor potential is closely related to the rate of action potentials in the second-order neurons. The information contained in these impulses is carried to the central nervous system for further processing.

Receptor cells within sensory organs transduce the energy associated with a stimulus into a receptor potential. Upon transmission to second-order neurons, the graded receptor potential is translated into a series of impulses.

Elucidation of neuronal connections

One important task of neuroethological research is to reveal the **connections** within the neural network involved in the control of a particular behavior. This is most commonly achieved by neuroanatomical means, especially by the application of **neuronal tract-tracing techniques**.

Some principles that we need to consider when interpreting the results of tract-tracing experiments will be discussed using the hypothetical neural circuitry shown in Fig. 3.8. Suppose we know the location of a cluster of neuronal cell bodies, defining Nucleus A, in the brain. We are now interested in identifying the target site(s) to which the cell bodies of Nucleus A send their axons. In other words, we would like to elucidate the **projection pattern** of neurons comprising Nucleus A.

We can get a first hint about the location and nature of these target sites by applying an **anterograde tracer substance** into Nucleus A. Anterograde tracer substances are taken up by neuronal somata and transported in anterograde direction (i.e. in the direction of the normal information flow) within the axon to the terminal region. Today, a large variety of such tracer substances is available. Some of them, such as fluorescently labeled dextrans, can be visualized directly using fluorescence microscopy after cutting thin sections through the tissue of interest. Other substances, such as cobalt chloride or horseradish peroxidase, require a histochemical or enzymatic reaction to visualize labeled neurons.

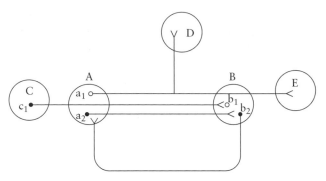

Figure 3.8 Hypothetical neuronal connections of two nuclei (denoted A and B). Both Nucleus A and Nucleus B contain subpopulations of neurons (labeled a_1 and a_2, as well as b_1 and b_2, respectively). Axons and their collaterals are represented by lines connecting the cell body with the terminal region(s). For further explanation, see text. (Courtesy: G. K. H. Zupanc.)

In our example, application of an anterograde tracer substance to Nucleus A would reveal labeled axonal terminals at sites B, D, and E. Is this result sufficient to get a definite picture of the projection pattern of the neurons of Nucleus A? Not necessarily, since not only axons originating from the cell bodies in Nucleus A may have taken up the label, but also axons running through Nucleus A (so-called **fibers-of-passage**). Obviously, the latter axons are not directly associated with the somata of Nucleus A, as can be seen from the figure. In our example, they arise from somata in Nucleus C.

To decide whether the labeled axonal terminals, indeed, originate from neurons in Nucleus A, we apply, in a second series of experiments, a **retrograde tracer substance** to the putative target areas B, D, and E. Retrograde tracers are taken up by axons and preferentially transported in retrograde direction from the axonal terminal region toward the cell body. Sometimes, the same substance as employed for anterograde tracing can be used, if it does not exhibit a significant preference for transport in anterograde direction. Also, after having reached the soma, some retrograde tracer substances enter the dendritic processes and may, thus, provide valuable information not only about the structure of the soma, but also about the dendritic morphology of the labeled neuron.

In our virtual experiment, application of a retrograde tracer substance to Nucleus B will show labeled somata not only in Nucleus A (defining the subpopulation a_2), but also in Nucleus C (defining the population c_1). As this analysis and a series of further anterograde and retrograde neuronal tract-tracing experiments will show, a subpopulation of cell bodies in A, referred to as a_1, project to Nucleus E, and, via collaterals, to Nucleus D, whereas a second subpopulation of neurons, called a_2, project to Nucleus B. Neurons of the subpopulation c_1 in Nucleus C also project to

Connections between different brain areas can be revealed using anterograde and retrograde tracer substances.

Nucleus B, but are not related to Nucleus A, except that their axons run through the latter nucleus. On the other hand, a subpopulation of neurons called b_2 exist in Nucleus B that project back to Nucleus A.

While the principles of neuronal tract-tracing are simple, in reality the researcher is often confronted with a number of problems requiring meticulous experimentation and careful interpretation of the results obtained. For example, the actual application of the tracer substance may not be restricted to the area of interest, but include adjacent areas. This could be due to the small size of the area of interest, making it difficult to precisely localize this brain region. Problems are also frequently caused by diffusion of the tracer substance from the application site into neighboring brain regions, especially if the tracer substance is highly hydrophilic. Thus, careful inspection of the application site after histological processing of the tissue is required to assess the validity of the labeling results.

Nowadays, several approaches are available to conduct neuronal tract-tracing experiments. The *in vivo* **approach** involves the administration of the tracer substance to the region of interest in the anesthetized animal, followed by killing the animal after a certain survival time. The length of this period depends on both the properties of the tracer substance (e.g. its molecular weight) and the distance over which the tracer has to be transported. *In vitro* **labeling** of neuronal connections is conducted on isolated pieces of neural tissue which are kept alive in oxygenated artificial cerebrospinal fluid. The advantage of this approach is that tracers can often be applied under visual control. On the other hand, such isolated pieces of tissue typically survive only for a few hours or a few days, thus limiting the distances over which connections can be traced. Neuronal connections can even be **traced in fixed tissue** using lipophilic dyes, which diffuse at very low velocity along the membranes of axons. All three approaches, available at both the light microscopic and ultrastructural level, can be combined with immunohistochemical or *in situ*-hybridization techniques (see below) to allow the neuroethologist to further characterize the neurons identified through tracing experiments.

Localization of tissue constituents

At the beginning of neuroethology, exploration of the structural correlates of behavior was often limited to the identification of individual neurons or neuronal assemblies involved in the control of a specific behavior. As we have seen above, this is commonly achieved through neuronal tract tracing, or combination of physiological approaches with intracellular labeling techniques. While the identification of the network components is still a major goal of neuroethological research, studies have been greatly extended since the 1980s to further characterize these components, especially in terms of their biochemical constituents.

Let us, for example, assume that we have obtained physiological and anatomical evidence that a certain neuronal cell cluster is crucially involved in the processing of sensory information relevant for the execution of a behavior. In the next step, we would like to know which neurotransmitters, neuromodulators, receptors, and second-messenger systems participate in this process. Two methods to address this issue are immunohistochemistry and *in situ* hybridization. The principles of these two techniques will be described in this and the following section.

Immunohistochemistry

Immunohistochemistry is based on the identification and localization of a cellular constituent (a so-called **antigen**) in a tissue section by labeling with a specific **antibody** (Fig. 3.9). The antibodies are generated by injecting the purified antigen (or a portion of the synthesized molecule) into a host animal (e.g. a rabbit). The immune system of this host animal recognizes the injected substance as a foreign compound and produces antibodies against it. When these antibodies are applied to a tissue section, they will bind to sites which contain the antigen.

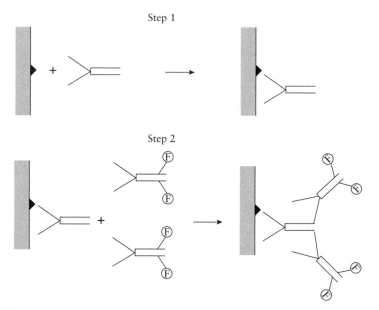

Figure 3.9 Immunohistochemical staining using secondary antibodies conjugated to a fluorophore. In the first step, the primary antibody binds specifically to a tissue site containing the antigen (symbolized by the filled triangle). In a second step, fluorescently labeled secondary antibodies bind to the constant fragment of the primary antibody. The fluorophore is indicated by the encircled 'F.' Since more than one secondary antibody can bind to a primary antibody, an enhancement of the signal results. (Courtesy: G. K. H. Zupanc.)

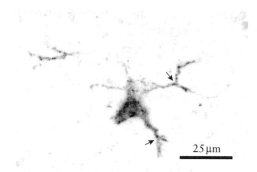

25 µm

Figure 3.10 Image taken of a cell (presumably an astrocyte) in the cerebellum of a teleost fish labeled with primary antibodies raised against the neuropeptide somatostatin, followed by binding of secondary antibodies conjugated to the fluorescent dye Cy3. The morphology of the labeled cell body and its processes, including the bifurcation pattern (arrows), is clearly visible. The cytoplasm of the cell contains packages of immunoreactive material. (Courtesy: G. K. H. Zupanc.)

The bound **primary antibodies** are characterized by two domains: one specific to the antigen and another one specific to the host animal. As a consequence of the latter feature, antibodies generated in a particular host animal, even if directed against different antigens, share a so-called constant fragment domain. To visualize the bound antibodies, **secondary antibodies** are used which bind to the host-specific constant fragment part of the primary antibody. If, for example, the primary antibody was raised in a rabbit, anti-rabbit secondary antibodies are required for binding to the primary antibody. Secondary antibodies are commonly conjugated to a **fluorescent label** (a fluorophore) or an **enzyme** which catalyzes an enzymatic histochemical reaction to generate a visible reaction product. When excited with a certain wavelength, the fluorophore emits a specific fluorescent color which can be viewed using fluorescence microscopy (Fig. 3.10). Reaction products of enzymatic reactions are directly visible under bright-field microscopes.

Demonstration of more than one antigen in the same tissue preparation can be achieved through **multiple immunostaining techniques**. Most commonly, these methods involve use of antibodies raised against different antigens in different host animals. Thus, for visualization, secondary antibodies conjugated to different chromophores (molecules that are responsible for the antibody's color) and directed against the host-specific constant portion of the respective antibodies can be employed. As a result, each of the different antigens is indicated by a different color.

To illustrate the multiple immunostaining technique, let us assume we are interested in demonstrating the existence of two neuropeptides—substance P and neuropeptide Y—in the same tissue section. We could

Immunohistochemical labeling involves binding of a primary antibody to an antigenic site in the tissue section, followed by secondary (and sometimes tertiary) reactions to visualize the binding site. It allows the researcher to study cellular constituents.

then apply to the tissue section a rabbit anti-substance P antiserum (i.e. antibodies directed against substance P, which were raised in rabbit) and a goat anti-neuropeptide Y antiserum (i.e. antibodies directed against neuropeptide Y, which were raised in goat). For visualization, we may use a sheep anti-rabbit FITC secondary antibody (i.e. an antibody raised in sheep against the constant fragment characteristic of rabbit antibodies and conjugated to the fluorophore fluorescein isothiocyanate, abbreviated FITC), and a horse anti-goat Cy3 secondary antibody (i.e. an antibody raised in horse against the constant fragment characteristic of goat antibodies and conjugated to the fluorophore Cy3). Using the proper excitation wavelengths, the FITC emits an apple-green fluorescence, and the Cy3 a red fluorescence. Therefore, green labeled structures in the tissue section indicate the presence of substance P, while red labeled structures point to the existence of neuropeptide Y.

In situ hybridization

In situ hybridization is the method of choice to study the distribution and density of particular sequences of nucleic acids, most commonly of messenger RNA (mRNA), in tissue. This technique was first employed in the late 1960s and early 1970s. It is based on the ability of single-stranded nucleic acid molecules to form a double-stranded duplex with any nucleic acid that contains a sufficient number of complementary base pairs (bp) (Fig. 3.11).

The three main classes of probes used for detection of the target RNA are complementary DNA (**cDNA**), RNA (**riboprobes**), and **oligonucleotides**. Riboprobes and cDNA probes are obtained through molecular cloning procedures, whereas oligonucleotides are produced by DNA synthesizers. Riboprobes and cDNA probes are typically very long, sometimes over 1000 bp. Oligonucleotides, by contrast, are rather short, generally consisting of 15–20 bp. They are targeted only at a portion of the entire RNA sequence.

Since the cDNA with its 'insert,' that is, the region complementary to the target mRNA, is usually contained within plasmid DNA and, thus, double-stranded, the insert DNA must be cut out of the plasmid DNA by means of restriction enzymes and denatured by boiling to obtain single-stranded DNA. Only half of the strands—those that are complementary to the mRNA—are able to hybridize. Denaturation is not necessary when using riboprobes, since they are already single-stranded. These RNA molecules are produced from cloned cDNA through reverse transcription procedures.

To detect the bound nucleic acid probes, they are labeled either with a **radioisotope** or an otherwise detectable molecule. The radioactively

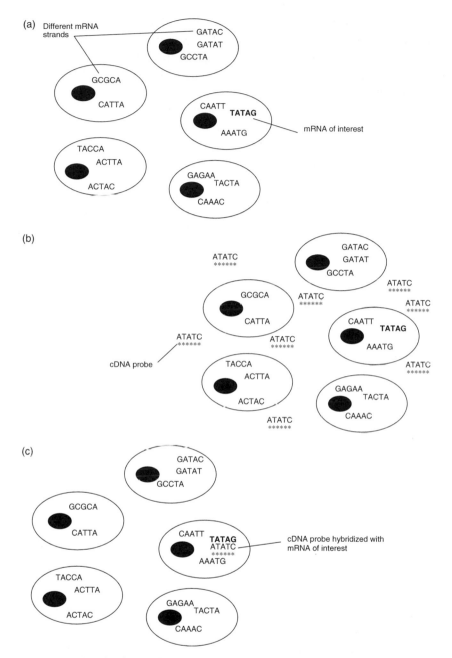

Figure 3.11 Principle of *in situ* hybridization using cDNA probes. (a) Whereas different cells may express different genes, and thus transcribe different mRNA molecules, the *in situ* hybridization technique enables the researcher to label specifically the mRNA of interest (indicated by **bold** typeface). (b) Fixed cells are permeabilized and incubated with a cDNA probe, whose sequence is complementary to the mRNA of interest. The probe is labeled (indicated by ***) for later identification. (c) The cells are washed to remove unbound probe. The remaining cDNA probe has hybridized to the mRNA of interest. This allows the researcher to identify and localize the cells expressing that mRNA. (Courtesy: J. Crowe.)

labeled nucleic acid bound to the target mRNA is visualized by X-ray film placed on top of the tissue section or by emulsion dipping of the microscopic slide with the tissue section. A 'foreign' molecule commonly used for non-radioactive labeling of nucleic acid probes is **biotin**. This glycoprotein binds, with high affinity, to **avidin**. The avidin is conjugated to rhodamine, for example, for detection by fluorescence microscopy. For detection by bright-field microscopy, the biotin can be conjugated to peroxidase, which is then employed in an enzymatic reaction.

The *in situ* hybridization technique allows the researcher to study the **expression of specific genes** by single cells. It provides information about the distribution of these cells and about the level of expression. Since mRNA is restricted to somatic regions of cells, only cell bodies are labeled by the *in situ* hybridization technique. The level of expression can be assessed, at least in semi-quantitative terms, by the density of label associated with the cell expressing the respective gene.

The information provided by *in situ* hybridization complements the results of immunohistochemical procedures. While *in situ* hybridization aims at detecting the message, immunohistochemistry is used to detect the resulting cellular protein. There are many instances in which the mRNA level does not correlate well with the protein level. In some extreme cases, high levels of message but no corresponding protein have been detected. This can be interpreted as an indicator of a high level of gene expression, paralleled by a rapid turnover of the resulting protein. On the other hand, presence of the protein but absence of the corresponding message suggests that the expression level is too low to be detected, whereas the protein—although produced in low numbers—is very stable, and thus accumulates over time in detectable amounts in the cell.

Selective elimination of cells

In many neurobiological investigations it is desirable to selectively kill a single cell, without destroying the surrounding tissue. The effect of the destruction can then be examined to determine the function of this cell.

In 1979, John Miller and Allen Selverston of the University of California, San Diego, introduced a technique which allows the researcher to achieve this goal. Single cells are filled with a fluorescent dye, such as Lucifer Yellow. This technique is commonly referred to as **photoinactivation**. Then, the dye-filled cell is killed by irradiating the tissue with light of a wavelength of maximal absorption, in the case of Lucifer Yellow with blue light. Within a few minutes, the cell is killed, as indicated by the absence of action potentials produced by the cell and postsynaptic potentials in the follower cells. Other cells in the vicinity of the dye-labeled cell are not affected by this treatment. The mechanisms by which the irradiation kills the dye-filled cell is unknown. It has been

In situ hybridization involves the hybridization of a labeled probe complementary to an mRNA sequence of interest. This technique allows the researcher to study the expression of specific genes in cells at the mRNA level.

speculated that the production of heat through photoabsorption or the generation of a toxic substance through photodecomposition of the dye causes the destruction of the cell.

Ethology: basic concepts and experimental approaches

Sign stimuli and releasing mechanisms

Our sensory organs are confronted every second with an enormous amount of information from the environment. However, only a tiny fraction of this information flow reaches our brain, and even less is consciously perceived. As a consequence of this tremendous reduction of the information flow, our perception of the world around us does not, by any means, reflect the 'true' and complete information present in the environment. Rather, what we perceive is a reflection of both the stimuli received and of the sensory and neural structures involved in the process of perception. This observation led biologist Jakob von Uexküll (see Chapter 2), at the beginning of the twentieth century, to the formulation of the *Umwelt* concept. Von Uexküll used this term not in the sense of its literal translation from German, namely 'environment,' but as the environment we perceive.

'Umwelt': that part of the environment which is perceived after sensory and central filtering.

> *Example: Many animals live in a sensory world very different from ours. In honey bees, the range of color vision is shifted toward the short-range end of the light spectrum, compared to ours—they can see ultraviolet light, but not red light. Using this sensory capability, they can see features of many flowers (so-called honey guides), on which they forage, that we are unable to perceive without technical means. Similarly, bats—in contrast to humans—produce ultrasonic sounds and analyze their echoes for the purpose of orientation. Weakly electric fish are able to perceive electric signals of such low amplitude that we have difficulty to detect, even with highly sophisticated technical apparatus. Each of these three examples demonstrates that the Umwelt of these animals is very different from ours, although the absolute physical environment is identical.*

The reduction in the flow of information has prompted ethologists to propose the existence of sensory and central filter mechanisms which select those stimuli from the environment that are biologically relevant, but ignore the others. These mechanisms are called **releasing mechanisms**. They determine the kind of stimuli to which the animal responds by producing an associated behavior. Thus, releasing mechanisms can be viewed as a sensory/central link between the stimulus and the resultant behavior. The component of the environment that triggers a given behavior is termed a **sign stimulus**. If the sign stimulus occurs in the context of social communication, it is often referred to as a **releaser**.

When early ethologists developed the concepts of releasing mechanisms and sign stimuli, they knew very little about the actual sensory and neural mechanisms responsible for stimulus filtering. What they studied was the relationship between stimuli and motor patterns purely from a behavioral point of view. The structural correlate of a releasing mechanism was treated as a 'black box.' Elucidation of the physiological nature of this black box remains the primary goal of sensory physiology and neurobiology in general, and neuroethology in particular.

How can the relevant features of a sign stimulus be identified at the behavioral level? Typically, careful observations of a behavioral situation lead to the formulation of a preliminary hypothesis. This hypothesis is, then, experimentally tested by the use of **dummies** or **models**.

Example: In European robins (Erithacus rubecula), adult males and females elicit aggressive behavior from the owner of a territory when trespassing its territory, while juveniles are not attacked. Adults are distinguished from juveniles by their red breast. This observation makes plausible the assumption that the red breast feathers are a key component of the stimulation regime triggering attacks by the owner of a territory. In his classic experiments, published in 1939, David Lack confirmed this hypothesis by placing various types of dummies into the territory of wild male robins, such as a stuffed adult robin, a stuffed juvenile robin (which has no red breast), a stuffed adult robin with the underparts painted over with brown ink, and a bunch of red breast feathers without the rest of the body (Fig. 3.12). The male robins threatened the bunch of breast feathers almost as heavily as the stuffed adult robbin. By contrast, the stuffed juvenile robin and the adult robin with the underparts stained brown elicited almost no threat responses and attacks, although these dummies look (to us!) more robin-like than the bunch of red feathers.

Supernormal stimuli

Frequently, it is possible to make models that produce a greater response from an animal than the natural object does. Such a model provides **supernormal stimuli**.

Examples: Experiments in several avian species have shown that, within certain limits, the parental bird prefers clutches with more eggs, or giant supernormal eggs over the smaller natural eggs, for incubation. Niko Tinbergen demonstrated this in a classic series of experiments

---

Sign stimulus: the component of the environment that triggers a specific behavior.

Figure 3.12 David Lack's classic experiment to examine the attack response of European robins. Among others, he placed one of these four dummies into the territories of male robins and monitored their behavior. While a mounted, naturally looking juvenile robin with its dull-brown breast (a), or an adult with the red breast and white abdomen stained brown (b), elicited either no attacks at all, or only a few threat responses, a simple bunch of red breast feathers (c) triggered similar attacks from the territorial owner, as does a stuffed adult (d). (After **Lack, D.** (1953).)

Figure 3.13 An oystercatcher incubates a 'supernormal' clutch of five eggs instead of the normal clutch consisting of only three eggs. (After **Tinbergen, N.** (1969).)

using oystercatchers (Haematopus ostralegus). If they have the choice, these birds prefer clutches with five eggs over normal clutches, which consist of only three eggs (Fig. 3.13). Similarly, when offered eggs of different sizes, they prefer supernormal eggs (Fig. 3.14).

A second classic example is provided by the pecking response of infant herring gulls (*Larus argentatus*) toward the parents' bill. The adult herring gull has a red spot near the end of the lower mandible, which contrasts with the yellow color of the bill. When the parent returns to the nest, the chick pecks at this spot, which in turn causes the parent to

Figure 3.14 An oystercatcher prefers a supernormal egg for incubation over her own egg (foreground) or a herring gull's egg (left). (After **Tinbergen, N.** (1969).)

regurgitate food. In a detailed study, Niko Tinbergen and Albert Perdeck examined the importance of the various features (such as color of the spot and color of the bill) of this stimulation regime for evoking the pecking response. While most variations elicited fewer responses than the natural stimulus, a thin red rod adorned at its tip with three sharply edged white bands had the opposite effect (Fig. 3.15). Now, the chicks pecked more vigorously than toward dummies imitating the parent's natural bill.

In humans, socially relevant features are often made supernormal by exaggerating them. One example is the use of lipsticks to create supernormal lips in women (Fig. 3.16). Another example is the enlargement of men's shoulders, in many societies, to emphasize the 'male' appearance (Fig. 3.17). The creation of supernormal shoulders is likely to have its roots in the behavior of human ancestors. When they ruffled up their hairs, for example, in aggressive situations, their shoulders seemingly increased in size.

> A supernormal stimulus produces a greater response from an animal than does the natural sign stimulus.

Law of heterogenous summation

A stimulation regime that elicits a specific behavioral response often consists not just of one stimulus feature. Rather, several independent and heterogenous features may be present. How do they influence the outcome of the behavioral response?

One possibility is that such independent and heterogenous features of a stimulus are additive in their effect upon the behavior. This is known as the **law of heterogenous summation** formulated in 1940 by Alfred Seitz, a student of Konrad Lorenz.

(a)

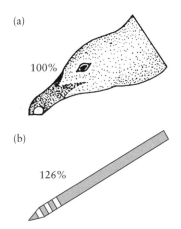

100%

(b)

126%

Figure 3.15 Pecking response toward a dummy imitating the natural appearance of an adult herring gull (a) and toward a thin red rod with three white bands near its tip (b). The numbers indicate the normalized responses elicited by each of the two models. (After **Tinbergen, N., and Perdeck, A. C.** (1950).)

An intriguing quantitative demonstration of this 'law' was performed by Daisy Leong, a student of Walter Heiligenberg, at the Max Planck Institute for Behavioral Physiology in Seewiesen, Germany. For her experiments, Leong used Burton's mouthbrooder *(Haplochromis burtoni)*, a cichlid fish from Lake Victoria in East Africa. Males of this species are highly territorial and extremely aggressive toward other fish, especially males. As Leong's experiments revealed, the following two features of the male coloration are crucial in eliciting aggressive responses from other males:

The law of heterogenous summation defines the additive effect of the independent and heterogenous features of a stimulus upon a behavior.

- A black eye bar that runs from the posterior end of the mouth to the eye.
- Bright orange spots in the pectoral region, as well as on the dorsal, caudal, and anal fins.

To test the effect of these two stimulation features quantitatively, Leong divided an aquarium into two parts using a glass partition. In one part, an adult male was placed together with a group of 10–15 young fish. In the second part, dummies were presented whose head pattern and pectoral coloration were systematically changed as follows (Fig. 3.18):

1. Dummy (1) had a black eye-bar, but no orange pectoral spots. Presentation of this dummy behind the glass partition increased the average bite rate toward the young fish by 2.81 bites/min.

Figure 3.16 Human lips made supernormal by the use of lipstick. (Courtesy: Jose Luis Pelaez, Inc./CORBIS.)

Figure 3.17 Enlargement of men's shoulders in various societies to emphasize the 'male' appearance. Top: Waika Indian from South America. Middle: Kabuki actor from Japan. Bottom: Alexander the Second of Russia. (After **Eibl-Eibesfeldt, I.** (1974).)

Dummy 1

Figure 3.18 Dummies used in the experiment by Leong. Dummy (1) with black eye-bar, but lacking an orange pectoral patch. Dummy (2) with orange patch, but without black eye-bar. Dummy (3) combines these two features—it incorporates both the black eye-bar and the orange patch. (After **Leong, C.-Y.** (1969).)

Dummy 2

Dummy 3

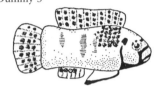

2. Dummy (2) had orange spots in the pectoral region, as well as on the dorsal, caudal, and anal fins, but no black eye-bar. Its presentation lowered the number of attack by 1.77 bites/min.

3. Dummy (3) combined both features by having the black eye-bar and the orange spots. This dummy elicited an average increase in the attack rate by 1.12 bites/min.

Application of the law of heterogenous summation predicts the following effect when Dummy (1) and Dummy (2) are combined: 2.81 attacks/min + (−1.77 attacks/min) = 1.04 attacks/min. Dummy (3), which incorporated the features of both Dummy (1) and Dummy (2) in one model, elicited 1.12 attacks/min. Thus, the predicted value (1.04 attacks/min) and the value observed in the experiments (1.12 attacks/min) are in remarkable agreement.

Gestalt principle

The effect of the orange spots and the black eye bar on the aggressive behavior of *Haplochromis burtoni* provides an excellent example that obviously follows the law of heterogenous summation. The total number of responses (here: attack rate toward the young fish) released by these two features of the stimulation regime is the same when presented successively, as when released by the whole. In general terms, the law of heterogenous summation states that the successive presentation of the individual components of a model elicits the same response as their simultaneous presentation in a combined model. In a simplified mathematical form, this could be summarized as follows:

$$Model_{sum} = Model_1 + Model_2 + ... + Model_n$$

However, while the law of heterogenous summation holds true in some instances, it does not do so in all. In the latter cases, the results can best be interpreted by applying the **Gestalt principle**. The German term *Gestalt* means 'configuration.' The Gestalt principle forms the central dogma of the Berlin school of psychology, an influential school of thought before the Second World War. It is also known as the **Gestalt school of**

psychology. The Gestalt principle states that **the whole is more than the sum of its parts:**

$$Model_{sum} >> Model_1 + Model_2 + \ldots + Model_n$$

In the most extreme case, the presentation of the individual components may not elicit any response. Rather, only the complete model which incorporates all the individual components will lead to a behavioral response.

Arguably, it has been suggested that the law of heterogenous summation applies predominantly to taxonomically 'lower' animals, whereas the Gestalt principle appears to underlie the mechanism of perception in more highly developed organisms. Moreover, there is some evidence that the mechanisms mediating the perception of behaviorally relevant signals change in the course of ontogenetic development. An intriguing study demonstrating such an alteration of the applicability of the law of heterogenous summation with age was published, in 1966, by T.G.R. Bower of Harvard University.

In this investigation, human infants were trained to turn their heads when a stimulus consisting of three individual components (resembling the face of an adult) was presented. The reinforcement consisted of a 15-sec long 'peek-a-boo' from the experimenter. Then, in the test situation, the combined stimulus and its individual component were presented in a counterbalanced order.

When these experiments were performed on 8-week-old infants, their response confirmed the applicability of the law of heterogenous summation (Fig. 3.19). On the other hand, when the identical experiments were conducted on 20-week old infants, they produced quite a different outcome. The whole was found to elicit more responses than did the sum of the responses elicited by its individual components.

Importance of condition of recipient

It is a well known fact that the effect of a sign stimulus not only depends on the stimulus presented, but also on the condition of the recipient.

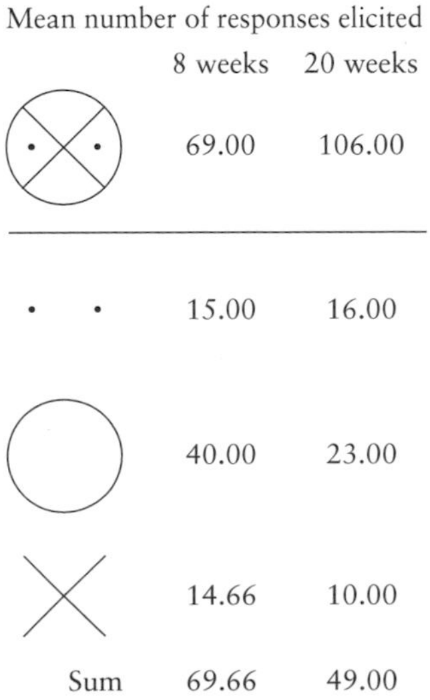

The Gestalt principle states that the effect of a stimulus upon a behavior is greater when presented as a whole than when presented sequentially in its individual components.

| | Mean number of responses elicited | |
	8 weeks	20 weeks
⊗	69.00	106.00
• •	15.00	16.00
◯	40.00	23.00
✕	14.66	10.00
Sum	69.66	49.00

Figure 3.19 Law of heterogenous summation versus Gestalt principle. Human infants were trained to make a head movement as a conditioned response when stimulated with a rough sketch of a human face (top row) or the three components of this face (lower three rows). In eight-week old infants, the sum of the mean number of responses elicited by the individual components is very similar to the mean number of responses evoked by the whole (69.66 vs. 69.00). By contrast, in 20-week-old infants, the whole stimulus elicits more responses than the sum of its parts (106.00 vs. 49.00). (Courtesy: G. K. H. Zupanc, based on data by **Bower, T. G. R.** (1966).)

For example, female frogs are strongly attracted by the calls of territorial males, if they are physiologically ready to lay eggs. If they have already laid eggs, or if they have not yet developed ripe eggs, the vocalizations of males elicit little approach behavior in females.

A quantitative demonstration of how the condition of the recipient of a stimulus can determine the type of a behavior elicited was published by Oskar Drees of the University of Kiel, Germany, in 1952. He used jumping spiders, a family (Salticidae) comprising approximately 2800 species. These spiders do not build webs, but capture prey by jumping onto them. Among their prey animals are insects, such as flies, that resemble, to a certain extent, potential sexual mates. It is, therefore, possible to construct dummies that are capable of evoking, in males, both attempted prey capture or courtship behavior. Which of the two types of behavior is elicited depends on the motivational state of the male. Satiated males respond even to rather crude dummies—black filled circles with one pair of legs—largely with courtship behavior (Fig. 3.20). If the males are left without food, the proportion of courtship behavior relative to attempted prey capture gradually decreases proportionally to the length of time the male has gone without food. However, it is possible again to increase the relative number of courtship behaviors displayed by presenting models with three pairs of legs, thus making the dummies more similar to potential mating partners.

The effectiveness of a sign stimulus on a behavior depends on both the stimulus presented and the condition of the recipient.

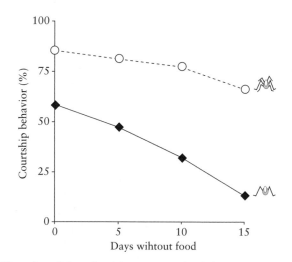

Figure 3.20 Effect of condition of recipient on courtship behavior in male jumping spiders. The number of courtship displays relative to the number of prey-capture activities depends on both the time the male is left without food and the complexity (three pairs of legs vs. one pair of legs) of the model presented. (Courtesy: G. K. H. Zupanc, based on data from, **Drees, O.** (1952).)

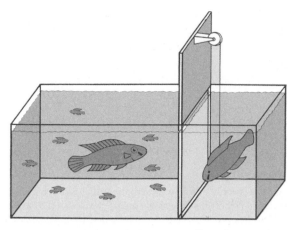

Figure 3.21 Test aquarium to examine the long-term effect of stimulation on the aggressiveness in the cichlid fish, *Haplochromis burtoni*. An adult male of this species was kept isolated in an aquarium with 10 small fish of the species *Tilapia mariae*. A dummy hanging on strings above the water, behind a screen, was invisible to the fish, until it was lowered, by remote control, into the water behind a glass partition. As an indication of its aggressiveness, the attack rate of the adult male toward the young fish was measured. (Courtesy: G. K. H. Zupanc.)

Motivational effect of releaser

The above examples may give the impression that stimuli exert only direct releasing effects. Although this is the most studied aspect of stimulation, a second behavioral consequence is probably of equal importance. Studies have shown that stimuli may also influence the **motivation** of the receiver. An elegant demonstration of this effect was achieved by Walter Heiligenberg together with his assistant Ursula Kramer. They used the same experimental set-up as employed by Daisy Leong in the experiments described above (Fig. 3.21). Initially, a male *Haplochromis burtoni* was kept with young fish in its part of the tank, without a dummy presented in the second part of the tank. Then, a male fish dummy was presented to the male on 10 consecutive days for 30 sec every 15 min from morning to evening. As a result, the initially low attack rate of the test fish toward the young increased markedly over time (Fig. 3.22). This has shown that the presentation of the male dummy not only has an immediate releasing effect upon the male's attack rate. It also leads to long-term changes in the motivation of the male to attack the young fish.

Releasers have not only an immediate releasing effect upon a behavior; they may also lead to long-term changes in the underlying motivation.

Communication

The concepts of releasers and releasing mechanism are intimately linked to that of **communication**. Communication, as Jack Bradbury and Sandra Vehrencamp, then at the University of California, San Diego (now at

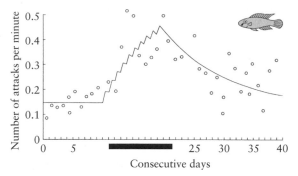

Figure 3.22 Long-term effect of repeated presentation of a male fish dummy (inset) on the attack rate of the cichlid fish *Haplochromis burtoni*. To establish the baseline level of aggression, the fish's attack rate was measured on the first 10 consecutive days. Then, starting with day 11, a model was presented every 15 min for 30 s over 8 h every day for 10 days (indicated by the bar underlying the abscissa). Upon completion of the stimulation, the attacks exhibited toward the young fish continued to be measured for another 20 days. The data points are the mean of six experiments. The line indicates the theoretical build-up and decay of the attack rate. (Courtesy: G. K. H. Zupanc. Based on data by **Heiligenberg, W. and Kramer, U.** (1972).)

Cornell University, Ithaca, New York), put it, 'is the glue that holds animal societies together.' It involves the production of information in the form of a **signal** by a **sender**, and its transmission to a **receiver**. Upon detection, the signal invokes a response in the receiver. In many instances, this is an immediate behavioral response. However, as it appears to be true especially in cases where a stereotyped social signal is repeatedly broadcasted, a tonic and motivational effect may also be exerted upon the receiver. An intriguing example discussed in the last section is the experiment performed by Walter Heiligenberg and Ursula Kramer to induce long-term changes in the motivation to attack young fish in a male cichlid fish. This was achieved by repeated presentation of a dummy incorporating releasing signals typical of territorial male conspecific.

Another feature of **true communication** is that both sender and receiver benefit from the information exchange. This criterion has been included in the definition of communication to exclude the quite frequent case that the stimulus produced by an animal is used by a receiver to the detriment of the sender. As we will discuss in detail in Chapter 6, the noise accidentally generated by mice is utilized by owls to localize and attack them. In this case, the provision of the acoustic information benefits the receiver, but certainly not the sender! To distinguish between signals that benefit the sender and information transfer that has no beneficial effects upon the sender, the term **cues** is used for the latter stimuli.

Communication: transfer of information from a sender to a receiver benefitting both partners.

Figure 3.23 Colorations of the cichlid fish *Hemichromis fasciatus* as indicators of aggressive motivation. The spotting pattern in the top drawing is typical of a neutral fish without a territory. The following three drawings represent fish of increasing aggressiveness. The coloration on the bottom is characteristic of a highly aggressive fish that is very likely to defend its territory against an intruder. (After **Wickler, W.** (1978).)

On the other hand, inclusion of the mutual benefit of the sender and the receiver as a criterion in the definition of communication does not necessarily restrict the information transfer to members of the same species. Foraging bees, for example, are guided to the center where the nectar is by the **honey guides**—the radiating lines on the petals of many flowers (see Chapter 2). This piece of information communicated by the flower to the bee results in mutual benefit of both partners, and thus is in agreement with the above definition of communication.

Based on the sensory modalities involved in their detection, different types of communication signals are distinguished. Many animals use **visual signals** for communication. For example, the cichlid fish *Hemichromis fasciatus* indicates its readiness to flee or attack through different spotting patterns in its coloration (Fig. 3.23). This is done by moving pigments within melanophores in the skin. Such signals allow an intruder to assess the likelihood of the owner of a nesting site to defend its site. This assessment, then, forms the basis for the intruder to decide whether or not it is economical to attack the owner and try to take over his territory.

Morphological structures involved in the provision of communicatory information are, characteristically, especially conspicuous and often exaggerated far beyond their need for normal function. In male fiddler crabs, one claw is enormously enlarged and frequently brightly colored (Fig. 3.24). To amplify the signal even more, this claw is waved rhythmically during courtship and aggressive encounters with other males.

Chemical signals (also known as **pheromones**) have been intensively studied in insects and mammals. The **trail substances** of ants are well-known examples of these signals (Fig. 3.25). When returning to the nest from a food source, a foraging ant intermittently secretes a tiny amount of the trail substance, thus defining a path between the nest and the food source. Other worker ants can, then, follow this trail to further exploit the food source.

Chemical signals are the only type of signals that may persist even in the absence of their producer. Moreover, since it is difficult to change chemical signals quickly (something relatively easy when employing acoustic signals), they often convey relatively stable messages, such as the sexual condition of a female or the ownership of a territory. Also, many chemical signals (referred to as **primer pheromones**, in contrast to **releaser**

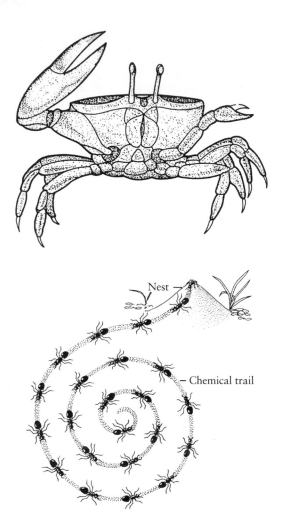

Figure 3.24 Male fiddler crab with one claw enormously enlarged. This claw is used as a signal in courtship and aggressive encounters. (After **Manning, A., and Stamp Dawkins, M.** (1992).)

Figure 3.25 Demonstration of the efficacy of trail substance as a chemical signal. This pheromone was laid down by an experimenter in a spiral-like fashion. Ants encountering this signal follow the spiral path. (After **Keeton, W. T.** (1980).)

pheromones) produce long-term alterations in the physiological condition of the receiver. They do not necessarily generate any immediate behavioral change. Rather, since they affect the receiver's motivation, the effect of primer pheromones becomes visible as measurable behavior only later, when encountering communication signals. It has been demonstrated, for example, that the estrous cycle of female mice can be synchronized by the odor of a male mouse—even in the absence of this mouse. Similarly, the odor of a strange male mouse can lead to the termination of the pregnancy of a newly impregnated female mouse.

A classic ethological example which demonstrates the complexity of the sensory cues involved in communication will be presented at the end of this chapter. The core element of this communication system is the **dance language** of honey bees (*Apis mellifera*), a behavior discovered through the pioneering research conducted over several decades by

Karl von Frisch (see Chapter 2). To understand this system, we need to take first a look at the life cycle of the honey bee.

Honey bees are believed to have evolved from tropical ancestors which colonized the temperate zones of the world by building combs within hollow trees. By massing together and generating metabolic heat in the insulated cavity, the combs offer the bee colony protection from the changeable weather.

A colony consists of up to 80 000 individuals, which can be grouped into three types of bees:

1. Drones, which are males.
2. Workers, which are females whose activity of the ovaries is suppressed, so that they are not able to lay eggs.
3. One queen, which is a fully developed female and the only mother in the colony.

Over her lifetime, each worker bee takes on a series of tasks. They include the following:

1. Within the first two weeks after metamorphosis, to act as a nurse bee, which incubates the brood and feeds the larvae.
2. In the next stage, to become a home bee, functioning as a store keeper, house cleaner, and guard.
3. When the wax glands become functional, to help to cap cells and build the comb.
4. Finally, for the last three weeks of her just six-weeks-long life, to become a field bee, foraging for nectar and pollen.

The field bee forages primarily upon flowers that produce pollen and secrete nectar. She scouts for food at distances of up to 10 km from the hive. Upon arrival back to the hive, she feeds other workers and attracts them by a so-called recruitment display. As a characteristic feature of this behavior, the field bee fans her wings and releases a recruitment pheromone. Then, she performs on the vertical sheet two types of dances in the hive, called round dance and waggle dance, respectively. The patterns of these dances are summarized in Figs 3.26 and 3.27.

The **round dance** consists of a circling—first to one side, then to the other—followed by numerous repetitions of this pattern. The dance excites other bees which begin to follow the dancing bee and keep their antennae close to her. The round dance indicates neither distance nor direction of the food source in relation to the hive. The dancing bee simply communicates that she found a food source in the vicinity of the hive, at a distance of up to 80 m. However, upon leaving the hive, the recruited bees remember the scent they smelled when they touched with their antennae the body of the dancer and when they were fed droplets of material from

Figure 3.26 Round dance. This type of dance is characterized by a circling of the dancer and the close contact via the antennae of the other bees. (After **Frisch, K. v.** (1977).)

Figure 3.27 Waggle dance. The dancing bee runs a short distance in a straight line, while waggling the abdomen from side to side. This waggle run is repeated after circling back to the beginning of this straight portion of the dance. (After **Frisch, K. v.** (1977).)

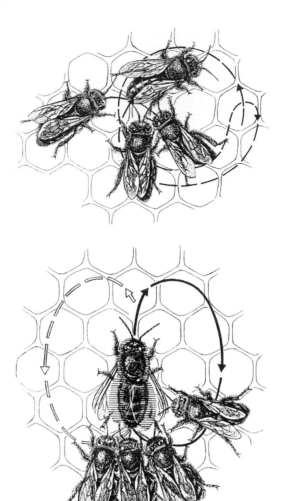

the newly discovered food source. It is this scent they look for in the vicinity of the hive.

The second type of dance, the **waggle dance**, is displayed when the food source is located more distantly to the hive. When displaying this dancing behavior, the bee runs a short distance in a straight line, while waggling the abdomen from side to side very rapidly. This part is called the waggle run. After circling back to her starting point (referred to as the return run), she performs another waggle run, followed by circling back on the other side. This pattern is repeated several times.

In contrast to the round dance, in the waggle dance distance and direction of the new food source are communicated. The **distance** is correlated with several components of the dance, such as the duration of the waggle run, the number of waggles, and the frequency of the sound. The diagram in Fig. 3.28 illustrates this correlation, showing an inverse relationship between the distance of the food source and the number of

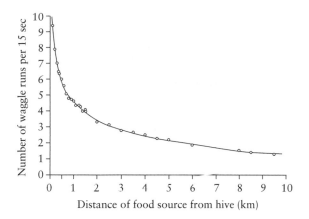

Figure 3.28 Relationship between distance of the food source from the hive and number of waggle runs per 15 sec performed during the waggle dance. (After **Frisch, K. v.** (1977).)

waggle runs per unit time. The waggling is sensed in the darkness of the hive by the other workers crowding around the scout bee. This is probably done by perceiving oscillatory air currents around the dancer. These acoustic near-fields are produced by the dancing bee through vibrations of her wings with a frequency of 200–300 Hz.

The second parameter communicated through the waggle dance is the **direction** of the food source. The location of the food source relative to the position of the sun is indicated by the direction of the waggle run. This is illustrated by the three examples shown in Fig. 3.29.

- A waggle run straight up the vertical comb indicates that the food source is located in the direction of the sun (Fig. 3.29(a)).
- A waggle run straight down the vertical comb indicates that the food source is located in the opposite direction of the sun (Fig. 3.29(b)).
- A waggle run at a certain angle indicates that the food source is at that angle to the sun. In the given example, a run 60° to the left of the vertical indicates that the food source is 60° to the left of the sun (Fig. 3.29(c)).
- A run 120° to the right of the vertical indicates that the food source is 120° to the right of the sun (Fig. 3.29(d)).

Similar dancing patterns could be evoked in numerous experiments in which bees were trained to communicate to other bees in the hive the location of an artificial food source consisting of a petri dish of sugar water. Such experiments demonstrated that the bees use the force of

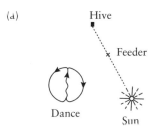
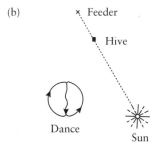
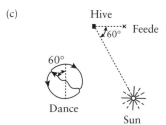
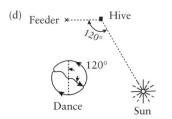

(a)
(b)
(c)
(d)

Figure 3.29 Communication of direction through the waggle dance. The feeders with sugar solution differ in their positions from the hive relative to the sun. (a) The feeder is placed straight in the direction of the sun. On the vertical comb, the waggle run is straight up. (b) The feeder is straight opposite to the sun. The waggle run is directed straight down. (c) The feeder is at 60° left to the sun. The waggle run is directed upward, with an angle of 60° to the vertical (dashed line). (d) The feeder is at 120° right to the sun. The waggle run is directed downward, with an angle of 120° to the vertical (dashed line). (After **Frisch, K. v.** (1977).)

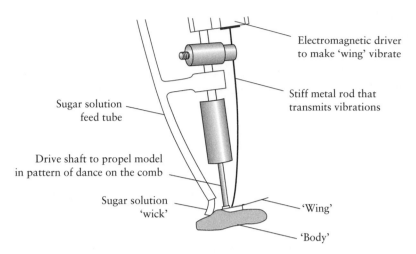

Figure 3.30 Artificial dancing bee. The mechanical model consists of brass covered with beeswax ('body') and a piece of razor blade ('wing'). At its 'head,' the model offers, via a feed tube, a sugar solution to other bees. The wing can be vibrated to produce sound similar to the one generated by dancing bees. (After **Manning, A., and Stamp Dawkins, M.** (1998).)

gravity as a symbol for the sun when they perform the waggle dance inside the dark hive.

The importance of the individual sensory and behavioral elements involved in the dances for the communication process has been demonstrated by the use of dummies. While earlier attempts had failed to elicit foraging of worker bees by dummies performing round and waggle dances, Axel Michelsen of the University of Southern Denmark, Odense (Denmark) and Martin Lindauer of the University of Würzburg (Germany), together with their coworkers, succeeded in the late 1980s in constructing such an artificial bee (Fig. 3.30). This model was made of brass covered with a thin layer of beeswax. It had artificial wings that could be vibrated so as to produce sound similar to the one of real dancers. The dummy was left in the hive for several hours to acquire the odor of the colony.

This model was, indeed, able to recruit worker bees to fly to food dishes they had not visited previously. Detailed experiments in which different directions and distances were indicated by the dance pattern proved the original hypothesis of Karl von Frisch that these parameters can be communicated through the dances. The experiments also demonstrated the importance of the acoustic near-field produced by the dancer and explained why previous experiments using models had failed—without sound, bees could not be recruited.

Summary

■ The elementary unit of the nervous system is the neuron. In most neurons, the cell body (also referred to as soma) gives rise to two types of processes, dendrites and axons.

■ Axons serve to conduct signals over wide distances. The projection of a neuron describes the route of an axon from its origin to its target region.

■ The contact point between two neurons is known as the synapse. Electrical synapses are identical to gap junctions in terms of their structure. Chemical synapses are polarized in that a transmitter substance is released from the presynaptic bouton into the synaptic cleft. After diffusion to the postsynaptic element of the recipient cell, the transmitter molecules bind to specific receptors, which leads to the generation of—excitatory or inhibitory—postsynaptic potentials.

■ Important types of transmitter substances are acetylcholine; amino acid transmitters, such as glutamate, glycine, and GABA; biogenic amines, including catecholamines, serotonin, and histamine; and neuropeptides.

■ Glia are non-neuronal cells in the nervous system. They have many important functions; for example, they provide myelin sheaths to axons. Such myelinated axons are found in many parts of the nervous system of vertebrates and, to a lesser extent, in invertebrates.

■ The electrical properties of neurons can be studied by intracellular and extracellular recordings.

■ At rest, the inside of a neuron is negatively charged compared to the outside, with the difference in the range of -60 to -80 mV. This is caused by a differential distribution of ions across the plasma membrane. Depolarization brings this membrane potential closer to zero. Hyperpolarization increases the membrane potential.

■ Small increases in current in a neuron lead to graded potentials. However, as soon as a critical threshold is reached through depolarization of the membrane, action potentials (also referred to as impulses or spikes) are generated. They are all-or-nothing events and are caused through influx of Na^+ ions into the cell, followed by efflux of K^+ ions out of the cell.

■ Sensory organs act as interfaces between an animal's environment and its central nervous system. The result of the transduction process is the generation of a receptor potential, which is translated into a series of impulses ('frequency coding').

- Neuronal projections can be elucidated through neuronal tract-tracing methods. These techniques employ both anterograde and retrograde tracer substances.

- The two major techniques available to reveal tissue constituents are immunohistochemistry and *in situ* hybridization. They provide information about tissue-specific antigens and expression of specific genes, respectively.

- A method to selectively kill single cells consists of filling the cell with a fluorescent dye and irradiating the cell with light of the absorption maximum of the dye.

- The information flow arising from the environment of an animal is dramatically reduced as a result of sensory and central filters called releasing mechanisms. Thus, the *Umwelt* of an animal reflects only a tiny portion of the absolute physical environment.

- Releasing mechanisms act as sensory/central links between a stimulus originating from the environment and the resultant behavior. The component of the environment that triggers a given behavior is called a sign stimulus or, if it occurs in the context of social communication, a releaser.

- The ethologically relevant features of a sign stimulus can be examined through the use of dummies or models.

- Models providing supernormal stimuli elicit a greater response from an animal than does the natural object. This property is frequently exploited in human societies to make more attractive certain features used for communication.

- The relationship between the individual components of a sign stimulus when evoking a behavioral response can be described by two principles: (i) the law of heterogenous summation states that two or more separately effective stimulus properties are additive in their partial effects when combined with one another; (ii) the Gestalt principle applies to situations in which the combined stimulation regime is more effective than the sum of its parts.

- The effectiveness of a sign stimulus depends not only on the features of the stimulus, but also on the condition of the recipient.

- Besides having an immediate releasing effect, sign stimuli may also produce long-term changes in the receiver's motivation to generate the corresponding behavioral patterns.

- Communication involves the transfer of information from a sender to a receiver to the mutual benefit of both partners. Information is conveyed through presentation of signals.

■ In one of the best studied communication systems, the dance language of honey bees, field bees communicate information about new food sources to other worker bees. This communication process is mediated by a variety of different signals, including olfactory, tactile, and acoustic stimuli.

Recommended reading

Alcock, J. (1998). *Animal Behavior: an evolutionary approach*. Sixth edition. Sinauer Associates, Sunderland, Massachusetts.

The classic introduction to animal behavior from an evolutionary point of view.

Bradbury, J. W., and Vehrencamp, S. L. (1998). *Principles of animal communication*. Sinauer Associates, Sunderland, Massachusetts.

The best and most comprehensive text available on animal communication. Can be used both as a textbook and as a reference book.

Goodenough, J., McGuire, B., and Wallace, R. A. (2001). *Perspectives on animal behavior*. Second edition. John Wiley & Sons, New York.

If there were a single text to be recommended for an animal behavior course, it would be this book. Comprehensive, balanced, and well-written, the textbook not only gives an excellent overview of historically important ideas, but also covers many current issues associated with the mechanisms and the evolution of animal behavior.

Kandel, E. R., Schwartz, J. H., and Jessell, T. M. (2000). *Principles of neural science*. Fourth edition. McGraw-Hill, New York.

Classic textbook providing comprehensive information at a more advanced level.

Manning, A., and Stamp Dawkins, M. (1998). *An introduction to animal behaviour*. Fifth edition. Cambridge University Press, Cambridge/ New York/Oakleigh.

A classic introduction to the study of animal behavior giving an overview of the major ethological concepts.

McFarland, D. (1993). *Animal behaviour: psychobiology, ethology, and evolution*. Second edition. Addison Wesley Longman, Harlow.

A highly recommended textbook on animal behavior. Easy to read, it conveys much of the excitement associated with the subject.

Polak, J. M., and Van Noorden, S. (1997). *Introduction to immunocytochemistry*. Second edition. BIOS Scientific Publishers, Oxford.

Although brief, this excellent introductory text covers all major theoretical and practical aspects of immunocytochemistry.

Shepherd, G. M. (1994). *Neurobiology*. Third edition. Oxford University Press, New York/Oxford.

A classic introductory textbook covering all major aspects of neurobiology. Especially recommended to readers with little prior knowledge in the neurosciences, and to those who would like to get a broad overview of neuroscience.

Zigmond, M. J., Bloom, F. E., Landis, S. C., Roberts, J. L., and Squire, L. R. (ed.) (1999). *Fundamental neuroscience*. Academic Press, San Diego/London.

With more than 1500 pages more a reference work than an introductory textbook. Major focus on mammalian, including human, systems.

Questions

3.1 Some species of weakly electric fish produce certain types of discharges almost exclusively during the breeding season. Based on this observation, scientists have proposed that these signals act as releasers in communication between males and females. Suggest experiments to verify this hypothesis.

3.2 Suppose a certain brain region called 'nucleus controllaris' in an animal is suspected to be involved in motor control of a specific behavior. Design experiments through which you could verify this hypothesis.

3.3 The experiments you suggested in your answer to Question 3.2 have, indeed, supported the hypothesis that nucleus controllaris is crucially involved in control of the specific behavioral pattern. How could you elucidate possible interactions of this brain region with other areas of the central nervous system? How could you identify transmitters and neuromodulators mediating input to nucleus controllaris?

Spatial orientation and sensory guidance

4

■ Introduction
■ Classification of orienting movements
■ Orienting behavior without a nervous system
■ Geotaxis in vertebrates
■ Echolocation in bats
■ Summary
■ Recommended reading
■ Questions

Introduction

Animals orient their body in a specific way relative to the environment. To maintain orientation after disturbance, or to change position in certain behavioral situations, the animal needs to continuously collect and process relevant stimuli from the environment and translate them into proper behavioral action.

In this chapter, we will examine how animals manage to perform this process of translation. We will first have a look at the diversity of orienting movement in animals and summarize the rules defining how to classify and name these movements. Then, we will discuss in more detail the cellular and neural mechanisms governing spatial orientation. In the first part of this discussion, we will use a particularly remarkable example—an animal that, although completely lacking a nervous system, is nevertheless able to produce, with high sophistication, orienting movements in response to environmental stimuli. In the second part of this discussion, we will have a closer look at how vertebrates detect and maintain equilibrium position using the force of gravity as a frame of reference. Finally, in the third and last part, we will describe how bats succeed to orient in the dark using echolocation.

Classification of orienting movements

The orienting responses of animals are extremely diverse. A first major attempt to classify the different types of these behaviors was undertaken by Alfred Kühn which is described in his book *Die Orientierung der Tiere im Raum* (Spatial Orientation of Animals), published in 1919. His system was extended by Gottfried Fraenkel and Donald Gunn in their book *The Orientation of Animals*, published in 1940. A number of aspects of the classification system of these three authors are still widely applied, while others have been modified since then. Particularly, it has become increasingly evident over the past decades that orientation movements involve not just one sense but multisensory guidance.

In general, scientists distinguish between primary and sensory orientation. **Primary** (or **positional**) **orientation** involves control of body posture, whereas **secondary orientation** controls the response of an animal with respect to a particular stimulus from the environment. Secondary orientation is, then, subdivided into the following two major types of orienting responses:

1. **Taxis** defines an orienting movement in freely moving animals (in sessile animals, such a response is referred to as **tropism**). The body of the animal is oriented, and possibly also moved, in a particular direction with respect to the source of stimulation.

2. **Kinesis** involves a behavioral response in which the animal's body is not oriented with respect to the source of stimulation. The animal's main response involves a change in speed of movement and/or in rate of turning. This undirected movement alters the position of the animal in relation to the source of the stimulus.

Taxis: orienting reaction or movement in freely moving organisms directed in relation to a stimulus.

Example: Woodlice (Porcelia scaber) *are small isopods that cannot tolerate prolonged periods of dryness. They, therefore, tend to aggregate in moist places, where they are rather motionless. On the other hand, in dry areas they exhibit increased locomotor activity, which enhances the probability of their finally reaching damper sites. Thus, and as first suggested by Donald Gunn and associates of the University of Birmingham, England, in the 1930s, the tendency of woodlice to aggregate in moist places can solely be explained by proposing that locomotor activity is controlled by relative humidity. The assumption of a directed movement, that is, a taxis behavior of the animal toward areas of higher humidity, is not necessary. Rather, we can classify the response of woodlice as a kinesis behavior.*

When applying the term 'taxis' to describe the directed movement of an animal, a prefix is added to define the nature of the stimulus. For example, 'phototaxis' indicates a reaction to light (Greek *photos* = light).

The possible addition of the modifiers 'positive' and 'negative' indicates the animal's mode of reaction to the stimulus. 'Positive' means an orienting movement toward the source of stimulation; 'negative' points to an orienting movement in opposite direction.

Example: Positive phototaxis indicates an orienting movement of an animal toward light.

Other taxis movements include the following:

- **Anemotaxis:** orienting movement in relation to the wind direction
- **Chemotaxis:** orienting movement triggered by a chemical stimulus
- **Galvanotaxis:** orienting movement with electric current acting as a directing stimulus
- **Geotaxis:** orienting movement in reference to gravity
- **Phonotaxis:** orienting movement elicited by an acoustic stimulus
- **Rheotaxis:** orienting movement directed by the current of water
- **Thermotaxis:** orienting movement induced by heat
- **Thigmotaxis:** orienting movement with a rigid surface acting as a directing force.

A certain degree of deviation from this terminology occurs in some other terms used to describe taxis movements, such as phobotaxis. This latter term indicates an orienting response of an animal induced by an aversive stimulus.

Orienting behavior without a nervous system

The response of an organism to certain environmental stimuli does not necessarily require highly developed sensory organs, such as eyes or ears. Even the presence of a nervous system is not necessary. A thorough analysis of the behavior of such 'lower' organisms started at the beginning of the twentieth century by a number of researchers, and has become particularly known through the publications of **Herbert Spencer Jennings** (see Box 4.1). The animal-like unicellular protist *Paramecium* sp., for example, exhibits a wide range of orienting behaviors—despite the fact that this ciliate protozoan lacks sensory cells and neurons. Chemotaxis is one of these orienting behaviors. This reaction can be observed when paramecians encounter a bubble of carbon dioxide (CO_2) or a drop of diluted acetic acid (Fig. 4.1). Due to diffusion, there are circular zones surrounding the CO_2 bubble or the drop of acetic acid that differ in acidity. Near the center of the bubble or the drop, the acidity is rather high, that is, the pH is low. With increasing distance from the

pH: method to quantitatively express the concentration of H_3O^+ ions in a solution. The pH is defined as the negative logarithm to the base 10 of the hydrogen ion concentration. Pure water is neutral and has a pH of 7. A pH of less than 7 indicates an acidic solution, a pH of greater than 7, a basic (alkaline) solution.

BOX 4.1   Herbert Spencer Jennings

Herbert Spencer Jennings. (Courtesy: Bettman/CORBIS.)

While his name has become largely forgotten by the beginning of the twenty-first century, the central principle that guided his work—to perform an objective, experimental analysis of behavior—has laid the foundation for modern behavioral research. It is the merit of the American **Herbert Spencer Jennings** to have established this principle at a time when many behavioral studies were still dominated by subjective and rather anecdotal approaches.

Jennings lived from 1868 to 1947. After graduating from highschool, he taught for some time in country schools and at a small college. Jennings received his undergraduate training at the University of Michigan, where he also started to study rotifers and protozoans—the organisms that remained the subject of choice in most of his investigations. Following a year of graduate study at Michigan, he attended Harvard, where he received his Ph.D. in 1896 with a descriptive study of the development of rotifers. In 1897, he went to Europe to work with the then well-known German physiologist Max Verworn in Jena, and to spend some time in Italy at the Naples Biological Station. After a decade of temporary positions, Jennings joined in 1906 the faculty of Johns Hopkins University, where he remained until his retirement in 1938. Soon after arrival at Johns Hopkins, he stopped his behavioral work and devoted the rest of his life to genetics, particularly the study of inheritance in protozoans.

Jennings behavioral studies and theoretical concepts, summarized, among others, in his classic book *Behavior of the Lower Organisms* (published in 1906), had a profound influence not only in biology, but also on the development of behavioristic psychology. Among his students was John B. Watson (see Chapter 2), the founder of the School of Behaviorism, who extended many of the principles developed by Jennings, through the study of lower organisms, to the behavioral analysis of man. However, in contrast to many prominent behaviorists, Jennings denied the general applicability of the hypothesis that external stimuli are the sole determinants of behavior. Instead, he emphasized the complexity and variability of behavior and the importance of internal factors. This zoological way of thinking, combined with the objective approach he applied toward the study of behavior, put Herbert Spencer Jennings many years ahead of most of his contemporaries.

center, the pH gradually increases. If the paramecian gets too close to the CO_2 bubble, or the drop of acetic acid, it reacts by stopping, backing away, and swimming forward again. These movements are accompanied by gyrations, revolutions around the longitudinal axis of the cell. On the other hand, the paramecian also shows a negative taxis response to zones of low acidity. Through avoidance of zones of high acidity and zones of low acidity, the paramecian is finally 'trapped' in a zone of optimal pH.

It is thought that this chemotaxic response is an adaptation to the feeding behavior of paramecians. One of their major food sources is decay bacteria, which generally lower the pH of the surrounding water.

Gyration: revolution around longitudinal axis.

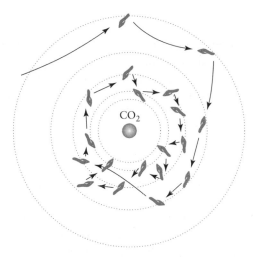

Figure 4.1 Chemotaxis response of a paramecian to a bubble of carbon dioxide. The bubble of CO_2 is surrounded by a pH gradient gradually increasing with increasing distance from the center. Zones of equal acidity are indicated by dotted concentric rings. Due to avoidance of both zones of high acidity and zones of low acidity, the paramecian tends to remain in a zone of optimal pH. (After **Keeton, W. T.** (1980).)

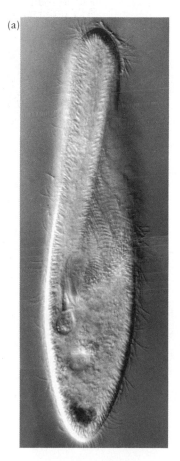

A positive chemotaxic reaction toward zones of mildly acidic pH, therefore, helps the paramecians to localize these bacteria. On the other hand, by exhibiting a negative chemotaxic response to zones of high acidity, they protect themselves from being damaged.

Cellular mechanisms of taxis behavior in paramecians

The cellular mechanisms mediating taxis behavior in ciliated protozoans is relatively well examined for two orienting movements, phobotaxis and galvanotaxis. The motor responses observed in both behaviors are mediated by cilia, subcellular organelles formed by microtubules and of similar structure as those of other eukaryotes. Their light microscopic and electron microscopic appearances are shown in Fig. 4.2. Much of the information on how proper stimulation of paramecians leads to a corresponding taxis behavior we owe to Hans Machemer of the Ruhr University in Bochum, Germany.

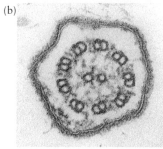

Figure 4.2 Light microscopic and electron microscopic appearance of cilia of a paramecian. The cilia are formed by numerous outward projections of the cell membrane. (a) At the light microscopic level, the cilia are visible as hair-like structures covering the whole surface of the paramecian. (b) An electron micrograph of a cross-section through a cilium reveals that the core structure is formed by nine doublet microtubules arranged in a ring around a pair of single microtubules. This array of microtubules is surrounded by the plasma membrane. (After **Machemer, H.** (1988).)

Phobotaxis

Cilia (singular: cilium): Hairlike appendages of many kinds of cells with a bundle of microtubules at their core. The microtubules are arranged such that nine doublet microtubules are located in a ring around a pair of single microtubules. This arrangement gives them the distinctive appearance of a '9 + 2' array in electron micrographs of cross-sections.

Phobotaxis can be elicited in paramecians, as well as in many other ciliates, by mechanical stimulation, for example, through touching of the cell membrane with an object. Without stimulation, the cilia move effectively in posterior and right direction, thus resulting in a left forward gyration of the paramecian. Mechanical deformation of the posterior cell end augments the normal ciliary rate and reorients the beat direction clockwise toward the posterior pole; this leads to a straight forward movement. On the other hand, mechanical deformation of the anterior cell end depresses the normal beat rate and reorients the beat direction counterclockwise close to the anterior pole. Thus, the paramecian shows a backward locomotor response, often accompanied by a right gyration. These movements lead to coordinated phobotaxic responses away from the source of potentially harmful mechanical impact.

These behavioral responses are initiated by local activation of mechanosensitive channels permeable only for certain ions; they are concentrated at the anterior and posterior poles of paramecians. The following cellular events, schematically shown in Fig. 4.3, can be distinguished depending on the site of stimulation:

- Mechanical deformation of the posterior end of the cell activates mechanosensitive potassium channels, leading to an efflux of K^+ ions, and thus to a K^+-dependent hyperpolarizing receptor potential.

- Mechanical deformation of the anterior end of the cell activates mechanosensitive calcium channels, leading to an influx of Ca^{2+} ions, and thus to a Ca^{2+}-dependent depolarizing receptor potential.

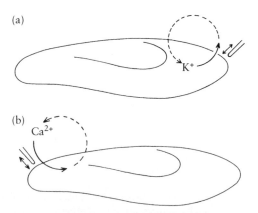

Figure 4.3 Changes in membrane potential induced by mechanical stimulation of *Paramecium* sp. (a) Mechanical stimulation of the posterior end activates mechanosensitive K^+ channels. This leads to an efflux of K^+ ions, and thus to an hyperpolarization of the membrane. (b) Mechanical stimulation of the anterior end activates mechanosensitive Ca^{2+} channels, which are concentrated in this region. The resulting influx of Ca^{2+} ions causes a depolarization of the membrane. (After **Machemer, H.** (1988).)

Electrophysiological experiments in a number of ciliates have shown that the local receptor potentials spread rapidly and almost without any decay across the somatic membrane, so that the membranes of almost all cilia are instantly hyperpolarized or depolarized, respectively. In the ciliary membrane, the depolarizing receptor potential activates voltage-sensitive Ca^{2+} channels, leading to an influx of Ca^{2+} ions into the cilium. This results in increased depolarization which causes an opening of further Ca^{2+} channels and an influx of Ca^{2+} ions through positive feedback. Due to this mechanism, the intraciliary Ca^{2+} concentration transiently increases from resting levels near 10^{-7} M to approximately 10^{-4} M. This elevation of intraciliary Ca^{2+} concentration triggers, at a subcellular level, a reversal of the ciliary activity. Theoretical considerations have shown that an increment by just 100 Ca^{2+} ions per 10 μm length of a cilium is sufficient to cause a rise of the Ca^{2+} concentration from 10^{-7} to 10^{-6} M.

In contrast to the depolarization-induced ciliary activation, the coupling between mechanosensitive ion channels in the membrane and the hyperpolarization-induced ciliary activation is less well understood. At present, experimental data favor the hypothesis that the hyperpolarizing receptor potential causes, through a yet unknown mechanism, a down-regulation of the intraciliary Ca^{2+} concentration, which then leads to a reorientation of the ciliary power stroke. Thus, according to this hypothesis, Ca^{2+} acts as a universal messenger of ciliary electromotor coupling, mediating the fast phobotaxic responses of paramecians by up- and down-regulation of intraciliary calcium.

Galvanotaxis

If direct current is flowing in a solution, paramecians will show a characteristic orienting behavior in relation to the resulting gradient: they will swim, with their anterior pole first, toward the cathode. Reversal of the polarity of the electric field causes a reversal of the swimming direction of the paramecians. This orienting movement toward the cathode is referred to as **galvanotaxis**.

Although the galvanotaxic behavior hardly ever occurs under non-experimental conditions, its detailed analysis has provided an excellent insight into how the transduction of an external stimulus is coupled to a motor response. For the sake of simplicity, we will assume that a paramecian swims between the two metal plates of a parallel plate capacitor, as shown in Fig. 4.4. The polarity of the capacitor is indicated by the '+' (anode) and the '−' (cathode) symbols. Between the two electrodes, separated by a distance of 1 cm, a voltage of 1 V is applied. Then, 1 V drops over 1 cm or, assuming homogeneous conductivity within the solution, 0.02 V (or 20 mV) over the 200-μm length of a paramecian. In other words, we have applied an electric field with an

Cathode: negative pole of a voltage source.
Anode: positive pole of a voltage source.

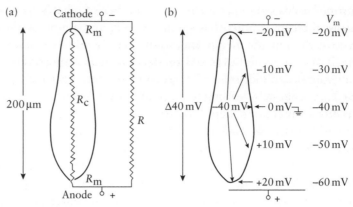

Figure 4.4 (a) Transcellular resistance of a paramecian. The resistance across the cell is equivalent to the sum of the resistances of the anterior half of the cell membrane, $R_{m\text{-anterior}}$, of the cytoplasm, R_c, and of the posterior half of the cell membrane, $R_{m\text{-posterior}}$. If a paramecian is brought into the electric field of a parallel plate capacitor such that its longitudinal axis is oriented parallel to the field lines, the transmembrane resistance and the resistance of the medium will be in parallel. (b) Polarization of a paramecian in an electric field. The transmembrane potential is equivalent to the difference of the potential within the cell and the potential of the external voltage gradient. As a result, the membrane potential is reduced (depolarized) toward the end closer to the cathode, and increased (hyperpolarized) toward the end closer to the anode. (After **Machemer, H.** (1988).)

Parallel plate capacitor: composed of two parallel metal plates isolated against each other. They have equal, but opposite, charges. By definition, the electric field lines arise from the positively charged plate (anode) and run, perpendicular to the plates, to the negatively charged plate (cathode).

intensity of 1 V/cm; the orientation of the electric field lines is perpendicular to the plates.

In such an electric field, a paramecian is subject to electrical stimulation, because it represents a spatially extended resistor arranged in parallel to the resistance of the medium. If the longitudinal axis of the paramecian is oriented parallel to the field lines, then the transmembrane resistance and the resistance of the medium are in parallel over the entire length of the cell. The resistance across the cell is the sum of the following three partial resistances arranged in series:

1. the resistance of the anterior half of the cell membrane, $R_{m\text{-anterior}}$ (approximately $8 \cdot 10^7$ ohm)

2. the resistance of the cytoplasm, R_c (approximately $1.5 \cdot 10^5$ ohm)

3. the resistance of the posterior half of the cell membrane, $R_{m\text{-posterior}}$ ($8 \cdot 10^7$ ohm).

These figures demonstrate that the resistance of the membrane of each of the two halves is roughly 500 times higher than the resistance of the cytoplasm. Thus, any voltage drop across the length of the cell is largely caused by the voltage drop across the resistances of the anterior and the posterior cell membrane, but only to a negligible extent by the voltage drop across the resistance of the cytoplasm. As a consequence, the entire cytoplasm is more or less at the same potential. In the case of a parallel orientation of the longitudinal axis of the paramecian and the electric field lines, roughly half of the assumed total voltage drop of 20 mV across the 200-μm length takes place at the resistances of each of the two membrane portions of the anterior and the posterior half of the paramecian.

To examine the effect of the voltage gradient on the polarization pattern of the membrane of the paramecian, we arbitrarily set the potential of the medium at the level halfway between the anterior and posterior pole at a reference value of 0 mV. Then, relative to this reference point, the potential of the medium at the level of the cell pole closer to the cathode is −10 mV, and at the level of the cell pole closer to the anode +10 mV. Thus, the voltage difference between these two points is, as expected, 20 mV. The potential within the cell at the level of the reference point is −20 mV, as the potential drops by this value across the two halves of the cell membrane. On the other hand, the same potential of the cytoplasm is found at the end of the cell closer to the cathode and the anode, respectively, because, as demonstrated above, the entire cytoplasm is more or less at the same potential. This has, however, an important consequence for the potential drop across different parts of the membrane: while at the cathodal end the membrane potential is only −10 V [−20 mV − (−10 mV)], at the anodal end the membrane potential is −30 mV [(−20 mV − (+10 mV)]. Thus, the part of the cell membrane closer to the cathode is depolarized, whereas the membrane portion closer to the anode is hyperpolarized.

The polarization of the cell membrane of the paramecian bears consequences for the beat direction of the cilia. As we have seen above, depolarization of the membrane results in the cilia beating toward the anterior pole of the cell, while hyperpolarization causes the cilia to beat toward the posterior pole. Thus, the cilia in the segments of the cell that are depolarized or hyperpolarized, respectively, beat in opposite direction. Since the beating toward the anterior end takes place only at depolarization values above approximately 5 mV, the portion of the cell that beats toward the posterior end is larger than the portion beating toward the anterior end. Overall, this causes the paramecian to adopt a position with its anterior end oriented toward the cathode (**homodromic orientation**), as any deviation from this orientation will automatically be corrected by the asymmetric number of cilia on either side of the depolarized segment beating in anterior direction, or of the hyperpolarized segment beating in posterior direction, respectively (Fig. 4.5). Simultaneously, due to the larger number of cilia beating in posterior direction, the paramecian will swim, with its anterior end first, toward the cathode.

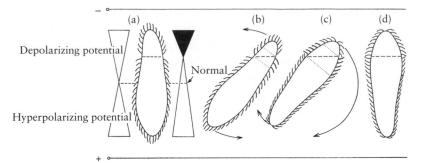

Figure 4.5 Orientation of a paramecian in an homogenous electric field of a parallel plate capacitor. (a) Homodromic orientation. The longitudinal axis is oriented parallel to the field lines traveling from the anodal plate in perpendicular direction to the cathode. Also, the anterior end of the paramecian is oriented toward the cathode. Those cilia that are subject to a depolarizing potential greater than approximately 5 mV beat toward the anterior pole. The remaining cilia beat in opposite direction. This orientation is stable because an equal number of cilia on either side of the longitudinal axis are activated; it results in movement of the paramecian toward the cathode. (b) and (c) Deviation from this homodromic orientation causes an asymmetry in the beat orientation of the cilia within the zone marked by the dotted lines, and thus to a turning movement indicated by the arrows, until the stable homodromic orientation of (a) is adopted. (d) Antidromic orientation, with the posterior end of the paramecian closer to the cathode. This orientation is unstable, since through an intermediary position similar to the one shown in (c), any deviation leads automatically to the homodromic orientation of (a). (After **Machemer, H.** (1988).)

A possible **antidromic orientation**, that is, an orientation of the paramecian with its longitudinal axis parallel to the field lines but with the posterior end positioned closer to the cathode and the anterior end closer to the anode would seemingly result in the opposite movement. However, this orientation is—in contrast to the homodromic orientation—unstable, so that any deviation automatically leads to a homodromic orientation and a 'correct' galvanotaxic response of the cell.

Geotaxis in vertebrates

The vast majority of animals assume a preferential position relative to the earth's force of gravity. An obvious prerequisite of this **geotaxis** response is that the animal is capable of **gravireception**. In vertebrates, the sensory organ mediating this sensory ability is the **otolith organ** of the inner ear. Together with the **semicircular canals** (which detects rotational movements of the head), it forms the **vestibular organ**. In mammals, the otolith organ gives rise to the **cochlea**, which is the sensory organ involved

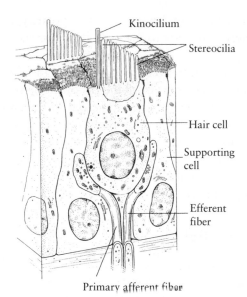

Figure 4.6 Hair cells of the vestibular system. The characteristic feature of these cells is the presence of hair-like projections at their apical surface, which are formed by several rows of stereocilia and a single kinocilium. The hair cells are embedded into a matrix of supporting cells covered with microvilli at their apical surface. At its base, each hair cell is innervated by an afferent process of a vestibular ganglion cell and an efferent process originating from cells in the brain stem. The efferent process mediates input to regulate the activity of the hair cell, whereas the primary afferent fiber provides an output channel to convey sensory information to the brain. (After **Kandel, E. R., and Schwartz, J. H.** (eds.) (1985).)

in hearing. All three organs—the otolith organ, the semicircular canals, and the cochlea—form together the **labyrinth**.

Labyrinth: otolith organ plus semicircular canals plus cochlea.

The otolith organ, whose structure and function will be examined in more detail below, consists of a patch (called **macula**) of sensory cells and a covering matrix. Because the sensory cells have at their tips fine 'hairs', they are commonly referred to as **hair cells**. Hair cells are also a central component of the semicircular canals and the cochlea, as well as of the lateral line system of fishes and aquatic amphibia. As shown in Fig. 4.6, the 'hairs' of each hair cell are composed of two types of cellular projections: one **kinocilium**, which, in mammals, disappears soon after birth, and a few dozens of **stereocilia**. Whereas the kinocilium is a true cilium exhibiting the typical 9 + 2 microtubuli structure in cross-sections, the stereocilia are, contrary to what their name implies, not cilia, but microvilli composed of an actin cytoskeleton ensheathed by a tube of plasma membrane.

The stereocilia are hexagonally packed and increase in length with increasing proximity to the kinocilium, which, if present, is always found at the tall extreme of the bundle of stereocilia. This arrangement allows the researcher to arbitrarily define an axis of polarity, running from the row of the shortest stereocilia to the kinocilium (or the row of the tallest stereocilia).

The stereocilia are extensively cross-linked. One type of connection known as the **tip link** emerges from the tip of the stereocilium and runs almost vertically along the extended long axis of the hair cell to join the side-wall of the adjacent taller stereocilium. In the macula, there is a covering of **otoliths**. Their main component are calcium carbonate crystals glued together by a jelly-like matrix. At the basal region of the hair cell body, synaptic contact is made with the ending of the **afferent (vestibular or eighth) nerve**.

The 'hair' bundle of the hair cell is composed of stereocilia (which are actually microvilli) and a kinocilium (which is a true cilium).

Effective physiological stimulus

For a long time, it remained unclear whether the mode of operation of the otolith organ depends on the **pressure** exerted by the otolith on the hair cells of the macula, or whether a **shearing force** provides an effective physiological stimulus. This question was addressed, and finally answered, by a series of experiments published by Erich von Holst (see Chapter 2) in 1950.

These experiments are based on the idea of increasing the weight of the otoliths. This raises both the pressure and the shearing force, but, as we will see below, not necessarily in equal amounts. If an animal tries to hold the shearing force constant, then the experiment would demonstrate that this parameter serves as an effective physiological stimulus. If, on the other hand, an animal tries to compensate for the pressure force, then the animal uses the latter parameter to define its position relative to gravity.

To provide the experimenter with precise measurements, Erich von Holst used two popular aquarium fishes, the tetra (*Gymnocorymbus* sp.) and angel fish (*Pterophyllum* sp.). The body of these two species is enormously laterally compressed, which greatly facilitates measurements of the angle the fish assumes relative to gravity. Moreover, both fishes are not more than a couple of centimeters long, which makes it possible to place the aquarium with the fish into a centrifuge and, thus, to experimentally increase the weight of the otolith.

The gravitational field to which the otolith organ is exposed can be manipulated by placing the animal into a centrifuge and modifying the speed of rotation.

In addition to gravity, the incident angle of light also determines the position of the fish. Normally, light comes straight from above, so that both the system mediating the **dorsal light reaction** and the mechanism underlying the response to the gravitational field tell the fish to adopt a vertical position. When light is presented from the side, the two positional mechanisms provide conflicting information. The result is a compromise, as shown in Fig. 4.7. Now, the fish assumes a tilted posture, with the dorsal part of its body oriented somewhat toward the source of light. The exact angle at which the fish deviates from the vertical position, which is

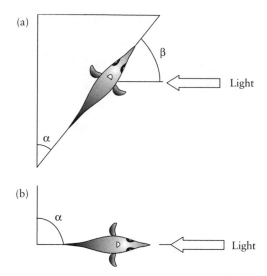

Figure 4.7 The orientation of the body of an angelfish is largely determined by gravity and the incident angle of light. (a) When light is presented to an intact fish from the side, the fish assumes a tilted position, with the dorsal part of its body oriented at a certain angle toward the source of light. (b) When the otoliths on both sides of the head are removed and identical stimulus is presented, the fish orients exclusively toward the light source. (After **Holst, E. v.** (1950).)

suggested by the otolith organ, depends, among other factors, on the intensity of the light. When both otoliths are removed, the fish relies solely on the light information and completely ignores the gravitational field.

To exclude the effect of light, von Holst kept the incident angle of light constant throughout the experiments. When he doubled the weight of the otoliths by increasing gravity from 1 to 2 g in the centrifuge, the fish adopted a position which showed that it attempted to hold the shearing force constant (Fig. 4.8). Thus, as suggested by these whole animal experiments, it is the shearing force component which provides the effective physiological stimulus to the otolith mechanism.

Physiological properties of hair cells

The finding of von Holst that the shearing force provides the effective physiological stimulus is reflected by the physiological properties of the hair cells. Due to the weight the otoliths exerted upon them, any tilting of the otolith organ causes a bending of the 'hairs'. The bending is sensed by the hair cells and leads to a characteristic change of the physiological activity exhibited at rest. The pattern of change is illustrated in Fig. 4.9,

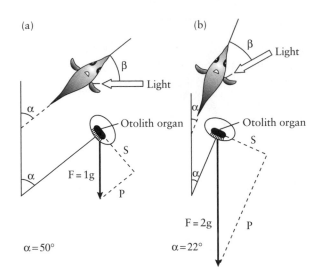

Figure 4.8 Experimental design employed by Erich von Holst to decide whether shearing or pressure force is used by a fish to detect gravity. (a) When light is presented from the side, the fish assumes a tilted position, with its back oriented somewhat toward the light source. Below the fish, the corresponding orientation of the otolith is drawn. The gravitational force, F, acting upon the otolith has two components—a shearing force, S, and a pressure force, P. (b) To increase the gravitational force from 1 to 2 g, the fish was placed into a centrifuge. Now the angle α that the fish maintains relative to the force of gravity F reveals that the shearing force S is held constant, while the pressure force P increases. Note that the incident angle of light is adjusted in the second experiment such that it is identical to the angle in the first experiment. (After **Holst, E. v.** (1950).)

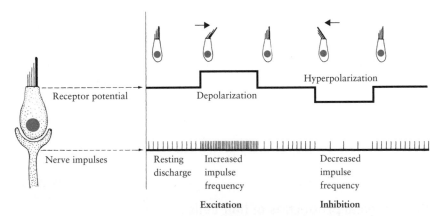

Figure 4.9 Physiological response of a hair cell to stimulation. The receptor potential of the cell is measured intracellularly, the rate of firing in the afferent fibers of the eighth nerve by extracellular techniques. Bending of the stereocilia toward the row of the tallest stereocilia causes depolarization in the hair cell and an increase in spike frequency in the vestibular nerve compared to the resting discharge. Conversely, bending of the stereocilia away from the row of the tallest stereocilia results in hyperpolarization and a decreased impulse frequency. (After **Flock, Å.** (1965).)

and can be summarized as follows:

1. Bending of the stereocilia toward the row of the tallest stereocilia results in depolarization of the hair cells and an increase in firing in the afferent fibers of the vestibular nerve.
2. Bending of the stereocilia away from the row of the tallest stereocilia results in hyperpolarization of the hair cells and a decrease in firing in the afferent fibers of the vestibular nerve.

Transduction mechanism

The **transduction** of the mechanical stimulation of the stereocilia into an electrical response has largely been elucidated through the work of James Hudspeth (now at Rockefeller University in New York) and David Corey (now at Harvard Medical School in Boston/Massachusetts). As their micromanipulation experiments performed on isolated sensory epithelium have demonstrated, the transduction is mediated directly by **mechanically sensitive channels**. This mode is in contrast to many other sensory receptor cells, such as photoreceptors and olfactory neurons, which employ cyclic nucleotides or other second messengers. The latter strategy has the advantage that the stimulus signal can be amplified. Furthermore, feedback within the metabolic pathway provides the opportunity for gain control, resulting, for example, in adaptation and desensitization. On the other hand, the transduction without the intervention of a second messenger results in a much higher speed of response. Hair cells operate not only more quickly than other sensory receptor cells, but even faster than neurons themselves. In the fastest hair cells found, the delay between the deflection of the hair bundle and the onset of a receptor current has been estimated to be only 10 μsec! Another remarkable feature of hair cells is the high sensitivity of their response. Studies have shown that a deflection of the tip of the stereocilia by as little as ± 0.3 nm, corresponding to $\pm 0.003°$, may be sufficient to trigger a response.

1 nanometer (nm) = 10^{-9} m.

Most transduction channels are concentrated near the tip of the stereocilia, rather than being located at the bases. The number of channels is probably quite low. In the bullfrog (*Rana catesbeiana*), the hair bundles of the sacculus have 100–250 transduction channels each. Since each of these hair cells has about 60 stereocilia, assuming even distribution through the bundle, each stereocilium would have only 2–5 channels.

The transduction channels are relatively non-selective for cations, but, due to the high concentration of potassium in the ionic milieu around the hair bundle, the major carrier for the transduction current is K^+. The channel gate is connected to the adjacent taller stereocilium by tip links, elastic filaments that exhibit spring-like properties (Fig. 4.10). During displacement toward the taller stereocilium, the tip links provide the

(a) Ion channel opens when
stereocilium bends

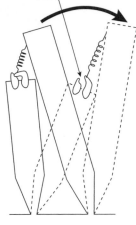

(b)

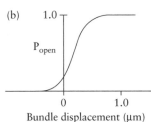

Bundle displacement (µm)

Figure 4.10 (a) Tip-link model for transduction. A mechanical stimulus originating from the bending of the 'hairs' of the hair cells acts directly on a mechanically sensitive ion channel in the stereocilium by modifying the tension in the 'gating spring' attached to the channel gate. Displacement of the bundle toward the tallest stereocilia stretches these tip links and, thus, increases tension. (b) Sensitivity curve relating displacement of the hair bundle to probability of opening of the transduction channels. The diagrams also demonstrate the enormous sensitivity of hair cells. Displacement of only one-third of a micrometer, corresponding roughly to the diameter of one stereocilium, already provides a saturating stimulus. Note that a certain fraction of channels are open at rest. (After **Pickles, J. O., and Corey, D. P. (1992).**)

tension to open the transduction channels. When deflection occurs in the opposite direction, the tip link tension slackens, thus increasing the probability that the channels close.

In an unstimulated hair cell, a small portion (approximately 10–25%) of the transduction channels are already open. The resulting inward, positively charged flux contributes, partially, to the depolarizing resting potential of about −60 mV. Deflection of the hair bundle toward its tall edge opens additional channels, and the influx of cation produces a depolarizing voltage charge. This initial event activates a variety of voltage-dependent channels across the basolateral membrane, thereby setting up the receptor potential. As the ultimate result, the release of transmitter is triggered at the basal region, where the hair cell synapses onto the ending of the afferent fibers, which, in turn, causes excitation in the vestibular nerve. Conversely, deflection of the stereocilia toward the short edge of the hair bundle closes transduction channels which are open at rest, thus leading to a reduced influx of cations into the cell. As a consequence, hyperpolarization occurs, resulting in a decrease in transmitter release and inhibition of firing in the fibers of the vestibular nerve.

Echolocation in bats

As mentioned at the beginning of the last section, hair cells play a pivotal role not only in maintaining an animal's equilibrium, but also in the processing of auditory information. How audition is involved in the orientation of animals has been particularly well examined in one of neuroethology's prime model systems—the echolocation in bats.

Bats are well known for their ability to emit high-pitch sound, the frequency of which is typically beyond the range of human hearing. In the dark, bats make use of echoes of this so-called **ultrasound** for orientation while flying, and for localization of prey animals during hunting. Ultrasound is particularly well suited for the detection of small objects because of its short wavelength.

Ultrasound: sound in the frequency range above that of human hearing, that is above approximately 20 kHz.

Exercise: Calculate the wavelength of sound of 80 kHz.

Solution: The wavelength is calculated according to the following formula:

$$s = f \cdot \lambda$$

where s, speed of sound (343 m/sec); f, frequency of sound; λ, wavelength of sound. Thus, the wavelength of sound of 80 kHz is approximately 4.3 mm.

Such wavelengths are short enough for sound to be reflected even by tiny objects. On the other hand, ultrasound exhibits a much stronger atmospheric attenuation than sound in the frequency range of human audition. For instance, sound of 20 kHz is attenuated by the atmosphere by 0.5 decibel per meter (dB/m). At 40 kHz, the attenuation is 1.2 dB/m, and at 100 kHz 3 dB/m. This attenuation effect limits echolocation to frequencies below 150 kHz, and to rather small distances between bat and object.

Bats form a large order, Chiroptera, among mammals, with approximately 950 different species. The two suborders are the Megachiroptera and the Microchiroptera. The Megachiroptera comprise about 150 species; they are, as the Greek prefix (*mega* = big) indicates, large bats. By contrast, the second suborder, Microchiroptera, are, as again the Greek prefix suggests (*micro* = small), small compared to the Megachiroptera. The Microchiroptera consist of approximately 800 species.

All but one genus of the Megachiroptera, which are frugivorous, do not echolocate at all. Instead, they have developed big eyes and rather small ears, and possess excellent night vision.

The beginnings of echolocation research

The beginnings of the study of bat echolocation date back to the end of the eighteenth century when several scientists conducted a series of experiments on how bats orient in the dark. Among them were the Italian Lazaro Spallanzani (1729–1799), a professor at the Universities of Reggio, Modena, and Pavia, and the Swiss Charles Jurine, a member of the Geneva Natural History Society. Jurine had found that, if he tightly sealed the ears of a bat with candle wax, the animal helplessly collided with obstacles in its flight path. Spallanzani confirmed these results and extended them by demonstrating that blinded bats could avoid obstacles perfectly well in closed rooms both during the day and at night. Based on these and many other experiments, Spallanzani concluded that the ears, rather than the eyes, of bats serve to orient the bat during flight.

The proposal of Spallanzani contrasted with the hypothesis of an influential contemporary, the French naturalist Georges Cuvier

(1769–1832), who explained the bat's capability through the presence of a nerve net on its wings that, he thought, acts as a highly sensitive sense of touch. Although Cuvier's explanation lacked experimental support, it was his hypothesis that was generally accepted for almost 150 years.

The situation changed only in 1938 when a Harvard undergraduate, Donald Griffin, published the results of a study in which he had examined the bat's orientation capabilities. In his investigation, Griffin used a novel device, the so-called sonic detector, which had been invented by the physicist George Pierce to detect ultrasound. What Griffin discovered laid the foundation for thousands of new research projects. He found that bats kept in a room emit high-energy ultrasonic pulses when flying around. The rate of production of these pulses increased as the bats approached objects. Similar to the experiments conducted by Spallanzani, Griffin demonstrated that plugging of the bat's ears severely interfered with its ability to avoid obstacles. The same kind of impairment occurred when he taped the mouth of the bat, thereby preventing the animal from emitting sound. Griffin concluded that bats scan their surroundings with their own-generated ultrasound, and that they use information contained in the reflected signals for localization of objects and orientation in the environment. He coined the term **echolocation** for this ability.

Classification of bat ultrasound

Based on their frequency spectrum, the sound produced by bats can be divided into three categories:

1. frequency-modulated (**FM**) signals
2. quasi-constant-frequency (**QCF**) signals
3. signals consisting of both constant-frequency components and frequency-modulated components (so-called **CF–FM** signals).

FM signals consist of short pulses, typically lasting for less than 5 msec, that quickly sweep downward in frequency during the course of the pulse. Since they cover a wide range of frequencies, these sounds are also called **broadband signals**. The North American big brown bat (*Eptesicus fuscus*) is such an 'FM bat'. A sonogram of the chirp-like pulses emitted by this species is shown in Fig. 4.11(a). As this plot demonstrates, the pulses last between 0.5 and 3 msec and sweep downward by about one octave. The figure shows, in addition, that, when the bat gets closer to a target, the pulses become shorter, and the frequency range covered by the individual pulses is shifted towards lower frequencies.

QCF signals dominate in a rather narrow frequency range, and are, thus, often referred to as **narrowband signals**. Changes in frequency, if they occur, are relatively slow. QCF signals last much longer than FM signals, typically between 10 and 100 msec.

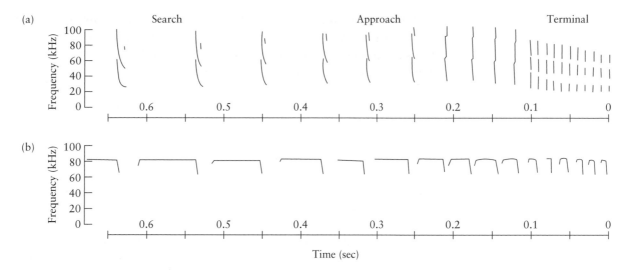

Figure 4.11 Sonograms of echolocation signals emitted by (a) the big brown bat (*Eptesicus fuscus*) and (b) the greater horseshoe bat (*Rhinolophus ferrumequinum*). The signals were recorded while the bats were approaching an insect in flight. Each of the two sequences includes pulses produced during the search, approach, and terminal capture stage. The time scale is plotted such that it counts down to the time of capture. The sonograms illustrate the different types of signals produced and the distinct signal patterns occurring in the different stages of prey hunting. (After **Simmons, J. A., Fenton, M. B., and O'Farrell, M. J. (1979).)**

Combined CF–FM signals consist of a long, constant tone, which is followed by a downward frequency-modulated sweep. A well-studied example of this type of signal is the sound of the greater horseshoe bat (*Rhinolophus ferrumequinum*). This signal is comprised of a constant-frequency component of approximately 83 kHz, followed by a brief downward frequency-modulated sweep. Such CF–FM pulses of *Rhinolophus*, recorded during pursuit of an insect, are shown in Fig. 4.11(b). As also revealed by the recording, the calling pattern changes in the course of pursuit. The closer the bat gets to the prey, the more CF–FM pulses it produces, and the shorter the individual pulses become.

The differences between the different types of calls reflect adaptations to specific foraging areas and hunting behaviors. Long narrowband signals are typically found in species that forage in open spaces, where the bats use them for long-range echolocation. Brief broadband signals are predominantly used by species that hunt near the ground or in an environment rich of denser vegetation. The bats employ the latter category of signals to distinguish prey from background clutter, and for measuring echo travel times. The type of signal emitted also depends on the behavioral situation. 'FM species,' for example, may include long QCF components during flight in open spaces.

In the following, we will describe in detail two aspects of the neuroethology of echolocation in bats: the mechanisms underlying distance estimation and the neural substrate of the so-called Doppler shift analysis.

FM signals: distance estimation

Theoretical considerations have suggested that FM signals are particularly well suited to measure the time delay between the emitted pulse and the return of the echo, because they allow a precise measurement of the timing of the signal. This information can be used to estimate the distance of the bat from the target.

> *Exercise: Assuming a velocity of sound in air of 344 m/sec (at a temperature of 20 °C), what distance does an pulse-echo delay of 1 msec correspond to?*
>
> *Solution: Within 1 msec, the sound wave travels 0.344 m or 344 mm. Upon emission from the bat's mouth and reflection by an object, the sound wave has to travel back over a similar distance to reach the bat's ear. Hence, the distance between bat and object is approximately 344/2 or 172 mm.*

Behavioral experiments by James Simmons, now at Brown University in Providence, Rhode Island, on *Eptesicus fuscus* have shown that the animals use, indeed, signal-to-echo delays to estimate target distance. Simmons initially trained bats to discriminate targets placed at different distances. Then, he replaced the natural echo caused by reflection of the sound from the target by electronically controlled 'phantom echoes'. This experimental manipulation allowed Simmons to vary the delay between the emitted sound and the returning 'echo,' so as to minimize the acoustic distance between the two phantom targets. These experiments demonstrated that the threshold above which the bats could still discriminate two targets based on signal-to-echo delays was 60 μsec, corresponding to a difference in target distance of 10–15 mm.

Further experiments revealed that the bats can even discriminate between a target fixed at a certain distance and a second jittering target with a temporal resolution of as good as 1 μsec. This is equivalent to 1/1000 the duration of a 'typical' action potential, and corresponds to a forward and backward movement of the second target of just 200 μm—a more than astonishing ability! How the central nervous system of the bat achieves this enormous temporal resolution is unknown.

Neurophysiological recordings have revealed neurons within the bat brain whose properties make them well suited to measure the delay between the emission of a pulse and the arrival of the returning echo. One

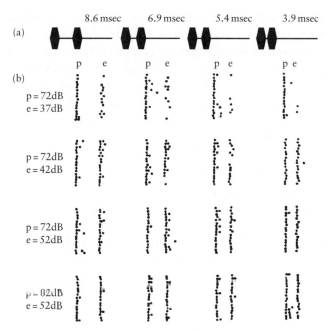

Figure 4.12 Properties of neurons in the inferior colliculus encoding echo delays. The neurons were stimulated with pairs of signals mimicking the emitted pulse (p) and the returning echo (e) at four different pulse-to-echo delay times (a). The response of these neurons at four different pulse/echo intensities, defined in decibel (dB), is shown in (b). Each line of dots represents the spike response following stimulation with one pulse/echo stimulus. For each echo delay and intensity combination, the results of 16 individual recordings are shown. (After **Pollak, G. D., Marsh, D. S., Bodenhamer, R., and Souther, A.** (1977).)

such neuronal class was identified in a midbrain structure, the inferior colliculus, by George Pollak (now at the University of Texas at Austin) in the late 1970s. These neurons respond best to pairs of FM pulses by encoding for the time interval between the emitted pulse and the returning echo. They, thus, serve as accurate time markers. Figure 4.12 summarizes this response to paired stimulus pulses simulating the emitted pulse and the returning echo. At the top of this figure, four different stimulus pairs with different pulse-to-echo delays are shown. These stimulus pairs were applied at different intensities, indicated on the left of the three rows below. In the first row, the response of the neurons to a pulse presented at an intensity of 72 dB and the simulated echo presented at 37 dB is shown. In the other experiments summarized in the second, third, and fourth row, the pulse/echo intensities were 72/42, 72/52, and 82/52 dB, respectively. Each black dot represents an action potential produced in response to the respective stimulus pair.

Taken together, this plot reveals two important features of these neurons. First, they copy accurately the different pulse-to-echo delays by generating one or two action potential in a phase-locked manner. Second, the timing of these action potentials is largely independent of the intensity of the pulse and the echo. The latter feature is important, since the perceived loudness of the pulse or echo should not affect the encoding of the timing. This is very much in contrast to many neurons in other sensory systems, which tend to respond to stimuli of higher intensity with shorter latency.

The delay-encoding property of the neurons is due both to a sharp tuning of each of the delay-sensitive neurons to a particular frequency within the FM sweep, and to a low threshold for the production of an action potential in response to a stimulus presented at that frequency. Since different neurons lock to the emitted pulse and the returning echo of different frequency components within one FM sweep, multiple measurements of the time delay are taken. This increases the precision with which the distance between bat and target is estimated.

The information encoded by the time-marker neurons in the inferior colliculus is then, together with other pieces of information collected in lower auditory sensory structures, conveyed to the auditory cortex. There, different parameters of auditory signals are encoded in different, anatomically separated areas. One of these regions, the **FM–FM area**, processes information related to echo delays. The properties of the neurons in this area have been analyzed in great detail by Nobuo Suga, one of the pioneers of the neuroethology of bats. Suga, who has been on the faculty of Washington University in St Louis (Missouri) since 1969, and his associates have carried out most of their studies on the mustached bat, *Pteronotus parnellii*, which emits CF–FM signals.

Similar to the neurons in the inferior colliculus, neurons in the FM–FM area of the auditory cortex of *Pteronotus parnellii* respond poorly when a pulse, echo, QCF signal, or FM sweep is presented alone, but do so vigorously when a sound pulse is followed by an echo at a particular delay time. Detailed analysis has shown that they compare the emitted pulse with the delayed echo. These delay-sensitive neurons are arranged in a topographic fashion, with the delay time increasing along one axis. The range of delays represented along this axis varies from 0.4 to 18 msec, which corresponds to target distances between 7 and 310 cm. This is in excellent agreement with the range over which the bats react to prey under natural conditions.

CF signals: Doppler shift analysis

As mentioned above, the echolocation signals of *Rhinolophus ferrumequinum* consist of a rather long (typically 10–100 msec) CF component with a frequency of 83 kHz, followed by a short downward-sweeping FM component (typically lasting a few milliseconds). Theoretical considerations, similar to the ones used in the interpretation of radar signals, suggest that the long CF part of the signal could be well suited for a so-called **Doppler shift analysis**. Doppler shifts occur whenever the source of a sound and the receiver of this sound are in relative motion toward one another. The bat experiences such Doppler shifts under two behaviorally relevant situations. The first situation arises when the bat emits a pulse of 83 kHz, while it is flying. Then, the echo returning from an object that the bat is approaching is higher than 83 kHz, usually between 83 and 87 kHz. This causes a problem, since the bat's acoustic system is most sensitive to sound at 83 kHz, and the range of tuning to this frequency is quite narrow. To compensate for this loss in sensitivity, the bat lowers the frequency of the sound emitted as soon as it detects a positive Doppler shift, that is, an increase in frequency of the returning echo compared to the internal reference frequency. The adjustment in emitted frequency is made according to the magnitude of the Doppler shift. This behavioral response, called **Doppler shift compensation**, was discovered by Hans-Ulrich Schnitzler, now at the University of Tübingen, Germany, in 1968. It ensures that the frequency of the echo is kept in the range of most sensitive hearing. In addition, the lowering of the frequency of the emitted sound also helps *Rhinolophus* to protect itself from deafening with its own very intense sounds, since the frequency region just below 83 kHz is distinguished by a remarkably high hearing threshold.

Doppler shifts were discovered by the Austrian physicist Christian Johann Doppler (1803–1853) in 1842. They involve a shift in frequency and occur whenever the source of any kind of propagating energy, such as light or sound, moves relative to the receiver of this energy. For example, the siren of an ambulance, which is perceived as a constant pitch by a listener as long as both ambulance and listener are stationary, changes in pitch (= frequency) when the ambulance goes by. As the ambulance approaches, the frequency increases, since sound waves from the siren are compressed toward the observer. As the ambulance moves away, the frequency decreases, since the sound waves are stretched relative to the listener.

A second situation in which *Rhinolophus* makes use of Doppler shift analysis occurs during prey detection. *Rhinolophus* usually hunts for insects, mainly flying moths and beetles. The fluttering of the wings of the prey animal produces Doppler shifts in the echo frequency, which, despite their small size, can be detected by the bat. Figure 4.13 shows an oscillogram and real-time spectra of echoes returning from a flying noctuid moth. Comparison of the oscillogram and the spectra with those of a stationary moth demonstrates that the movement of the wings causes strong rhythmical changes in amplitude and spectral composition, including

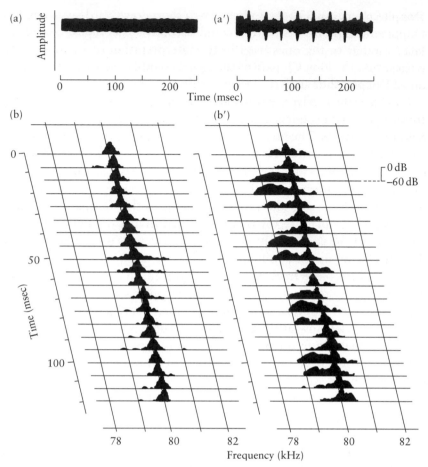

Figure 4.13 Frequency and amplitude modulations of the echolocation signals of the greater horseshoe bat caused by flying moths. For the analysis, a noctuid moth was mounted in the acoustic beam of a loudspeaker transmitting a continuous sinusoidal signal at a frequency of 80 kHz. (a) Oscillogram of the echo returning from the moth *Phytographa gamma* at rest. (a′) Oscillogram of the echo returning from the flying moth. The clearly visible amplitude modulations are caused by the fluctuations of the reflection diameter of the moth, which originate from the movements of the wings. (b) Real time spectra of the echoes returning from the moth at rest. The echoes exhibit single peaks at the transmitted frequency, 80 kHz. (b′) Real time spectra of the echoes returning from the flying moth. The spectra are asymmetrically broadened toward lower frequencies. These broadenings occur in the rhythm of the wing beat of the moth. (After **Schnitzler, H.-U., and Ostwald, J. (1983).**)

an asymmetrical broadening of the frequency of the echo compared to the emitted sound.

Unlike the Doppler shifts caused by the relative motion of an object in the flight path of the bat, *Rhinolophus* is unable to compensate for the

Doppler shifts produced by the fluttering of targets. The latter types of Doppler shifts occur at each wing beat of the insect, thus making them too fast for the vocal system of the bat to follow. The frequency of the echo returning from a flying insect is, therefore, continuously oscillating up and down around 83 kHz.

Can *Rhinolophus* detect the rather small Doppler shifts originating from the flutter of an insect's wings? In order to answer this question, Schnitzler and his group trained bats to discriminate between a sinusoidally oscillating target and a similar but motionless target. By controlling amplitude and frequency of the target's oscillations, defined modulations of the returning echo could be produced. After a training phase, the bats were able to discriminate between an oscillating and a motionless target. Then, the magnitude of the oscillations were reduced to determine the discrimination threshold. The results of these experiments demonstrated that the bat's detection system is extremely sensitive. A target oscillating at a rate of 35/s (which is typical of the wing movements of many insects) causes Doppler shifts of ±30 Hz, that is, a 83 000-Hz signal emitted by the bat is modulated such that the echo ranges from 82 970 to 83 030-Hz—a maximum deviation from the emitted frequency of less than 0.04%! This is an order of magnitude better than the frequency discrimination ability of humans in the best frequency range.

That the bat is, indeed, able to discriminate with this tremendous sensitivity is further demonstrated by a second behavior of *Rhinolophus*. The bats always lengthen the CF component of the sound, while reducing the length of the FM sweep, when oscillations of targets are imitated. This lengthening of the CF component of the signals provides the bat with more time to analyze a fluttering target.

Adaptations of the auditory system

The behavioral experiments have shown that the sound-transmitting structures of *Rhinolophus* are specialized so as to keep the frequency of the returning echo in an 'expectation' window of 83 kHz, as long as the Doppler shifts do not occur at too high a rate. Is this specialization of the sound-transmitting structures paralleled by a specialization of the sensory and neural structures receiving and processing the echoes?

Like in any other mammal, in bats sound pressure waves are transmitted down the external auditory canal of the outer ear to the eardrum or tympanum. There, the sound pressure waves cause the tympanum to vibrate at the same frequency (in case of a pure tone) or frequencies (in case of more complex sound) as the sound wave. These vibrations are transmitted from the outer to the inner ear via the three middle-ear ossicles—malleus, incus, and stapes. The footplate of the stapes adheres to the oval window, a patch of membrane that vibrates in concert with the vibrations of the tympanum and the three ossicles. Vibrations of the oval window cause vibrations in the cochlear fluid of the scala vestibuli, one of the three chambers of the cochlea. These vibrations travel along the coils of the cochlea in the scala vestibuli to the apex, where they enter the scala tympani. In the latter structure, the vibrational waves are propagated back down the coils of the cochlea. The third cochlear duct, the scala media, is positioned between the scala vestibuli and the scala tympani. It contains the auditory receptor cells—the inner and outer hair cells.

The hair cells are arranged in several rows on the so-called basilar membrane. Their stereocilia project upward, where at least some of them make contact with the tectorial membrane. Vibrations in the cochlear fluid induce displacement of the basilar and tectorial membranes, which, in turn, produce a shearing force on the hair cells, including their stereocilia. This deflection leads, like in any other hair cell and as described in detail above, to a systematic change in the membrane potential of the hair cells.

How are the various parameters of sound, particularly frequency, encoded by the cochlea? Studies of the Hungarian scientist and engineer Georg von Békésy (1899–1972), in the 1930s and 1940s, showed that different frequencies of sound exert their maximal effect at different points along the basilar membrane. High frequencies lead to a maximal displacement of the basilar membrane near the oval window, low frequencies closer to the apex. This encoding of sound frequency by the position of the hair cell along the basilar membrane is referred to as the **place theory of hearing**.

In most mammals, the frequency decreases on a logarithmic scale with increasing distance from the oval window. However, in *Rhinolophus*, similar to other bats that use a long CF component in their echolocation signals, this representation of frequencies on the basilar membrane exhibits a remarkable specialization. Around the place where 83 kHz are represented (roughly in the region of 80–86 kHz), the basilar membrane is greatly expanded in terms of both length and thickness compared to lower frequencies. This structural expansion of the basilar membrane is, by analogy with the fovea of the retina, frequently referred to as the **acoustic fovea**. The highest frequency expansion factor, expressed as length of basilar membrane over one octave, is found just above the resting frequency. In this range, the frequency expansion factor is more than 40 times higher than the corresponding factor in the frequency range below 70 kHz.

This over-representation of frequencies in the range of the CF component is accompanied by a high density of innervation of the cochlea by first-order neurons. It has been shown that the patch of hair cells representing the frequency range from 3 kHz below to 3 kHz above the resting frequency is innervated by 21% of all first-order ganglion neurons that make synaptic contact with the cochlear hair cells. These figures underline the importance of this rather small frequency range for sensory processing of acoustic signals.

A similar over-representation of frequencies relevant for echolocation, as found in the sensory structures and first-order neurons, occurs in higher-order brain structures. In the auditory cortex of *Rhinolophus*, the number of neurons tuned to frequencies in the range of echolocation signals is highly over-represented compared to neurons that are most sensitive to other frequencies. As Hans-Ulrich Schnitzler and his associate Joachim Ostwald found, these neurons can be divided into two groups: one that encodes frequencies within the 'expectation window' of the CF component, and the other with a frequency representation in the range between the resting frequency of the CF component and approximately 70 kHz, which covers the bandwidth of the FM sweep. Schnitzler and Ostwald called these regions 'CF area' and 'FM area', respectively. This finding shows that the CF part and the FM part of the echoes are represented by two distinct areas of the auditory cortex. Such a distinct representation suggests that these two areas of the auditory cortex analyze different types of information contained in the different parts of the echo.

Further physiological experiments revealed neurons within the CF area of *Rhinolophus* that are selectively responsive to signals mimicking the echo caused by a flying moth. Figure 4.14 shows such a response. As a control signal, a pure sine-wave tone was used. No response of the neurons was evident to control signals (Fig. 4.14(a)). However, as soon as the pure tone was modulated so as to reproduce moth echoes, a clear response became apparent (Fig. 4.14(b)). Each wing beat in the moth echo caused a phase-locked discharge of the moth-echo-selective neurons.

Counter sonics: the prey's adaptations

As the two examples discussed in detail above demonstrate, the peripheral and central receiver structures of the bats exhibit a number of adaptations for the analysis of behaviorally relevant parameters conveyed in the echoes. However, do the adaptations of the bats to the moth echoes tell the entire story of the bat–prey relationship? The answer is a definite 'no,' because noctuid moths, in turn, have developed a number of counter-adaptations to avoid predation by the bats.

These counter-strategies were examined in great detail by Kenneth Roeder of Tufts University in Massachusetts more than half a century ago in what can be considered one of the first neuroethological studies. Roeder found that the moth's adaptations include the development of ears that are, in general, most sensitive between 20 and 60 kHz—the

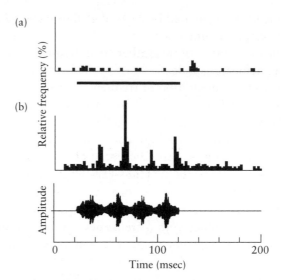

Figure 4.14 Response of a moth-echo-selective neuron in the CF area of the auditory cortex of *Rhinolophus*. (a) Stimulation with a pure tone. The duration of the stimulus is indicated by the horizontal bar beneath the histogram. (b) Stimulation with a mimicked echo of a flying moth. The oscillogram of the stimulation echo is shown beneath the histogram. Each of the four wing beats, evident from the amplitude peaks, causes a phase-locked discharge of the neuron. This response is evident from the pronounced increase of the relative number of spikes immediately after the wing beat. (After **Schnitzler, H.-U., and Ostwald, J.** (1983).)

bandwidth used by most species of echolocating bats. The ear of the moth consists of a tympanic membrane stretched over an enclosed ear sac located on the metathorax. Two sensory neurons, designated A1 and A2, contact the tympanic membrane. Each of these acoustic neurons is most sensitive to ultrasound. However, A1 is the more sensitive neuron. It starts to respond when the bat is still quite far away, that is, the intensity of the sound is low. A2 is the less sensitive neuron; it begins to respond only when the bat is close to a few meters.

This physiological response characteristic of the A1 and A2 neurons is paralleled by a differential behavioral response. If the moth hears the bat in the search phase of hunting, it hides or flies away before the bat catches it. If the moth detects the bat sound only in the terminal phase of hunting (when both sensory neurons are saturated), it starts flying in loops to confuse the bat, or it power-dives to the ground by folding its wings.

Certain tiger moths (family Arctiidae) have even gone one step further by producing their own ultrasonic sounds. They emit high-frequency clicks during the terminal phase of attack. These clicks are thought to either signal that the moths are distasteful, or to cause a jamming effect by generating 'phantom echoes', thereby preventing the bat from predicting the position of the moving moth.

Summary

- Among orienting responses, taxes form an important category. They involve an orienting reaction or movement in freely moving organisms directed in relation to a stimulus.

- Taxis responses are commonly specified by a prefix indicating the nature of the stimulus, and by the modifiers 'positive' and 'negative' indicating orienting movements toward or away from the source of stimulation, respectively.

- Taxis responses occur even in organisms without a nervous system, such as the unicellular protist *Paramecium* sp. This ciliate protozoan exhibits, among others, chemotaxis, phobotaxis, and galvanotaxis.

- The taxis behaviors in paramecians are mediated by cilia, whose stroke pattern is largely determined by the intraciliary Ca^{2+} concentration.

- Mechanical stimulation of the posterior end of a paramecian leads to a hyperpolarizing receptor potential, whereas such stimulation of the anterior end results in a depolarizing receptor potential. These differential receptor potentials cause transient up- and down-regulation, respectively, of the intraciliary Ca^{2+} concentration, which, in turn, lead to opposite ciliary activities.

- In an electric gradient, paramecians exhibit galvanotaxic behavior by swimming, with their anterior pole first, toward the cathode. This behavior is triggered by a differential polarization of those parts of the cell that are closer to the cathode and the anode, respectively, and, thus, by the generation of a differential beat pattern of the cilia in different parts of the cell.

- Geotaxis involves orientation of the body relative to gravity. This behavioral response requires that the animal be capable of perceiving the earth's gravitation field, a sensory capability referred to as gravireception.

- In vertebrates, gravireception is mediated by the otolith organ of the inner ear. This organ consists of a macula of sensory cells called hair cells. The 'hairs' at the apical region of each hair cell are composed of two types of projections, one kinocilium and a few dozens of stereocilia. The stereocilia are hexagonally packed and increase in length with increasing proximity to the kinocilium, which, if present, is found at the tall extreme of the bundle of stereocilia. A mechanically sensitive channel at the tip of each stereocilium is connected with the side-wall of its adjacent taller neighbor via a tip link. The hairs of the hair cells are covered by otoliths.

■ Behavioral physiological experiments have shown that it is the shearing force, and not the pressure force, exerted by the otoliths on the hair cells of the macula that provides an effective physiological stimulus to signal positional changes of the head.

■ Any tilting of the head, and thus of the otolith organ, causes a bending of the hairs, which leads to a characteristic physiological response such that (i) bending of the stereocilia toward the row of the tallest stereocilia results in depolarization of the hair cells and an increase in firing in the afferent fiber making synaptic contact with these cells; (ii) bending of the stereocilia away from the row of the tallest stereocilia results in hyperpolarization of the hair cells and a decrease in firing in the afferent fibers.

■ The transduction of the mechanical stimulation of the stereocilia into an electrical response is mediated by mechanically sensitive transduction channels concentrated near the tip of the stereocilia. During displacement of the stereocilium toward the adjacent taller stereocilium, the tip link provides the tension to open the transduction channel, leading to an influx of cations, and thus to a depolarization of the membrane of the hair cells, which is followed by an increase in firing in the afferent fiber. Conversely, deflection of the stereocilia toward the shorter edge of the hair bundle closes transduction channels open at rest, leading to a reduced influx of cations into the cell, and thus, to a hyperpolarization, which is then followed by a decrease in firing in the afferent fiber.

■ Hair cells play also a central role in processing of auditory information, including information used for acoustic orientation. The neural basis of this ability has been particularly well examined in echolocating bats, which typically emit signals in the ultrasound frequency range

■ Based on the frequency spectrum, the ultrasound produced by bats can be divided into frequency-modulated (FM) signals; quasi-constant-frequency (QCF) signals; and signals consisting of a rather long constant-frequency component, followed by a downward frequency-modulated sweep (CF–FM signals)

■ FM signals are, among others, used to estimate the distance of the bat from an object. This is achieved by measuring the time delay between the emitted signal and the returning echo

■ In the inferior colliculus of bats, neurons have been identified that serve as time markers by encoding for the time interval between the emitted pulse and the returning echo. In the auditory cortex, neurons exist that respond best to a particular pulse-echo-delay time

■ The CF component of bat ultrasound is particularly well suited for the so-called Doppler shift analysis, which is performed during flight and when localizing flying insects. During flight movements of the bat, this analysis forms the basis for Doppler shift compensation to keep the frequency of the returning echo in the range of maximal sensitivity of the auditory system. During prey hunting, Doppler shift analysis is used to detect the fluttering of wings of the flying insects

- Both sensory structures in the ear and neural structures in the brain of the bat are specialized to maximize sensitivity to frequencies in the range of the CF component of the returning echo. This includes an over-representation of such frequencies in the cochlea and the auditory cortex

- To counteract bat echolocation, noctuid moths have developed a number of adaptations. They include the development of ears sensitive in the ultrasound range and, in some species, the production of high-frequency clicks.

Recommended reading

Adler, J. (1987). How motile bacteria are attracted and repelled by chemicals: an approach to neurobiology. *Biological Chemistry Hoppe-Seyler*, **368**, 163–173.

An excellent starting point to learn more about the genetic and biochemical basis of chemotaxis behavior of bacteria. Written by a biologist who originally wanted to study neural mechanisms of behavior in animals. However, due to the lack of proper approaches available for the study of higher organisms at the time when he started his career, Adler turned to bacteriology, where he laid the foundation for today's research in this fascinating field.

Griffin, D. R. (1958). *Listening in the dark. The acoustic orientation of bats and men.* Yale University Press, New Haven.

Although published half a century ago, still a superb book to read. Written by the discoverer of the bats' ultrasonic sounds, who has been not only an outstanding scientist, but also an extremely stimulating writer.

Hudspeth, A. J. (1989). How the ear's works work. *Science*, **341**, 397–404.

A stimulating review on the biophysics and physiology of hair cells, written by one of the pioneers in this area of research.

Pickles, J. O., and Corey, D. P. (1992). Mechanical transduction by hair cells. *Trends in Neurosciences*, **15**, 254–259.

This well-written review article focuses on the transduction of the mechanical stimulus into an electrical signal by the hair cells.

Questions

4.1 Describe how mechanical stimulation of paramecians leads, through proper activation of the cilia, to a phobotaxis response.

4.2 What is galvanotaxis? How is this behavioral response mediated at the cellular level in paramecians?

4.3 The postural reaction of vertebrates is largely determined by the action of the vestibular system. What is the effective physiological stimulus? How is this stimulus transduced into a physiological response of the vestibular organ?

4.4 How do bats that emit FM signals gauge distance to a prey animal? What neural substrate underlies this behavioral ability?

4.5 How does Doppler shift analysis help bats to detect prey? What adaptations of the bat's auditory system make this analysis possible?

Neuronal control of motor output: swimming in toad tadpoles

5

■ Introduction
■ The behavior
■ A physiological approach to study swimming behavior
■ The spinal circuitry controlling swimming
■ Operation of the swimming circuitry
■ Coordination of oscillator activity along the spinal cord
■ Summary
■ Recommended reading
■ Questions

Introduction

Although simple reflexes in mammals were first described as early as in 1906 by Charles Sherrington, we still cannot fully explain, at a cellular level, how the contractions of the muscles are achieved upon perception of a stimulus in the skin. Similarly, we largely lack a mechanistic understanding of how simple rhythmic motor patterns, such as walking, are generated in adult mammalian species.

To avoid the difficulties associated with the complexity of the adult mammalian nervous system, behavioral neurobiologists have searched for model systems in which the neural circuits controlling motor patterns are simple in terms of their structure and their function, and are readily accessible to experimental analysis. One of these model systems will be described in this chapter. It centers around a simple behavior, swimming, which appears to be controlled by just three neural cell types in the spinal cord of hatchling clawed-toad tadpoles. This model system was 'discovered' by Alan Roberts (see Box 5.1), who, together with his group at the

---

BOX 5.1   Alan Roberts

Alan Roberts in his laboratory at the University of Bristol. (Courtesy: A. Roberts.)

Alan Roberts, born 1941 in Rugby (UK), is Professor of Zoology at the School of Biological Sciences of the University of Bristol (UK). From 1960 to 1963, he studied zoology at the University of Cambridge, where his scientific hero was Hans Lissmann (see Chapter 7). Roberts then went to the USA to do his Ph.D. on crayfish escape behavior with Theodore Bullock (see Chapter 2), first at the University of California, Los Angeles, and later at the University of California, San Diego. In 1967, Roberts returned to the United Kingdom to become a

Research Fellow in the Department of Zoology of the University of Bristol, where he subsequently progressed through the faculty ranks, until he was appointed to a Personal Chair in 1991.

When commencing his first position at the University of Bristol, Alan Roberts was looking for an easily available animal species with simple behavior and a nervous system composed of only a limited number of neurons. Although he had done his Ph.D. on crayfish, Roberts, nevertheless, decided that the central nervous system of most adult invertebrates is still too complex and causes intractable problems. He, therefore, turned to the developing nervous system of 'simple' vertebrates. His final decision to use hatchling *Xenopus* embryos as a new model system was, as he says, 'inspired by George E. Coghill's book on *Anatomy and the Problem of Behavior*, published in 1929, and by the fact that a breeding colony of *Xenopus* was already established by other colleagues at Bristol at the time of [his] arrival.' Since then, Alan Roberts' choice has, indeed, proven to be an excellent one, having led to important breakthroughs in the analysis of the origin of behavior at the level of individual neurons.

University of Bristol (UK), has also made major contributions to the system's behavioral and neuronal analysis.

The behavior

Embryos of the clawed-toad (*Xenopus laevis*) hatch after two days of development. The first day out of the egg, the 5–6 mm long tadpoles spend most of the time hanging from a strand of mucus secreted by a cement gland located on the head. During this developmental stage, which lasts one day or so, the digestive system matures and the mouth opens. Then, the next phase of development, characterized by the constant swimming and filter-feeding of the tadpole, starts.

Although the tadpole is at rest most of the time during the first day of free life, it can initiate swimming upon stimulation by dimming the light or by touching any part of its body. The sequence of this behavior is

4 mm

Figure 5.1 Tracing of swimming movements of hatching *Xenopus* embryos. The sequence shows the lateral undulations of the body during swimming at 3.3 msec intervals. The alternate contractions of the antagonistic segmented trunk muscles on the two sides of the body produce waves of bending that continuously progress in caudal direction. The arrowheads indicate the points of maximal curvature. (After **Arshavsky, Y. I., Orlovsky, G. N., Panchin, Y. V., Roberts, A., and Soffe, S. R.** (1993).)

shown at 3.3 msec intervals in Fig. 5.1. When touched, the tadpole bends to the opposite side and swims away by producing **lateral undulations** of the body at a frequency of 10–25 Hz. These undulating movements are the result of alternate contractions of the antagonistic segmented trunk muscles on the left and right sides. The contractions produce waves of bending spreading from the head to the tail at a speed of approximately 15 cm/sec. This drives the tadpole forward at about 5 cm/sec. Under natural conditions, swimming ceases when the tadpole bumps into objects in the water, at which point it attaches itself to substrate using secreted mucus. Under experimental conditions, swimming can be stopped by applying gentle pressure to the head or the cement gland. In the free-swimming tadpole, the stimulus-induced swimming helps the animal to escape from predators and can, thus, be regarded as an **escape behavior**.

Swimming in the hatching tadpole is performed by lateral undulations of the body and forms part of the escape behavior.

A physiological approach to study swimming behavior

The small size of the hatching *Xenopus* tadpoles bears both advantages and disadvantages for experimentation. In the embryonic stage just before hatching, the spinal cord is small, 100 μm in diameter and a few millimeters in length. At this stage of development, there are only approximately 1000 neurons on each side of the cord. Based on results of neuronal tract-tracing experiments (see Chapter 3), these neurons form just eight different anatomical categories. This makes it much easier to relate neuronal activity to certain behavioral patterns than is the case in more complex systems, such as the spinal cord of adult tetrapodes. Furthermore, the small size of the hatching *Xenopus* tadpoles eases pharmacological manipulations, as there is a rapid diffusion of drugs to spinal neurons. Also, in wholemount preparations all neurons can be seen from the surface, thus making it unnecessary to cut sections of the central nervous system to study the morphology of neurons.

On the other hand, the small size makes electrophysiological recordings from the whole free-swimming animal impossible. However, Alan Roberts and associates discovered that this limitation can be overcome by studying **fictive swimming**, instead of real swimming behavior. This is achieved by immobilizing the animal by blocking synaptic transmission at the neuromuscular junction with α-bungarotoxin. Then, the tadpole can be held on its side in a small dish, perfused with a physiological saline solution, and dissected to expose muscles and spinal cord. **Extracellular recordings** can, for example, be made by placing electrodes near motor neuron axons innervating the swimming muscles. **Intracellular recordings** are possible from individual neurons of the spinal cord using glass capillary microelectrodes inserted across the cell membrane. Figure 5.2(a) sketches the preparation for such physiological experiments. The top trace of Fig. 5.2(b) shows the neural activity of a single motor neuron recorded with an intracellular microelectrode during fictive swimming, which can be evoked by similar stimuli as actual swimming in the intact animal. The two traces below are corresponding extracelluar recordings

'Fictive behaviors' are sequences of motor neuron activity occurring without the production of actual movement or muscular contraction. Such behaviors can be observed in immobilized whole animals or in isolated preparations of the nervous system, in which muscles are removed.

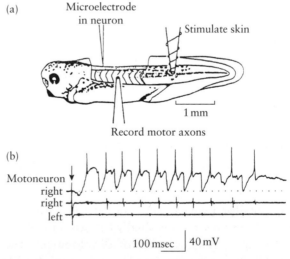

Figure 5.2 (a) Preparation for monitoring physiological activity of spinal neurons during fictive swimming in a tadpole with the brain removed. Fictive swimming is evoked by stimulation of the skin. (b) Activity of motoneurons during fictive swimming. The lower two traces are the results of extracellular recordings and show the alternating motor discharge traveling to the swimming muscles on either side of the body. Note the alternation of discharges of the motoneurons on the left and right side. The top trace, obtained by using intracellular techniques, shows the activity of a single motoneuron. (After **Roberts, A.** (1990).)

of the alternating motor discharge traveling to the swimming muscles on either side of the body. They demonstrate the alternation of firing of the motoneurons on the left and the right side of the spinal cord.

α-bungarotoxin is a constituent protein of the venom of the Southeast Asian krait (*Bungarus multicinctus*). By binding to nicotinic postsynaptic receptor sites, this neurotoxin irreversibly blocks cholinergic transmission at the neuromuscular junction, thus producing muscle paralysis.

The spinal circuitry controlling swimming

Out of the eight different types of neurons in the spinal cord of hatching tadpoles, only three appear to play a role in generating the basic swimming motor pattern. They are referred to as motoneurons, descending interneurons, and commissural interneurons, and together form the **central pattern generator** (see Box 5.2) that controls swimming. The central

---

BOX 5.2   Central pattern generator

The production of **rhythmic motor patterns** is a universal phenomenon in animals. Examples are not only basic physiological activities, such as ventilation and chewing, but also more obvious behaviors, such as swimming and walking.

At the beginning of the twentieth century, two models were proposed to explain the generation of such rhythmic movements. One model, whose main proponent was Charles Sherrington, is based on the assumption that sensory receptors continuously monitor the activity of the muscles. This information is then fed back into the motor system to activate reflexes that trigger the next step in the sequence of the individual repeated actions. Thus, this model proposes that rhythmic motor patterns are the result of a **chain of reflexes**.

The alternative model assumes the existence of an 'oscillator' within the central nervous system. This oscillator is based on a network of neurons that provide the timing to rhythmically activate the muscles. This concept was first proposed by Graham Brown of the University of Liverpool; the neuronal oscillator is nowadays commonly referred to as the **central pattern generator**.

A central pattern generator may, thus, be defined as a neuronal network that can produce a rhythmic motor pattern in the absence of any sensory input, or input arising from a higher center within the central nervous system and conveying timing cues. The number of neurons encompassing central pattern generators ranges between 10–30 and thousands, depending on the system and the animal species.

The requirement that the rhythmicity be generated in the absence of sensory input does not exclude the possibility that such input may shape the rhythmically generated motor pattern to adjust the output produced at a given time to the other behaviors performed by the animal. Moreover, the neuronal network underlying the central pattern generator is subject to **modulation** by neuromodulators, such as catecholamines or neuropeptides (see Chapter 8). This modulation may lead to alterations in the frequency or the intensity of the rhythmic motor output.

Two major mechanisms of rhythm generation have been identified. In some central pattern generators, the individual neurons themselves exhibit oscillatory activity and, thus act as **pacemakers**. In other types of central pattern generators, the rhythmicity is the result of **synaptic interactions** among neurons that are themselves not rhythmically active. In other words, in the latter types, the rhythmicity is a property of the circuit, but not of the individual neuron.

The rhythmic activity of central pattern generators can be maintained even in the absence of real motor activity, for example after isolation of neuronal tissue containing the central pattern generator, or in immobilized animals. This allows the researcher to study the physiological properties of the central pattern generator circuit, while it produces **fictive behaviors**, instead of real motor activities.

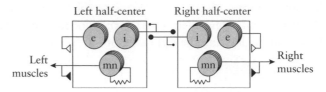

Left half-center   Right half-center

Left
muscles

Right
muscles

Figure 5.3 Basic circuitry of neurons in the spinal cord controlling swimming in the *Xenopus* embryo. As this wiring diagram shows, the central pattern generator consists of two half-centers, one on each side of the spinal cord. Each circle represents a small population of neurons. The motoneurons (mn) innervate the swimming muscles on the ipsilateral side. The individual motoneurons in one half-center are electrotonically coupled. The excitatory interneurons (e) provide premotor excitation and a positive feedback; their connections are restricted within each half-center. The inhibitory interneurons (i) have strong reciprocal inhibitory connections between half-centers and weak recurrent inhibitory connections within half-centers. The excitatory connections are indicated by filled and open triangles, the inhibitory connections by filled circles. The zig-zag line symbolizes electrotonic coupling. (After **Roberts, A., Soffe, S. R., and Perrins, R. (1997).**)

pattern generator in the spinal cord consists of two parts, one on the left side and the other on the right side. They are called **half-centers**. The pattern of connectivity within the underlying neural network is shown in Fig. 5.3.

The peripheral axons of the **motoneurons** excite the swimming muscles in the trunk, which are segmentally organized, by releasing the transmitter substance acetylcholine. Descending longitudinal axons originating from these neurons make excitatory synaptic contact with other, more caudally located motoneurons. This excitation is mediated by nicotinic acetylcholine receptors and provides positive feedback to motoneurons in other segments. In addition to these cholinergic synaptic contacts, the motoneurons within one segment also make electrical connections with each other.

Based on their physiological properties, the **descending interneurons** are also called **excitatory interneurons**. They excite not only the motoneurons, but also the commissural interneurons, as well as other excitatory interneurons, on the same side of the cord. The latter action of the excitatory interneurons can be regarded as a positive feedback. The excitatory interneurons release glutamate as a transmitter at their synapses, which then binds to postsynaptic receptors. The excitatory postsynaptic potentials, EPSPs (see Chapter 3), produced in the target cells consist of two components: a slow component mediated by glutamate receptors of the N-methyl-D-aspartate (NMDA) type, and a fast component mediated by glutamate receptors of the α-amino-3-hydroxy-5-methylisoxazole-propionic acid (AMPA) type. The fast AMPA excitation helps more caudal neurons to reach firing threshold, whereas the slow NMDA excitation sustains the next cycle of activity on the same side of the spinal cord.

In contrast to the descending interneurons, the **commissural interneurons** exert an inhibitory action upon their target cells; hence, they are also called **inhibitory interneurons**. They project to the opposite side of the cord, where they produce inhibitory postsynaptic potentials, IPSPs (see Chapter 3), in all three cell types. Some of the inhibitory interneurons also send out an axon on the same side. This projection mediates inhibition of the cells on the ipsilateral side, although the corresponding IPSPs are much smaller than those on the contralateral side. The inhibitory interneurons release the amino acid glycine. Upon binding of this transmitter to receptor molecules in the postsynaptic neuron, channels are opened that result in an influx of chloride ions into the cell. This current drives the potential in a negative direction.

The central pattern generator controlling the rhythmic muscle contractions during swimming in hatching tadpoles consists of two half-centers in the spinal cord and three types of neurons—motoneurons, excitatory interneurons, and inhibitory interneurons.

Operation of the swimming circuitry

During fictive swimming, the three types of neurons in one given half-center of the central pattern generator produce only one spike per swim cycle. Due to the electrical coupling of the motoneurons, the spikes generated by the neurons on one side appear almost simultaneously. However, the spikes produced by the left and the right half-centers alternate.

The production of spikes by the motoneurons is caused by the EPSPs evoked by input originating from the ipsilateral excitatory interneurons. After the firing, the motoneurons remain depolarized relative to the resting potential before swimming is initiated. When neurons on one side of the central pattern generator fire, all the neurons on the other side receive, in the middle of their activity cycle, IPSPs mediated by the contralateral projection of the inhibitory interneurons. Thus, overall the two half-centers of the swimming central pattern generator exert an inhibitory action on each other.

How is the rhythmic activity of the central pattern generator produced? Experimental evidence suggests that it is primarily inhibition, rather than excitation, of the spinal neurons that is responsible for **rhythmogenesis**. As illustrated in Fig. 5.4, in the absence of rhythmical drive from the central pattern generator, injection of a suprathreshold depolarizing current into a motor neuron elicits a single spike, but no repeated firing. On the other hand, injection of a pulse of hyperpolarizing current into a cell evokes **rebound firing**. However, the latter effect is observed only if the cell is steadily depolarized, as is the case in swimming. This observation has led Alan Roberts and his associates to propose that rhythmogenesis is

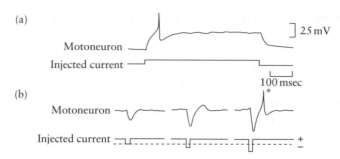

Figure 5.4 Effect of depolarization and hyperpolarization of spinal motoneurons on spike generation. (a) Injection of a suprathreshold depolarizing current elicits a single spike, but no repeated firing. (b) Injection of pulses of hyperpolarizing current into a steadily depolarized cell elicits rebound firing (indicated by asterisk). (After **Arshavsky, Y. I., Orlovsky, G. N., Panchin, Y. V., Roberts, A., and Soffe, S. R.** (1993).)

Rebound firing: the production of action potential(s) by a neuron after it has been hyperpolarized.

based primarily on the rebound phenomenon occurring in response to the inhibitory input received from the contralateral half-center of he central pattern generator. It is unclear whether the pacemaker properties of the interneurons play any role in generating the rhythm, but if they do, that role is then secondary. Therefore, it appears that the main factor determining the cycle period of the swimming rhythm is the duration of the reciprocal IPSPs. Indeed, the cycle period during swimming (40–100 msec) is effectively the sum of the IPSP durations in each half-center, namely two times 25–40 msec.

Coordination of oscillator activity along the spinal cord

During swimming, motor activity produced in the half-centers of each central pattern generator starts at the head and progresses toward the tail. The resulting rostrocaudal delay in motor output, which assumes values between 1.5 and 5.5 msec/mm, appears to be controlled by a combination of several mechanisms. One mechanism is based on the projection pattern of the excitatory interneurons. These neurons make not only synaptic contact with motoneurons and interneurons within their own half-center, but they also activate, via descending axons, half-centers in more caudal segments. This pattern of connectivity promotes a progression of activation of the swimming central pattern generators in rostrocaudal direction. Another factor causally involved in controlling the longitudinal progression of motor activity is a reduction in the number of excitatory interneurons and inhibitory interneurons found per segment in rostrocaudal direction—regions located closer to the head have more interneurons than those closer to the tail. Related to this rostrocaudal reduction in the number of interneurons is a rostrocaudal gradient in both the excitatory and the inhibitory synaptic input received by the motoneurons. The importance of this gradient in setting up the longitudinal delay in motor activity can be demonstrated by pharmacological manipulations: reduction of the gradient in excitatory input by application of the glutamate agonist NMDA to the caudal spinal cord reduces the delay time, whereas application of an NMDA antagonist to caudal segments increases the delay by increasing the rostrocaudal excitatory gradient. Taken together, these mechanisms promote the establishment of a 'leading' oscillator in spinal segments near the head.

Summary

■ Central pattern generators play an eminent role in the production of rhythmic motor patterns. The neural network underlying these oscillators can fulfill their function even in the absence of sensory input, or input from higher central sites conveying timing cues.

■ The central pattern generator controlling swimming in hatching embryos of the clawed-toad (*Xenopus laevis*) is readily accessible to physiological experimentation, because its rhythmic activity can be initiated even after blocking synaptic transmission at the neuromuscular junction, when the immobilized tadpoles perform 'fictive', instead of real, swimming.

■ Such physiological experiments, combined with anatomical studies, have suggested that three types of neurons in the spinal cord are sufficient to explain the generation of the rhythmic pattern during the contractions of the trunk muscles during swimming. They are the motoneurons that innervate the swimming muscles; the excitatory interneurons that make excitatory synaptic contacts with the motoneurons on the ipsilateral side of the spinal cord; and the inhibitory interneurons that project to the contralateral half-center of the central pattern generator, where they exert an inhibitory effect upon all three types of neurons.

■ During fictive swimming, a given half-center of the central pattern generator produces only one spike per swim cycle. The spikes produced by the left and the right half-centers alternate.

■ The spikes produced by the motoneurons are caused by EPSPs evoked by input originating from the ipsilateral excitatory interneuron. When the neurons on one side fire, all the neurons in the contralateral half-center receive, in the middle of their activity cycle, IPSPs mediated by the contralateral projection of the inhibitory interneurons. It is primarily this inhibition, rather than the excitation, that is responsible for the rhythmic pattern produced by the central pattern generator.

■ During swimming, the rhythmic activity produced in the half-centers of the central pattern generators progresses from head to tail along the spinal cord. The resulting rostrocaudal delay in motor output is caused by (i) synaptic contact of excitatory interneurons with half-centers in more caudal segments of the spinal cord; (ii) rostrocaudal reduction in the number of excitatory interneurons and inhibitory interneurons; (iii) rostrocaudal gradient in both excitatory and inhibitory input received by motoneurons.

Recommended reading

Arshavsky, Y. I., Orlovsky, G. N., Panchin, Y. V., Roberts, A., and Soffe, S. R. (1993). Neuronal control of swimming locomotion: analysis of the pteropod mollusc *Clione* and embryos of the amphibian *Xenopus*. *Trends in Neurosciences*, **16**, 227–233.

This comparative analysis of the operation of the central pattern generators underlying swimming in the marine mollusk Clione and the amphibian embryo Xenopus suggests intriguing common mechanisms, despite the differences in neuronal hardware.

Orlovsky, G. N., Deliagina, T. G., and Grillner, S. (1999). *Neuronal control of locomotion: from mollusc to man*. Oxford University Press, Oxford.

A successful attempt to integrate our current knowledge of the organization and operation of locomotor control systems in a variety of different species, both invertebrates and vertebrates.

Questions

5.1 What are central pattern generators? How do they function? What approaches are available to study their physiological properties?

5.2 Describe the structural organization and physiological operation of the central pattern generator controlling swimming in hatching toad tadpoles.

Neuronal processing of sensory information

6

- Introduction
- Recognition of prey and predators in the toad
- Directional localization of sound in the barn owl
- Summary
- Recommended reading
- Questions

Introduction

Our sense organs are continuously hit by an enormous amount of information originating from numerous biological and non-biological sources within the environment. Yet, only a tiny fraction of this information is perceived by the sensory receptors, and further processed in the brain. This observation, as already mentioned in Chapters 2 and 3, has led Jakob von Uexküll to the formulation of his *Umwelt* concept. *Umwelt*, in von Uexküll's terminology, is not the absolute environment. Rather this term denotes the world around us, as we—or animals—perceive it.

The reduction of the information flow is accomplished by sensory and central filters. These filters form an integral part of the structures determining the animal's behavioral response to a stimulus. Their function is to extract the biologically relevant information and ensure its further sensory processing, so that, finally, a certain stimulus is linked to the initiation of the appropriate motor action. The sum of the structures participating in the selective release of a behavior is often referred to as **releasing mechanism**, although some ethologists limit this definition to sensory structures and exclude motor structures.

The aim of this chapter is to discuss how this processing of sensory information is implemented at the neuronal level. We will use two particularly well-examined model systems to illustrate this.

The first example centers around the toad's ability to visually recognize prey objects and distinguish them from predators. How is this achieved? In 1959, J. Y. Lettvin, H. R. Maturana, W. S. McCulloch, and W. H. Pitts from the Massachusetts Institute of Technology in Cambridge, Massachusetts, proposed, in a paper entitled *What the frog's eye tells the frog's brain*, that much of this recognition process takes place at quite an early stage of sensory processing. They described 'bug detectors' in the retina of the frog that respond best 'when a dark object, smaller than a receptive field, enters that field, stops, and moves about intermittently thereafter.' Although, as we will see below, subsequent investigations have shown that the real situation is far more complex, this work has stimulated an intensive search for **feature detectors,** neurons that respond selectively to rather specific features of a sensory stimulus.

The search for feature detectors has been greatly facilitated from the late 1950s on by the advances made in the development of intracellular recording techniques (see Chapter 3). This method has enabled researchers to correlate perception data, commonly obtained through analysis of the animal's response upon presentation of a stimulus, with single-cell responses. Among the early pioneers of this approach were Otto-Joachim Grüsser, Otto Creutzfeldt, and Günter Baumgartner in the laboratory of Richard Jung at the University of Freiburg (Germany). This trio of German neurobiologists performed a quantitative analysis of the properties of single neurons in the cortex of cats. One of the first scientists who applied this approach to more neuroethologically oriented studies was Ursula ('Ulla') Grüsser-Cornehls, the wife of Otto-Joachim Grüsser. She initiated this work using the frog retina during a research visit to the laboratory of Ted Bullock at the University of California Los Angeles (see Chapter 2).

The second example discussed in detail in this chapter focuses on the ability of owls to localize prey solely on the basis of sound generated by the prey animal. As it is the case with the toad, the well-defined behavior exhibited by the owls in response to such sound offers an excellent opportunity to explore how physical parameters associated with the sound are centrally processed to solve a problem immensely important for the owl.

Feature detectors: neurons that respond selectively to specific features of a sensory stimulus.

Recognition of prey and predators in the toad

The model system

The true toads form the family Bufonidae, within which *Bufo* comprises more species than any other genus—approximately 250. They are widely

Figure 6.1 Prey-catching behavior of the common toad, evoked by a worm-like stimulus. When a rectangular stripe is moved in the direction of its long axis within the lateral visual field of the toad, leading to a monocular perception of the object, one of the first responses is a turning movement toward the prey (a). The following behaviors include a binocular fixation of the prey (b), snapping (c), swallowing (d), and wiping of the mouth (e). (After Ewert, J.-P. (1980).)

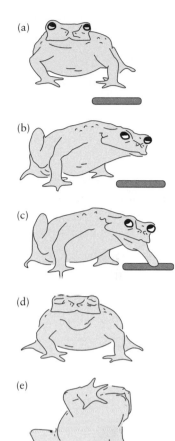

distributed all over the world, with the exception of a few major regions including Australia, Greenland, New Guinea, New Zealand, and Madagascar. Within Central Europe, the most abundant species is the **common toad** (*Bufo bufo*). Its ability to **recognize prey and enemies** has proven a particularly well-suited model system to investigate the behavioral and neural mechanisms of **feature detection**. Indeed, this system was among the first to be studied in great detail in neuroethology—an achievement mainly due to the work of **Jörg Peter Ewert** and his group of the University of Kassel in Germany.

The natural behavior

Before we discuss some of the key experiments performed by Ewert and his group, we will first have a look at the toad's natural behavior. When a common toad is motivated to catch prey, and a small moving prey object appears in its visual field, the toad responds with characteristic behaviors, which are shown in Fig. 6.1. One of the first responses is an **orienting movement toward the prey**. The following sequence is variable and depends on the exact stimulus situation, but typically includes (if the prey is close enough) stalking up the prey, binocular fixation, snapping, swallowing, and, finally, wiping of the mouth with the fore-limbs.

A first indication of what features may be used to distinguish between prey and enemy objects is given by observations of the toad's natural behavior. Among the enemies that prey on toads are snakes. Therefore, whenever possible, a toad tries to avoid an encounter with snakes. If this is not possible, the toad shows a characteristic avoidance response. It blows itself up, assumes a stiff-legged posture, and exhibits its flank, as shown in Fig. 6.2(a). A similar behavior can be evoked through presentation of a head-rump dummy, as illustrated in Fig. 6.2(b). On the other hand, there are also animals, such as leeches, that look somewhat similar to snakes, but are not dangerous to toads. Nevertheless, depending on the leech's posture and movement, the toad may confuse them with a snake. This happens if a leech lifts its sucker in the air (Fig. 6.2(c)). In this case, the toad takes it for an enemy. If, on the other hand, a leech walks jerkily along with its frontal sucker on the ground, it is regarded as prey (Fig. 6.2(d)).

The natural prey catching behavior of the toad consists of a series of well-defined individual behavioral patterns.

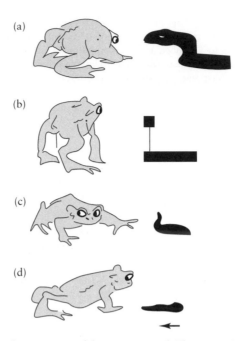

Figure 6.2 Enemy and prey images of the common toad. The perception of a snake elicits a characteristic avoidance response (a). Presentation of a head-rump dummy results in the same type of defense posture (b). Such a behavior can even be evoked by a leech with raised frontal sucker (c), implying that it fits the enemy image. On the other hand, as soon as the sucker is on the ground, and the leech walks along its long axis, the toad responds with an orienting movement typically exhibited toward prey (d). (After **Ewert, J.-P.** (1980).)

Dummy experiments

To examine what features are used to distinguish prey and enemy, toads are placed in a glass cylinder and presented with several models. This experimental set-up is shown in Fig. 6.3. The toad's behavior in response to these dummies is then measured by the rate of turning of the head toward the prey.

Dummy experiments in which the toad is stimulated with a small, narrow, dark stripe have shown that the following features characterize the key stimulus 'prey':

- direction of movement
- area dimensions relative to the direction of movement.

The importance of the direction of movement is illustrated by the results of the following experiments (Fig. 6.4): When the stripe moves in a worm-like fashion, that is, parallel to its long axis, it signals 'prey'. When, on the other hand, the long axis of the same stripe is oriented perpendicular to the direction of movement, the stimulus loses its key

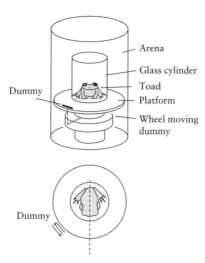

Figure 6.3 Experimental set-up to examine the toad's response toward prey objects. The toad is placed on a circular stage within a glass cylinder. Dummies, such as a black stripe, can be moved around the toad with various velocity by means of an electric drive, The wall of the arena, which in reality is opaque and not transparent as shown here, provides the background against which the dummy is viewed. (After **Ewert, J.-P. and Ewert, S. B.** (1981).)

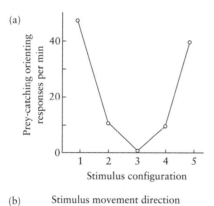

Figure 6.4 Importance of direction of movement of stripe dummies for the efficacy to release prey catching in the common toad. (a) The stripes are moved in different directions relative to their longitudinal axis, as indicated in (b) The rate of prey-catching activity is maximal when the stripe is moved parallel to its long axis ('worm configuration'), as done with dummies (1) and (5). The prey-catching response is minimal when the stripe is moved in the direction of its short axis ('anti-worm configuration'), as revealed by dummy experiments (3). Movement in directions intermediate between these two extreme configurations results in intermediate responses, as demonstrated by dummy experiments (2) and (4). (After **Ewert, J.-P.** (1980).)

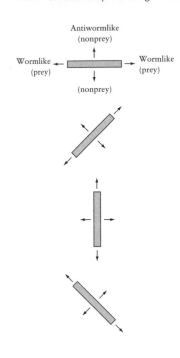

Figure 6.5 Illustration of the concept of worm configuration and anti-worm configuration. Movement of an elongated object, such as a rectangle, along its longitudinal axis signals 'prey.' Movement perpendicular to the long axis elicits no prey-catching response. Rather, this stimulus configuration commonly triggers enemy-avoidance behavior. Note that the 'worm configuration' and the 'anti-worm configuration' are solely defined by movement of the object relative to its long axis. Thus, these configurations are invariant to the actual direction of movement, as illustrated here in the two-dimensional plane. (After **Ewert, J.-P.** (1980).)

feature. Such models appear to be perceived as 'threat,' rather than as 'prey,' and elicit a threat reaction, or no response at all.

These results have led to the concept of 'worm' and 'anti-worm' configuration of dummies, which is illustrated in Fig. 6.5 and can be summarized as follows:

- a **worm configuration** is achieved by movement of a rectangle in direction of its long axis
- an **anti-worm configuration** results when a rectangle moves in direction of its short axis.

> In the toad's sensory world, 'prey' are elongated objects that move in direction of their long axis. By contrast, 'enemies' are objects that move in direction of their short axis.

Note that movement along the long axis does not necessarily imply movement in horizontal direction. Rather, a rectangle presented in an upright position could also move in vertical direction—and still elicit the same kind of response as do rectangles moving along their long axis in horizontal direction. This configuration may, for example, imitate a situation when a caterpillar climbs up a stalk of grass.

The importance of the area dimension relative to the direction of movement is demonstrated by experiments in which rectangles of various lengths but constant width are used. These rectangles are moved either in the direction of their long axis, or in the direction of their short axis. As anticipated, based on the results of the previous experiments, the best response is evoked by movement of the rectangles along their long axis. Moreover, within certain limits, the greater the length of the rectangle (the 'wormier' the object looks like), the greater the toad's response (Fig. 6.6).

By contrast, rectangles moving along the short axis have low releasing values. These values decrease even further, if the length of the long axis increases.

Small moving squares have releasing values similar to those of rectangles moved in a worm-like fashion. For longer squares, the releasing value decreases, until it reaches zero. Still longer squares evoke a new type of behavior. Instead of turning the head *toward* the stimulus, the toad turns *away* from the square—thus resembling the toad's response toward an approaching enemy.

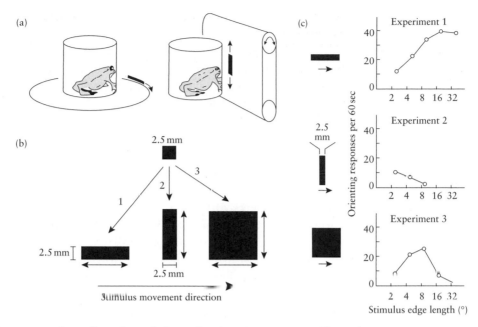

Figure 6.6 Importance of area dimensions relative to direction of movement. (a) The toad is placed in a glass cylinder and stimulated, at a constant distance of approximately 7 cm, with models moved in the horizontal or vertical plane. (b) Starting with a 2.5×2.5 mm² square, the area of the dummy is systematically increased. In the first set of experiments (1), the vertical edge is kept constant at 2.5 mm, while the horizontal edge is elongated stepwise in direction of the movement ('the worm becomes longer'). In the second set of experiments (2), the horizontal edge is kept constant at 2.5 mm, while the length of the vertical edge is stepwise increased ('the snake becomes taller'). In the third set of experiments (3), squares of increasing size are presented. (c) The number of orienting responses is plotted as a function of stimulus edge length, expressed as the size of the visual angle (in degrees). Using worm-like objects (experiment 1), an increase of the long axis evokes—within certain limits—an increasing number of prey-catching responses. When employing the anti-worm configuration (experiment 2), an increase of the long axis leads to a reduction of the already quite low rate of prey-catching activity. Presentation of squares of increasing size (experiment 3) leads initially to an increase in prey catching. However, further increase in size reduces the prey-catching activity down to zero. (After Ewert, J.-P. (1980).)

In search of feature detectors

How is recognition of prey and enemies achieved in the toad's brain? Are different features (such as size, contrast, motion, color, etc.) of an object processed by different neurons, and if so, at what brain level does convergence of these different information channels take place? Or are object categories, such as those for prey and enemies, already formed at an early stage of information processing, possibly even in the retina of the eye?

How selectively do neurons involved in the processing of visual information respond to specific configurations of an object?

To explore the ability of neurons to detect features relevant to object recognition, **recording experiments** are performed. During the recordings, the toad is immobilized by injection of a neuromuscular blocking agent. This treatment is necessary to avoid displacement of the electrode within the brain due to possible movements of the animal. To get access to the brain, a small piece of skull is removed under local anesthesia. Then, by means of a three-dimensionally movable micromanipulator, a recording electrode is advanced to the brain region of interest.

As in the behavioral experiments, the toad is stimulated during the recording experiments with black elongated stripes or squares moving in a worm-like or anti-worm-like fashion. To qualify as a recognition neuron, one expects a cell to exhibit a similar selectivity in discriminating worm- and anti-worm like stimuli, as shown by the whole animal in the behavioral tests. For example, presentation of worm-like stimuli with increasing length of the long axis should result in similar increases in neuronal activity by prey-recognition neurons. Moreover, only the respective recognition neuron should be responsible for initiating the corresponding motor response, and no other neuron should activate this motor pattern.

Feature detectors can be identified by recording from the brain, while exposing the animal to behaviorally relevant stimuli.

The visual system

The first processing station of the toad's visual system is the retina. The receptor cells are linked, via amacrine and bipolar cells, to the ganglion cells. Typically, one ganglion cell receives input from a number of receptor cells. The corresponding visual area defines the **receptive field** of a particular ganglion cell. The axons of the retinal ganglion cells form the optic nerve. This nerve innervates various locations of the brain. As we will see in detail below, two of these areas are particularly important for the recognition of prey and enemies: the **optic tectum**; and the posterior thalamus and pretectum, commonly called the **thalamic-pretectal area**. Figure 6.7 summarizes the major neural pathways involved in prey catching and escape response in the toad.

The optic tectum forms the roof of the midbrain. Its name comes from the Latin word for roof, *tectum*. The homologous structure in mammals is the superior colliculus.

The link between the eye and the tectum, mediated by the axons of the retinal ganglion cells, is referred to as the **retinotectal projection**. In amphibians, it is entirely contralateral—the left optic nerve terminates in the right optic tectum, the right optic nerve travels to the left optic tectum.

As a result of this contralateral projection, neurons located in the right tectum respond to stimulation within the receptive field of the corresponding

The major neural parts of the visual system of the toad are: the receptor cells and ganglion cells in the retina; the optic nerve formed by axons of the ganglion cells; and the optic tectum and the thalamic-pretectal area.

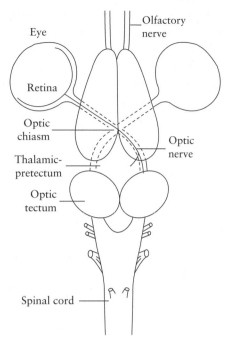

Figure 6.7 Diagram of the basic pattern of connectivity between the retina on the one side, and diencephalic and mesencephalic brain regions on the other, in the common toad. The axons of the ganglion cells form the optic nerve. These cells project contralaterally over wide, but specific areas within the thalamic–pretectum in the diencephalon and the optic tectum in the mesencephalon. For sake of simplicity, only one such projection is shown. Also, reciprocal connections between the thalamic–pretectum and telencephalic nuclei are not included in this simplified diagram. (After **Ewert, J.-P.** (1974).)

receptor and ganglion cells in the left retina. Moreover, retinal ganglion cells project to the optic tectum in a systematic point-to-point fashion. Adjacent receptive fields are, therefore, represented by adjacently located tectal neurons. This leads to the establishment of a topographic map of the visual world on the surface of the optic tectum. This map is, specifically, referred to as a **retinotopic map**.

Recording experiments

By employing the above approach, Ewert and his group conducted a long series of recording experiments at the different levels of the visual system. Such recordings have shown that there are at least four classes of ganglion cells in the retina, called R1 through R4. They differ in size of their receptive fields and in sensitivity to rather broadly defined features of objects, such as object size and contrast, movement, and ambient illumination. However, the response of these four types of cells reveals no

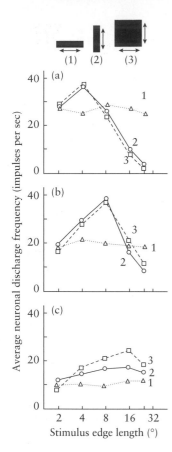

Figure 6.8 Stimulus–response profile of three classes of retinal ganglion cells in the common toad. For the stimulation, the same types of configurations were used as in the behavioral experiments (see Fig. 6.6): '1' denotes a rectangle moved in a worm-like fashion; '2' rectangle moved in an anti-worm-like fashion; '3' square moved along one of the stimulus edges. The length of the stimulus edge is expressed as the size of the visual angle (in degrees). Each data point indicates the mean response of 10 different neurons of the same class. (a) Response of retinal ganglion cells of class R2. (b) Response of retinal ganglion cells of class R3. (c) Response of retinal ganglion cells of class R4. None of the response profiles resembles those obtained in behavioral experiments. (After **Ewert, J.-P.** (1980).)

selectivity in distinguishing between worm- and anti-worm-like objects similar to the one observed in behavioral tests. Figure 6.8 summarizes the results of physiological experiments on three classes of these retinal cells, called R2, R3, and R4 neurons. Obviously, if more selective feature detectors exist, they must be located deeper in the brain.

At the next level, the thalamic pretectal area, there is a large number of different types of neurons. In their initial work, Ewert and his group identified at least 10 different classes, referred to as TP1 through TP11. These different types of neurons become activated by various visual stimuli, some of which are quite complex. One of these neuronal cell types, TP3, responds to large moving visual stimuli. Neurons of this class show only little response to stimulation when elongated objects move in a worm-like fashion. Rather, they respond best when such objects are moved in an anti-worm-like fashion, as shown in Fig. 6.9(a). A similar response profile is displayed by TP4 neurons.

Different categories of cells also exist in the optic tectum. Ewert and colleagues grouped them into eight classes, T1 through T8. Each of them is maximally activated by different visual stimulation regimes. One class, T5, gives strong responses to moving stimuli. Recordings from these cells have revealed two cellular populations defining the subclasses T5(1) and T5(2). T5(1) neurons encode for object extension parallel to the direction of movement, but fail to clearly differentiate between worm- and anti-worm-like objects (Fig. 6.9(b)). T5(2) cells show a similar response profile, but are, in addition, able to differentiate between object extensions parallel to and across the direction of movement—their response is strongest to worm-like stimuli and weakest to anti-worm-like stimuli. Moreover, comparison of the stimulus–response profile obtained through behavioral experiments and that defined by the spike frequency of the T5(2) neurons has demonstrated a nearly identical shape of the curves (Fig. 6.9(c)): elongation of the long axis of a rectangle moved in a worm-like fashion results, initially, in a gradual increase of the response, but in a decline after further elongation. By contrast, elongation of the long axis of anti-worms leads to a steady decrease of the stimulus–response curve.

T5(2) neurons in the optic tectum show many properties that qualify them as feature detectors responding best to worm-like stimuli.

Figure 6.9 Stimulus-response profile of TP3 (a) cells in the thalamic–pretectum and of T5(1) and T5(2) neurons (b and c, respectively) in the optic tectum of the common toad. For the stimulation, rectangles moved in a worm-like fashion ('1'), rectangles moved in an anti-worm-like fashion ('2'), and squares moved along one stimulus edge ('3') were employed. Each data point is the mean of 20 recording experiments. For interpretation of the diagrams, see text. (After **Ewert, J.-P.** (1980).)

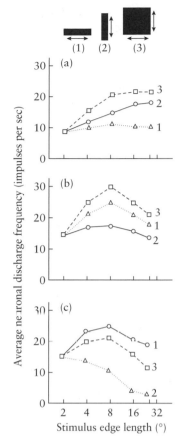

Stimulation experiments

Compelling evidence for the involvement of T5(2) neurons in prey recognition comes from stimulation experiments carried out in freely moving toads. For electrical stimulation, an electrode is permanently implanted in a small region of the tectum. Before stimulation, the same electrode is used to record the activity of cells located near its tip. By moving a small object across the contralateral visual field, the receptive field of this group of neurons is determined. If the object is moved within the receptive field, neuronal activity is recorded. Outside the receptive field, no activity, or highly reduced activity, is observed.

In the crucial part of the experiment, stimulating current is delivered through the implanted electrode. Depending on the exact position of the electrode, different responses are evoked. Stimulation in the thalamic-pretectal region elicits escape responses. Stimulation in the tectum, on the other hand, results in various behavioral patterns characteristic of prey catching, such as a turning of the head, snapping, and swallowing. When turning of the head is elicited, this orienting movement is directed to that part of the visual field which corresponds to the receptive field of the respective neurons. The toad, then, behaves as if a natural prey object were present.

Connections with other brain regions

Physiological and neuroanatomical experiments have shown that T5(2) cells project to appropriate motor centers involved in prey-catching behavior, providing further evidence for the involvement of these neurons in prey recognition. These motor centers are located in the medulla oblongata in the hindbrain and the spinal cord.

Moreover and most significantly, T5(2) cells receive pronounced inhibitory input from the thalamic-pretectal area, especially TP3 neurons. It is thought that this inhibition plays an important role in the toad's selectivity for certain configurational features of objects, including those relevant for prey–enemy discrimination.

This hypothesis is supported by experiments in which the thalamic-pretectal/tectal connection was lesioned. Such experimental manipulations produce a number of altered responses of tectal neurons (Fig. 6.10(a)).

Electrical stimulation of the optic tectum, believed to activate T5(2) neurons, elicits orienting movement in toads characteristic of prey-catching behavior.

T5(2) cells receive inhibitory input from the thalamic-pretectal area. This connection is thought to contribute to the enhancement of selectivity for features relevant for prey–enemy discrimination.

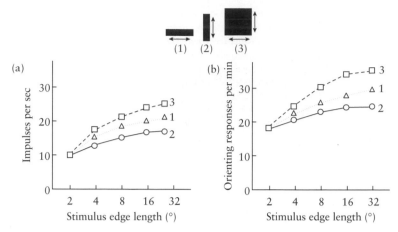

Figure 6.10 (a) Effect of thalamic–pretectal lesions on the response of T5(1,2) neurons to stimuli of different configuration ('1', rectangle moved in worm-like configuration; '2', rectangle moved in anti-worm-like configuration; '3', square moved along one stimulus edge). The neurons respond to a size increase of each of the three configurational objects with an increase in firing rate. Their ability to distinguish between the three configurations, normally observed in the intact animal, is largely lost in the lesioned animal. (b) Effect of thalamic-pretectal lesions on prey-catching behavior, as measured by the head-turning responses in the common toad. The animals respond to any moving object, independent of its configuration ('1', '2', and '3' as in (a)), with prey-catching activity. The data points are the means of recordings from 20 individual T5 neurons and 20 lesioned toads, respectively. (After **Ewert, J.-P.** (1980).)

First, a strong increase in visual responses of tectal neurons. Second, an impairment of the configurational selectivity of T5(2) neurons. Third, an increased sensitivity of T5(1) and T5(2) neurons to moving large objects—even to those that were ineffective prior to lesioning. Fourth, an inability to distinguish between moving objects and (through movement of the animal) self-induced moving retinal images. Fifth, a failure to estimate object distance.

Similar deficiencies are exhibited by toads with thalamic-pretectal/ tectal lesions when tested in behavioral experiments. Such toads are unable to discriminate between prey and enemy—they respond to *any* moving object with prey-catching behavior, but lack an enemy avoidance response (Fig. 6.10(b)).

Directional localization of sound in the barn owl

The behavior

In the second half of this chapter, we will turn to a different model system—the directional localization of sound in the barn owl—to see how

BOX 6.1   Masakazu Konishi

Mark Konishi in 2003. (Courtesy: M. Konishi.)

Today recognized as one of the leaders in neuroethology, Masakazu ('Mark') Konishi succeeded in combining an interesting ethological question—how owls localize prey using sound—with a rigorous neurobiological approach. In addition to providing an answer to this specific question, the research of his group has also revealed important general computational rules and neural principles of how the brain processes sensory information to produce behavior.

Konishi was born in Kyoto (Japan) in 1933. While still at school, he developed a keen interest in biology. As he puts it, "As a child, I enjoyed fooling animals. When I read *The Study of Instinct* by Niko Tinbergen in 1953, my junior year in college, I thought I found a perfect field for me. You get paid and praised for fooling animals with dummies." As an entry to such a career, he studied zoology at Hokkaido University in Sapporo (Japan). Since he felt greatly attracted by the American academic system, Konishi moved to the USA in 1958 to do his Ph.D. on birdsong under the renowned ethologist Peter Marler at the University of California, Berkeley. Upon receiving his doctorate in 1963, he went to Germany for his postdoctoral training. After a few months in the laboratory of the sensory physiologist Johann Schwartzkopff at the University of Tübingen, he joined the group of Otto Creutzfeld at the Max Planck Institute for Psychiatry in Munich, where he was involved in one of the first attempts to analyze visual receptive fields of neurons in the visual cortex of cats by single unit recordings. Although the project was only of limited success due to the enormous difficulties in recording long enough from neurons to map their receptive field, Konishi greatly benefitted from the expertise gained, as well as from the close vicinity of the Max Planck Institute for Behavioral Physiology in Seewiesen, where he met many of the leading ethologists.

After two years in Germany, Mark Konishi assumed an assistant professorship at the University of Wisconsin, Madison, but shortly thereafter moved to Princeton University. Fascinated by the behavioral studies of Roger Payne, he decided to work on sound localization in barn owls. A systematic physiological investigation of this problem began in 1976, a year after Konishi joined the faculty of the California Institute of Technology in Pasadena, to which he still belongs. Following an intensive search, a breakthrough was achieved when he found, together with his postdoc Eric Knudsen, a map of auditory space in the owl's midbrain. This contradicted the then widely held believe that the auditory system should have neither spatial receptive fields nor maps, because the first processing station of the auditory pathway, the inner ear, maps sound frequency instead of space. However, the discovery of Konishi and Knudsen is now not only generally accepted, but also marks the beginning of a systematic, and very successful, analysis by the laboratory of Mark Konishi of the neural steps involved in the processing of auditory information.

the various physical parameters of a stimulus are encoded in the brain. As in the toad, elucidation of the underlying neural mechanism in the barn owl has been greatly advanced through focusing on a relatively simple behavioral pattern.

Barn owls form a small family, Tytonidae, consisting of nine species. The most abundant member of this family is the barn owl, *Tyto alba*. This species has been intensively studied by Masakazu Konishi and his associates at the California Institute for Technology in Pasadena, since the 1970s (see Box 6.1). For his investigations, Konishi has capitalized on an extraordinary ability of barn owls: to precisely localize prey animals solely based on the noise produced by them. This sensory capability is likely to be an adaptation to the behavior of field mice, the main food source of barn owls. Field mice predominantly forage at night when they are barely visible. Moreover, they tend to move through tunnels in grass or snow, rather than across open areas. Thus, the visual system is of rather limited use in prey hunting.

While hunting in the dark, the barn owl visits a number of observation perches within its territory to survey the ground. Upon hearing the noise of a prey animal, the barn owl turns its head in a rapid flick so that it directly faces the source of the sound. Then, it swoops down to finally strike the mouse.

Barn owls can localize prey solely based on acoustic cues.

The fact that barn owls can find their prey solely based on the **noise** generated by the moving mouse was demonstrated by Roger Payne at Cornell University in Ithaca, New York, in a series of simple, though elegant, experiments. In one of these experiments, a barn owl was trained to strike mice released in leaves on the floor of a room in complete darkness. Then, with the lights turned off, a mouse-sized wad of paper was dragged through the leaves on the floor. The owl successfully struck the paper wad. This experiment demonstrates that the owl is not using olfaction, since the paper wad put out no mouse-like odor. Also, the owl's ability to precisely catch the mouse cannot be explained based on infrared sensitivity, as the paper wad and the leaves through which it was dragged were at the same temperature.

In a different experiment, also conducted in complete darkness, the floor of the experimental room was not covered with a layer of leaves, but with sand instead. This enabled the mouse to walk silently. To make the mouse's movements audible, nevertheless, the mouse was forced to tow a rustling leaf several centimeters behind its tail. Then, the barn owl struck the leaf, but ignored the mouse.

Azimuth: horizontal plane; Elevation: vertical plane.

Since the barn owl hunts from points above ground, it is not sufficient for it to locate the prey just in the horizontal plane (which is referred to as the **azimuth**). Rather, it must also determine the angle between itself and the prey in the vertical dimension (which is referred to as the **elevation**). How is this achieved?

Experimental approach

To answer this question, it was first necessary to obtain detailed information about the accuracy of prey localization. However, owls in free flight

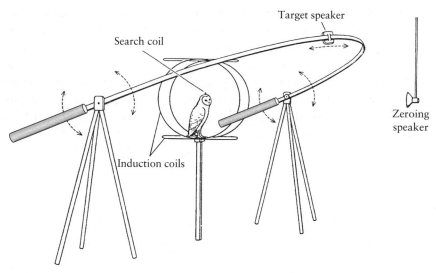

Figure 6.11 Simplified drawing of the electromagnetic angle detector system. This set-up is based on the natural response of hunting owls when hearing a noise. Then, the bird turns its head to face the source of sound. In the experiments, the owl is trained to remain on the perch, instead of approaching the presumptive prey. The movements of the owl's head are monitored by search coils mounted on top of the bird's head, and by induction coils between which the owl is positioned. Through this arrangement, any movement of the head induces, in the search coils, a current change that is recorded and analyzed by a computer. At the beginning of the experiments, the head is aligned by directing the owl's attention to a zeroing loudspeaker. Sound imitating the noise caused by a prey animal comes from a target speaker. This speaker can be moved around in both the horizontal and the vertical plane, at a constant distance from the owl's head. (After **Knudsen, E. I.** (1981).)

are of only limited use for this purpose. When the owl swoops down, it aligns its talons with the long axis of the mouse's body, just before striking the mouse. When the mouse turns to run in a different direction, the owl realigns its talons. This flexibility in the behavior of the owl during the free-flight phase of prey catching makes it difficult to employ a standardized behavioral paradigm. Therefore, owls are trained to remain on their perch during the experiment. Each time they turn their head in response to a sound stimulus, they are rewarded with a small piece of meat.

The experimental set-up is sketched in Fig. 6.11. At the beginning of the experiments, the head is aligned by attracting the owl's attention with a sound from a fixed source. This sound source is referred to as the **zeroing loudspeaker**. Stimulation takes place with sound from a second loudspeaker called the **target speaker**. This loudspeaker can be changed in both the horizontal and vertical plane, thus imitating the natural situation in which the angle between the prey and the owl also varies in these two dimensions.

The orientation of the response relative to the target speaker is monitored by means of an **electromagnetic angle-detector system**

The accuracy of sound localization, as indicated by a turning of the owl's head toward the source of the stimulus, can be monitored by means of an electromagnetic angle-detector system.

The barn owl can locate sound with high accuracy in both azimuth and elevation.

composed of two major components: First, two small coils of copper wire mounted on top of the owl's head and arranged perpendicularly to each other. They are called the **search coils**. The second part of the electromagnetic angle-detector system consists of two bigger coils between which the owl is positioned. These two coils carry electric current. They are called the **induction coils**, as they induce current flow in the search coils. Turns of the head cause predictable changes in current flow within the search coils. Analysis of these variations, therefore, enables the experimenter to determine both the azimuth and the elevation of head movements.

Accuracy of orientation response

Experiments employing the electromagnetic angle-detector system have revealed several important features of the barn owl's orientation system. Under optimal conditions (e.g. when the sound source is directly in front of the face), **the barn owl can locate sound within one or two degrees in both azimuth and elevation.** For comparison, humans are as accurate as owls in azimuth, but three times worse in elevation. The biological relevance of this accuracy will be demonstrated in the following exercise.

Exercise: A barn owl sits on its perch 5 m above ground. It hears a mouse directly underneath on the ground. Within what error range can it locate the mouse on the ground? How does this error range relate to the size of an adult field mouse?

Solution: The situation is sketched in Fig. 6.12. The angle α (1°) denotes the accuracy in azimuth, h (5 m) the height of the perch, c the hypotenuse, and x the error range. Therefore,

$$\tan \alpha = x/h$$

or

$$x = h \tan \alpha$$

By using the above values, we get

$$x = 5 \cdot 0.016 \text{ m}$$

Therefore, as a final result,

$$x = 0.08 \text{ m}$$

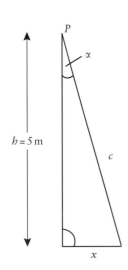

Figure 6.12 Sketch to illustrate the accuracy of sound localization in the barn owl. The owl sits on its perch (*P*) 5 m above ground. In the right triangle drawn, α denotes the accuracy in azimuth, *h* the height of the perch relative to ground, *c* the hypotenuse, and *x* the error range. The angle α is exaggerated. For explanation of the calculation of the error range, see text. (Courtesy: G.K.H. Zupanc.)

The error range of 0.08 m (or 8 cm) roughly corresponds to the length of a mouse. Thus, within the limits of its sensory accuracy, the barn owl is able to locate the mouse with sufficient precision. Moreover, after having started to swoop down to the mouse, the owl can still correct the course of its approach (e.g. if the mouse makes a turn), before it finally strikes the target. This improves the precision of localization even further.

The success rate with which an owl strikes a mouse is illustrated by results of behavioral experiments conducted by Roger Payne. A trained owl was kept in a completely darkened room, its floor covered with leaves. Then, single mice were released. In 16 trials, the owl made 16 strikes at the mouse that was at least 4 m away. It missed the mouse only 4 times, and never by more than 5 cm!

The accuracy of prey localization in both azimuth and elevation varies with several factors, among which the frequency range of the sound plays an important role. Although owls can hear over a broad range of frequencies (roughly from 100 to 12 000 Hz), experiments have shown that it is essential for accurate localization that sound contains relatively high frequencies between 5000 and 9000 Hz. The importance of this frequency range is underlined by a striking sensory adaptation discovered by Christine Köppl, Otto Gleich, and Geoffrey Manley of the Technical University Munich in Garching, Germany. They found that half of the owl's basilar membrane in the inner ear is devoted to the analysis of frequencies between 5000 and 10 000 Hz. A similar sensory specialization to behaviorally important frequencies has been found in bats, as we discussed in detail in Chapter 4.

Another important feature of the owl's orientation response is the rapidity with which the flick of the head occurs. This movement is initiated after a delay of only 60 msec after the onset of the sound. Maximum accuracy in sound localization can be achieved even if the sound ends before the head movement begins. This indicates that the owl does *not* use any feedback information while moving the head, and, thus, *no* iterative adjustment of the head's position relative to the source of sound occurs. Rather, the entire program controlling head movement is laid out before its initiation. In other words, the owl makes the computational decisions in regard to the head movement under **open-loop conditions**. On the other hand, mid-course corrections may be made when the owl flies toward a mouse in the dark, by listening to the noise generated by the moving mouse.

Physical parameters of sound involved in head orientation

Indications of what physical parameters are used by the owl in localizing sound are provided by head-orientation experiments in which one ear is blocked. The results of these experiments and their interpretations can be

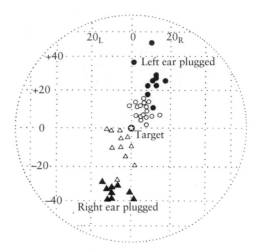

Figure 6.13 The functional significance of interaural intensity differences for localization of sound in elevation. A barn owl's auditory space is drawn in both azimuth and elevation. The target is presented in the origin of the bicoordinate system defined by these two parameters. Insertion of a loose ear plug (open symbols) into the left ear (circles) or the right ear (triangles) results in modest errors in elevation. Occlusion of the ears with tighter ear plugs (closed symbols), on the other hand, causes severe errors in elevation, while still affecting the accuracy of sound localization in azimuth, but only to a limited extent. (After **Knudsen, E. I., and Konishi, M.** (1979).)

summarized as follows:

1. If one ear is completely blocked, the owl makes large errors in localizing the source of sound. This demonstrates that the owl's ability to successfully locate prey depends on a comparison of the signals in the two ears.

2. If one ear is partially blocked, this results in significant errors in determining the elevation (Fig. 6.13). On the other hand, only slight errors occur in the accuracy of determining the azimuth. Partial blocking of one ear effectively reduces the intensity of the sound reaching the ear, but does not significantly alter the time of arrival of the sound at the two ears. These results, therefore, indicate that differences in intensity between the ears (referred to as **interaural intensity differences**) are the principal cues for locating sound elevation.

3. If the plug is inserted into the left ear, it causes the owl to direct its head above and a little to the right of the target (Fig. 6.13). Conversely, a plug in the right ear results in the owl facing below and a little to the left of the target. In either case, the tighter the ear plug, the greater the degree of error.

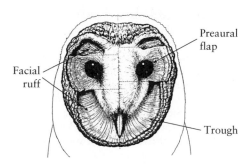

Figure 6.14 Facial ruff of the barn owl. This structure is formed by tightly packed feathers. The ear opening, located behind the preaural flaps, are connected to the surface of the ruff via two troughs running down the face to join below the beak. They collect high-frequency sound and funnel it into the ear canal. The preaural flap and ear opening is lower on the right than on the left side. Moreover, the right trough is tilted up, while the left trough is tilted down. These asymmetries lead to differently perceived intensities of sound by the left and right ear. For demonstration purposes, the owl's facial disc feathers, which normally cover the two troughs, were removed in this illustration. (After **Knudsen, E. I.** (1981) and **Knudsen, E. I., and Konishi, M.** (1979).)

The latter result reflects a vertical **asymmetry in the directional sensitivity of the two ears**. This can be attributed to a vertical displacement of the two ears. While the left ear is above the midpoint of the eyes and points downward, the right ear is below the midpoint of the eyes and points upward.

Moreover, there is a slight **asymmetry in the arrangement of the facial ruff**, which can be seen in Fig. 6.14. This structure is composed of stiff, dense feathers that are tightly packed. The surface of the facial ruff very efficiently reflects high-frequency sound. However, the ear openings are connected to the surface of the ruff via two troughs running through the ruff from the forehead to the lower jaw. The troughs serve a similar purpose as the external pinnae of the human ear: they collect high-frequency sounds and funnel them into the ear canals. Directional asymmetry of the two troughs is produced by the left trough being oriented downward, while the right trough is oriented upward.

Taken together, these asymmetries in the placement of the ears cause the left ear to be more sensitive to sounds coming from below, whereas the right ear is more sensitive to sound from above.

The importance of the facial ruff is underlined by experiments in which the feathers are removed. Then, the owl is unable to locate sound in the elevation. Upon acoustic stimulation, it always faces horizontally—regardless of the true elevation of the source.

Head-orientation experiments employing the electromagnetic angle-detector system have also shown that the most important cues for locating

(a)

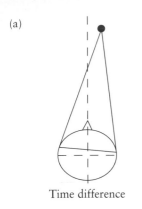

Time difference

Figure 6.15 Physical parameters used by the barn owl for sound localization. (a) Top view of a schematically drawn head of an owl with a sound source. Sound originating from a source not directly in front of the owl results in a difference in the length of the sound path between the two ears, thus causing an interaural time difference. This difference increases with increasing incidence of the sound relative to the midsagittal plane of the head. The interaural time difference is maximal when the sound comes directly from one of the owl's sides. It is used to determine the azimuth component of the location of the sound source. (b) Side view of a schematically drawn head of an owl with left (L) and right (R) ears. In the barn owl, the left ear is located higher than the right ear, relative to the eye level. Moreover, the two ears are sensitive in different vertical directions, an effect enhanced by the structure of the facial ruff. This directional asymmetry leads to interaural intensity differences that enable the owl to localize the sound source in elevation. (After **Konishi, M.** (1990).)

(b)

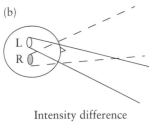

Intensity difference

Barn owls analyze interaural time differences and interaural intensity differences to locate sound in azimuth and elevation, respectively.

---

sound in the azimuth are differences in the arrival times of the two ears, technically referred to as **interaural time differences** (Fig. 6.15). Such differences occur whenever the distance from the source of sound to the two ears is different. Consequently, when the source is directly in front of the bird, there is no difference. When, on the other hand, the sound comes directly from one of the owl's sides, the difference is maximal. Within these two extremes, the interaural time difference varies with the incidence angle of the sound (measured relative to the midsagittal plane of the head) in a systematic way.

Additional support for the hypothesis that interaural time and intensity differences are used to locate sound in azimuth and elevation, respectively, comes from experiments in which miniature earphones were inserted into the owl's ear canal. Through this approach, acoustic stimuli can be generated that are equal in intensity, but differ in the time of arrival at the two ears; or signals that differ in intensity, but are equal in arrival time; or signals comprising even a third category in which different combinations of intensity and arrival time are used.

Such experiments evoke similar responses, as observed after the owl hears sound from outside sources (Fig. 6.16). If the intensity of the sound is held constant, but the sound is issued to one ear slightly before the other ear, the owl turns its head, mostly in the horizontal plane, in the direction of the leading ear. The longer the delay in delivering the sound to the second ear, the further the head turns.

Similarly, if the timing is held constant but the intensity is varied, the owl mostly moves its head in the vertical plane. Sound that differs in both the intensity and the arrival time between the two ears cause the owl to move the head in both horizontal and vertical direction. In each case, the degree to which the owl turns the head corresponds to the degree of the difference observed in these two parameters after generating sound at these imaginary locations. Thus, the owl appears to use, indeed, interaural

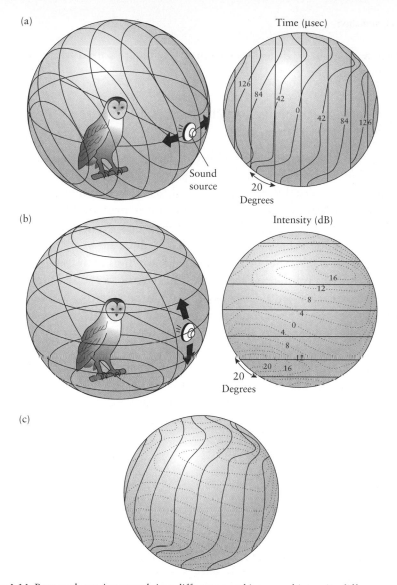

Figure 6.16 Barn owls use interaural time differences and interaural intensity differences to determine the azimuth and elevation components, respectively, of a source of sound within a bicoordinate system. For visualization, this bicoordinate system can be plotted along the surface of an imaginary globe encapsulating the owl's head. (a) An interaural time difference occurs when the sound source is not directly in front of the owl. This difference increases by 42 μsec for every 20° a sound source moves laterally. Thus, this parameter provides the horizontal (azimuth) component of the positional information of a sound source. (b) Interaural intensity differences determine the vertical (elevation) component of a source of sound. Sound originating from sources above eye level is perceived more intensely by the right ear. This difference increases with increasing vertical position of the sound source, as indicated by the respective decibel levels. On the other hand, sound originating from sources below eye level is perceived more intensely by the left ear. Again, the extent of this difference depends on the vertical position of the sound source, as indicated by the respective decibel levels. (c) Combination of the azimuth and elevation information defines each location in space. When the owl is stimulated by sound coming from such a location, it turns its head in the direction predicted by the model. (After **Konishi, M.** (1993).)

time and intensity differences to precisely locate sound in the horizontal and vertical plane.

The cochlear nucleus: parallel processing of time and intensity information

As we have seen above, the owl determines the azimuth and the elevation bicoordinate components of a sound source through analysis of interaural time and intensity differences. How is this achieved at the neuronal level?

The link between the ear and the brain is provided by the **auditory nerve**. Its axons originate from nerve cell bodies situated in the inner ear. Different sound frequencies are encoded by different fibers of the auditory nerve. By contrast, the codes for intensity and timing of the sound are not segregated within the auditory nerve. Rather, both sound parameters are carried by each fiber. Encoding of these parameters is achieved through variation of the rate and the timing of action potentials, respectively. With changing sound intensity, the fibers respond distinctly by changing their spike rate. In addition, each fiber fires at a particular phase angle of the spectral component to which it is tuned, a phenomenon referred to as **phase locking**. This physiological feature is illustrated in Fig. 6.17. Phase locking is used to encode the timing of the signal.

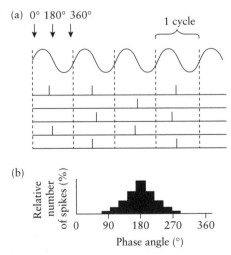

Figure 6.17 Schematic illustration of the principle of phase locking. (a) The stimulus provided is a sinusoidal sound wave running over four cycles. The auditory neuron depicted fires preferentially at or near 180°, as shown in the five traces. (b) The phase-locking property of a neuron can be verified by analysis of a period histogram. For such histograms, the relative frequency of spikes occurring at the various phase angles is determined. The plot shown here reveals that the vast majority of the neuron's spikes occur at or near 180°. Thus, the neuron analyzed is phase-locked to 180°. (After **Konishi, M.** (1992).)

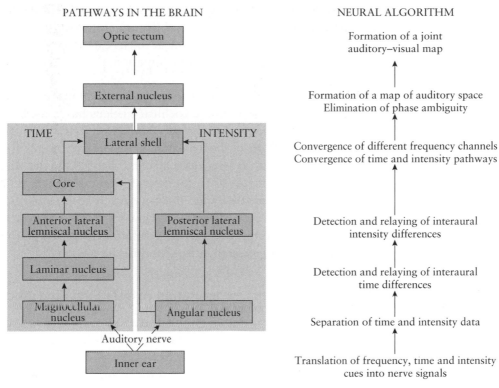

PATHWAYS IN THE BRAIN

NEURAL ALGORITHM

Figure 6.18 The neural circuit involved in sound localization. To the right, the corresponding steps of the neural algorithm are summarized. The neural pathway starts in the inner ear that is connected, via the auditory nerve, to two cochlear nuclei: the magnocellular nucleus and the angular nucleus. From this hierarchical level on, time and intensity data are processed in parallel pathways. After several computational steps, they converge again in the lateral shell. An auditory map is formed in the external nucleus. Projection of neurons of this nucleus to the optic tectum leads to the formation of a combined auditory–visual map. (After **Konishi, M.** (1992).)

Fibers of the auditory nerve innervate the **cochlear nucleus,** of which two are located in each cerebral hemisphere. This nucleus consists of two subpopulations of neurons differing in both morphology and physiology. Neurons of one subpopulation define the **magnocellular nucleus,** whereas neurons of the other subpopulation form the **angular nucleus.** Each fiber of the auditory nerve divides into the collaterals. One of them innervates the magnocellular nucleus, the other enters the angular nucleus (Fig. 6.18).

The branching of the auditory nerve fibers provides the structural basis to separate processing of phase and intensity information within the cochlear nucleus. A subpopulation of neurons of the angular nucleus are sensitive to variations in sound intensity over a large dynamic range, but do not exhibit phase locking to particular phase angles of the acoustic stimulus. By contrast, neurons of the magnocellular nucleus show phase locking, but, although sensitive to changes in intensity, exhibit a smaller dynamic range. Thus, the two cochlear nuclei can be viewed as filters: the angular neurons let pass intensity information, the magnocellular neurons timing information.

Support for the notion that intensity and timing information are processed in parallel within the cochlear nucleus has come from experiments in which Terry Takahashi, Andrew Moiseff and Masakazu Konishi prevented one of the two cochlear nuclei from firing. Injection of a local anesthetic into the magnocellular nucleus altered the selectivity of higher-order brain neurons to interaural time differences, but did not affect their response to interaural intensity differences. Similarly, inactivation of the angular nucleus resulted in changes in the selectivity of higher-order neurons to interaural intensity differences, without altering the selectivity of the response to interaural time differences.

Interaural time and intensity data are segregated in the first way stations processing auditory information: neurons of the angular nucleus process intensity information, whereas neurons of the magnocellular nucleus respond to timing information.

The laminar nucleus: computation of interaural time differences

The next way station after the magnocellular nucleus is the **laminar nucleus** in the brain stem. It receives input both from the magnocellular nucleus on the same (ipsilateral) side of the head and from its counterpart on the opposite (contralateral) side of the head. Thus, it is in the laminar nucleus that information from both sides of the ear converges for the first time within the brain.

As a detailed anatomical and physiological study conducted by Catherine Carr and Masakazu Konishi has shown, the laminar nucleus plays a crucial role in **measuring and encoding interaural time differences** through analysis of the corresponding interaural phase differences. Obviously, this requires the combination of timing data from both ears.

Based on such a convergence of input from the two ears, a theoretical model circuit for detection of interaural time differences was suggested as early as 1948 by Lloyd Jeffress, while he was a visiting professor at the California Institute of Technology. The two key elements of his model, which is shown in Fig. 6.19, are **delay lines** and **coincidence detectors**. Jeffress proposed that coincidence detectors receive input from both ears. However, the time of transmission of signals varies between the ears, due to the existence of the delay lines. They reflect the differences in arrival time at the two ears of an acoustic signal if the sound source is not placed directly alongside the midsagittal plane of the head. Thus, different coincidence detectors encode different transaural time differences.

Coincidence detectors fire most strongly if the phase-locked impulses generated at lower brain levels reach the detector simultaneously. By contrast, asynchronous arrival of the impulses at the coincidence detector results in relatively weak firing of the detector.

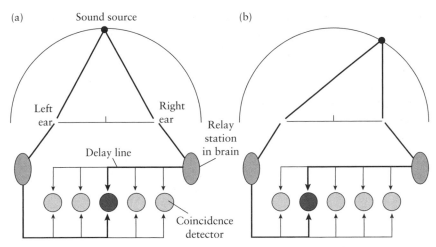

Figure 6.19 Generalized model circuit for detection of interaural time differences, as proposed by Lloyd Jeffress in 1948. A crucial element of this circuit is a series of coincidence detectors. They receive input from both ears, but only the one that receives impulses from the two sides, simultaneously, will fire (indicated by the dark circle). The fibers that connect the coincidence detectors with the relay station in the brain are constructed such that they function as delay lines. As the sound source moves from a position directly in front (a) to one located on one of the two sides (b), the coincidence detector that responds maximally changes as well. This differential responsiveness of the coincidence detectors forms the basis for the measurement of interaural time differences. (After **Konishi, M**. (1993).)

In the barn owl's brain, the axons of the magnocellular neurons serve as delay lines. Neurons within the laminar nucleus function as coincidence detectors. This basic neural circuitry is shown in Fig. 6.20. When two impulses locked to a certain phase angle of the respective frequency component arrive at roughly the same time, the input-receiving neuron of the laminar nucleus fires at its maximal rate. With decreasing degree of coincidence, its firing rate drops, until it finally ceases firing. As a result of this operation and of a systematic topographic arrangement of coincidence-detector neurons in the laminar nucleus, certain interaural time differences are encoded by certain laminar neurons defined by their location within the nucleus.

On the other hand, laminar neurons fire maximally not only when hit synchronously by impulses marking a particular interaural time difference. Rather, they also respond at maximum rate if this time difference is delayed or advanced by one full cycle (360°) or by integer multiples of this cycle of the sound wave, as shown by Fig. 6.21. This phenomenon is referred to as **phase ambiguity**. Obviously, phase ambiguity, as it persists in the laminar nucleus, imposes an enormous challenge upon the owl's ability to compute interaural time differences. However, as we will see below, this problem is overcome at higher levels of sensory processing.

Interaural time differences are computed in the laminar nucleus. This is achieved by neurons of this nucleus functioning as coincidence detectors and axons arising from the magnocellular neurons as delay lines.

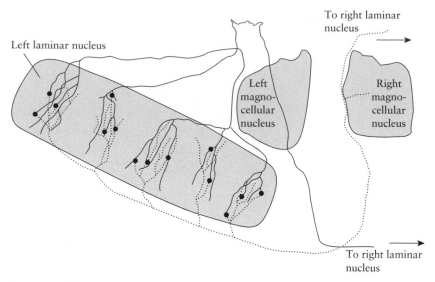

Figure 6.20 Schematic drawing of the dorsal brain stem area with the left laminar nucleus, as well as the left and right magnocellular nuclei. Upon this drawing, a reconstruction of an ipsilateral and contralateral magnocellular neuron is superimposed. This axonal arbor was labeled by using a neuronal tract-tracing technique. The fibers arising from the cell bodies in the magnocellular nucleus serve as delay lines. Neurons of the laminar nucleus (black dots) act as coincidence detectors. They fire at maximum rate when impulses carried by axons arising from the ipsilateral and contralateral magnocellular nucleus reach them simultaneously. For the sake of clarity, axons originating from the left and the right magnocellular nucleus are distinguished by closed and dotted lines, respectively. (After **Konishi, M**. (1993).)

Another complication arises from the fact that coincidence detectors, such as the neurons within the laminar nucleus of the barn owl, respond not only to **binaural stimuli**, but also to **monaural stimuli**. Discrimination between these two types of stimuli is observed only in higher-order neurons, the so-called space-specific neurons in the inferior colliculus (see below).

The posterior lateral nucleus: computation of interaural intensity differences

After processing in the laminar nucleus, timing information is relayed to higher brain centers on the opposite side of the head. These centers include the **anterior lateral lemniscal nucleus** and the **core** (Fig. 6.18).

Similarly, intensity information continues to be carried by a separate pathway after having been processed in the angular nucleus. Among these higher processing centers is the **posterior lateral lemniscal nucleus**.

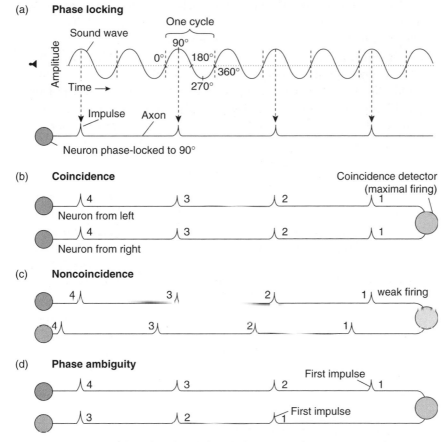

Figure 6.21 Illustration of the principles of phase locking, coincidence, non coincidence, and phase ambiguity. (a) As a stimulus, a sinusoidal sound wave of a specific frequency is used. An auditory neuron tuned to this frequency produces action potentials at a certain phase angle (here: 90°). In other words, the neuron is phase-locked to 90°. However, instead of firing every time the wave reaches this phase angle, the neuron depicted here fires only every other time. (b) In both ears, auditory neurons tuned to the same frequency lock to the same phase angle. If these impulses reach the coincidence detector simultaneously, this neuron will fire at maximum rate. (c) If the impulse trains reach the coincidence detector asynchronously, the neuron will fire rather weakly. (d) The coincidence detector is prone to phase ambiguity. In the case illustrated, the first impulse transmitted by the upper delay line is advanced by the time of two full cycles of the sound wave relative to the first impulse carried by the lower delay line. Despite this ambiguity, the coincidence detector will fire strongly. For the sake of illustration, the impulses have been numbered. (After **Konishi, M.** (1993).)

It receives excitatory input from the contralateral angular nucleus and inhibitory input from the contralateral posterior lateral lemniscal nucleus. Thus, the two posterior lateral lemniscal nuclei are reciprocally inhibitory.

The posterior lateral lemniscal nucleus is the brain region where intensity differences between the two ears are computed. The structural basis for this comparison of the intensity between the two ears is provided by the projection of each of the two posterior lateral lemniscal nuclei to their contralateral counterparts. At the synaptic level, this comparison is made by the two different types of input to neurons of the lemniscal nucleus: inhibitory input received, via the contralateral posterior lateral lemniscal nucleus, from the ipsilateral ear, and excitatory input from the contralateral ear. The difference between the strength of the inhibitory input and that of the excitatory input determines the rate at which neurons of the lemniscal nucleus fire.

However, as has been shown by Geoffrey Manley and Christine Köppl in collaboration with Masakazu Konishi, the threshold and the strength of the inhibition varies systematically along the dorsoventral axis of the posterior lateral lemniscal nucleus. This gradient results in neurons orderly arranged within the nucleus depending on their selectivity for different interaural intensity differences. In the left nucleus, ventral neurons respond maximally when sound is louder in the left ear; conversely, neurons in the dorsal portion fire most strongly when sound is louder in the right ear. Similarly, in the right nucleus, ventrally located neurons respond maximally when sound is louder in the right ear; neurons in the dorsal part exhibit maximal responsiveness when sound is louder in the left ear.

Interaural intensity differences are computed in the posterior lateral nucleus.

The lateral shell: convergence of timing and intensity information

Both the core and the posterior lateral lemniscal nucleus project to the **lateral shell** (Fig. 6.18), suggesting that **timing and intensity information converge** at this level of sensory processing. This hypothesis is strongly supported by recordings from neurons within this nucleus. Such experiments have demonstrated that most neurons respond both to interaural time differences and interaural intensity differences. However, localization of a sound source in space is not yet possible, as phase ambiguity still persists.

After parallel processing over several brain levels, time and intensity converge in the lateral shell.

External nucleus: formation of a map of auditory space

Phase ambiguity is overcome only at the next level of sensory processing, namely in the **external nucleus** (Fig. 6.18). This nucleus is located in a brain area equivalent to the inferior colliculus of mammals. The so-called **space-specific neurons** in the external nucleus respond to acoustic stimuli only if the sound originates from a restricted area in space. Sound coming from a source outside of this area either does not evoke a response at all, or does so only at a low level of firing, by the same neuron. The area in space to which neurons within a certain brain area respond is called the **receptive field**. Neurons of the left external nucleus have their corresponding receptive fields primarily in the right auditory space, whereas neurons of the right external nucleus process information predominantly of the left auditory space, although some overlap in regions near the midsagittal plane of the head does occur.

Another property of the external nucleus is that neighboring space-specific neurons have receptive fields representing neighboring regions in space. This leads to a systematic arrangement of these neurons such that sound azimuth is arrayed mediolaterally and sound elevation is mapped dorsoventrally. Overall, as a result of this arrangement, a **neural map of auditory space** is formed.

On each side of the brain, the neural map formed by neurons of the external nucleus represents an auditory region extending from 40° contralaterally to 15° ipsilaterally in azimuth and from 20° above to 20° below eye level in elevation. However, within this range, different space regions are represented to a different extent. For example, the region within 20° of the midpoint of the face in both azimuth and elevation is proportionally heavily represented. On the other hand, the number of neurons devoted to processing of information from regions near the lateral, dorsal, and ventral edges of the auditory space is relatively low. This causes the spatial resolution to be much better if sound comes from an area in front of the owl, than if sound originates from other regions. Similar results have been obtained through behavioral experiments.

An essential prerequisite to elicit a response from a specific space-specific neuron is that both the time difference *and* the intensity difference fall within the range to which the neuron is normally tuned. Stimulation by the correct time difference (or the intensity difference) alone is not sufficient for a response. These time and intensity difference combinations correspond to the horizontal and vertical location within the bicoordinate system of the neuron's receptive field. Stimulation of the owl through miniature earphones with correct combination of the parameters evokes not only a response in the corresponding space-specific neurons, but also triggers a turning of the owl's head toward the correct location in auditory space.

The final step in sensory processing: formation of an auditory–visual map

Under most natural conditions, or if not kept in complete darkness, the barn owl uses both the auditory and the visual system to localize prey. How is information shared between these two systems coordinated?

As investigations by Eric Knudsen (then at Stanford University) have shown, the auditory map provided by neurons of the external nucleus projects to the **optic tectum** (which is homologous to the superior colliculus

Space-specific neurons: respond only to acoustic stimuli originating from a restricted area in space.

Through systematic arrangement of space-specific neurons, a neural map of auditory space is formed in the external nucleus.

In the optic tectum, a joint auditory-visual map is formed.

of mammals). There, a joint **auditory–visual map** is formed. Each neuron of this map responds to both auditory and visual stimuli arising from the same point in space. In this map, the representation of the frontal region of space is greatly expanded. This corresponds to the behavioral observation that, in this region, the highest accuracy in localization of prey is achieved.

How is the joint auditory–visual map formed? Experiments by Eric Knudsen and Michael Brainard have shown that the alignment of the two sensory maps is controlled during ontogeny by visual instruction of the auditory spatial tuning of neurons in the optic tectum. To shift the visual field at a certain angle to the left or the right, the two scientists attached to barn owls displacing prisms mounted in spectacle frames in front of their eyes at 12–15 days of age, just as the eyes are beginning to open for the first time. After at least 70 days of prism experience, neurons of the optic tectum were found to be tuned to sound source locations corresponding to their optically displaced, rather than their normal, visual receptive field locations. Thus, these results demonstrate that, during development, visual experience fine-tunes the topography of the auditory map and aligns it with the visual space map.

The final behavioral action—the turning of the head toward the sound source—requires a link between the auditory–visual map and motor regions. This connection appears to be mediated by a projection of the auditory–visual map to a **motor map**. Stimulation of neurons within this map induces head movements in the direction of the corresponding receptive field represented by the auditory–visual map.

Summary

■ The first example discussed in this chapter is based on the ability of toads to recognize prey and predators and to respond with appropriate behavior patterns. In dummy experiments, such objects can be identified by using rectangular stripes. Movement of such rectangles in direction of their long axis ('worm configuration') elicits responses resembling those observed under natural conditions toward prey. Movement of the rectangles in direction of the short axis ('anti-worm configuration') results in responses typically exhibited toward predators.

■ Within certain limits, an increase in the length of the long axis of the rectangle leads to a greater response of the toad.

■ The major targets in the toad's brain of the retinal ganglion cells are the thalamic-pretectal area and the optic tectum. Electrical recordings from these brain regions, as well as the ganglion cells, have revealed a rather poor sensitivity of retinal ganglion cells to worm- and anti-worm-like stimulus configurations. By contrast, in both the thalamic-pretectal area and the tectum there are populations of neurons that are activated by more complex stimulus configurations. One cell type called TP3, in the thalamic-pretectum, is best activated by anti-worm-like stimuli. Conversely, a cell type termed T5(2) in the tectum, is activated by a worm-like stimulus configuration.

- In agreement with the results of the recording experiments, prey-catching behavior is released by electrical stimulation of the optic tectum, whereas stimulation of the thalamic-pretectal region activates escape behavior.

- The configurational selectivity of T5(2) neurons in the tectum appears to crucially depend on inhibitory input received from the thalamic-pretectal area. Lesioning of this connection results in the toad being unable to discriminate between prey and predator objects.

- The second example discussed in detail in this chapter centers around barn owls. As nocturnal hunters, they use the sense of hearing to localize prey, predominantly field mice. Upon hearing the noise generated by the prey animal, the owl turns its head in a rapid flick so that it directly faces the source of sound.

- The owl's head-turning response can be monitored with an electromagnetic angle detector system consisting of (i) search coils mounted on top of the owl's head, and (ii) induction coils between which the bird is positioned.

- The flick of the head is initiated approximately 100 msec after the onset of the sound. While moving the head, the owl does not use any feedback information. Thus, computational decisions regarding the head movement are made under open-loop conditions.

- Under optimal conditions, the barn owl can localize the source of sound within 1–2° in both the horizontal plane (azimuth) and the vertical plane (elevation).

- The owl's ability to precisely localize the source of sound is based on the analysis of interaural time differences and interaural intensity differences. Interaural time differences occur when the sound source is not directly in front of the owl; they define the location of the sound in the azimuth. Interaural intensity differences are the result of a directional asymmetry of the owl's ears and the facial ruff; they determine the elevation coordinate of a sound source.

- Sound intensity and timing are encoded in each fiber of the auditory nerve by variation of the rate of firing and the locking of the action potentials to a particular phase angle of the spectral component to which the respective fiber is tuned.

▪ Each fiber of the auditory nerve divides into two collaterals. One of them innervates the magnocellular nucleus, while the other enters the angular nucleus. Both nuclei are subdivisions of the cochlear nucleus. Whereas neurons of the angular nucleus are sensitive to intensity information, neurons of the magnocellular nucleus process timing data.

▪ The next way station is the laminar nucleus. It receives input from both the ipsilateral and the contralateral magnocellular nucleus. The main function of the laminar nucleus is to compute and encode interaural time differences. This is achieved by (i) the axons of the magnocellular nucleus serving as delay lines, and (ii) the neurons of the laminar nucleus functioning as coincidence detectors. However, timing information extracted is not unambiguous due to the existence of phase ambiguity at this level of sensory processing.

▪ Interaural intensity differences are computed in the posterior lateral lemniscal nucleus, which receives input both from its contralateral counterpart and from the contralateral angular nucleus.

▪ Timing and intensity information converge in the lateral shelf where neurons respond to both interaural time differences and interaural intensity differences.

▪ While phase ambiguity still persists in the lateral shell, it is overcome at the next level of sensory processing, namely in the external nucleus. In this brain region, space-specific neurons respond only to acoustic stimuli originating from their receptive field. Overall, these neurons form a neural map of the auditory space. Within this map, regions near the midpoint of the face are represented by significantly more neurons than are more lateral areas. This leads to high resolution, if sound comes from sources directly in front of the owl.

▪ The final step in sensory processing is the formation of an auditory–visual map achieved through projection of the neurons in the external nucleus to the optic tectum. This auditory–visual map projects to a motor map, stimulation of which induces head movements.

Recommended reading

Ewert, J.-P. (1974). The neural basis of visually guided behavior. *Scientific American*, **230**(3), 34–42.

Provides a good introduction to the research led by Jörg-Peter Ewert on the neural basis of object recognition in toads.

Ewert, J.-P. (1980). *Neuroethology: an introduction to the neurophysiological fundamentals of behavior*. Springer-Verlag, Berlin/Heidelberg/New York.

The first textbook of neuroethology. As many examples have been taken from Ewert's own research, this book is especially well suited for students who seek a comprehensive, yet easy-to-read, review of the neural mechanisms underlying object recognition in toads.

Ewert, J.-P. (1997). Neural correlates of key stimulus and releasing mechanism: a case study and two concepts. *Trends in Neurosciences*, **20**, 332–339.

An update of Ewert's work and an attempt to interpret the results of his classic work from a more modern point of view. Although sometimes difficult to read, useful as a supplement of the above two readings.

Knudsen, E. I. (1981). The hearing of the barn owl. *Scientific American*, **245**(6), 82–91.

A good starting point to be introduced to the work on behavioral mechanisms governing sound localization in barn owls.

Konishi, M. (1992). The neural algorithm for sound localization in the owl. *The Harvey Lectures*, **86**, 47–64.

Review of the neural mechanisms underlying sound localization in the owl. Suitable for more advanced readers.

Konishi, M. (1993). Listening with two ears. *Scientific American*, **268**(4), 34–41.

The best overview available to introduce students to the work of Mark Konishi and associates.

Questions

6.1 What signals do toads use to recognize prey and enemies? How is their response linked to the properties of cells in the retina and in specific parts of the brain?

6.2 Barn owls are able to localize prey solely based on noise generated by the prey animal. How is this achieved at the behavioral level? How are the physical parameters of sound essential for localization of the sound source processed and integrated in the brain?

Sensorimotor integration: the jamming avoidance response of the weakly electric fish, *Eigenmannia*

7

▨ Introduction

▨ The system and its components

▨ Behavioral experiments

▨ Neuronal implementation

▨ Reflections on the evolution of the jamming avoidance response

▨ Summary

▨ Recommended reading

▨ Questions

Introduction

One of the masterpieces of neuroethological research, which will be described in detail in this chapter, is the behavioral and neurobiological analysis of the **jamming avoidance response** of the knifefish *Eigenmannia* sp. This behavior was discovered by two Japanese scientists, Akira Watanabe and Kimihisa Takeda, both from the Tokyo Medical and Dental University. They described their findings in a paper published in the *Journal of Experimental Biology* in 1963.

Shortly after its discovery, Theodore ('Ted') Bullock of the Scripps Institution of Oceanography of the University of California, San Diego, began to study this seemingly exotic behavioral pattern. He assumed the function of this response to be maintenance of a private frequency of each individual fish for detecting objects, such as obstacles in the closer vicinity. He, therefore, named it jamming avoidance response.

BOX 7.1   Walter Heiligenberg

Walter Heiligenberg in his laboratory at the Scripps Institution of Oceanography. (Courtesy: G.K.H. Zupanc.)

One of the best and most complete case studies ever performed in neuroethology is the work of Walter Heiligenberg on the jamming avoidance response of weakly electric fish. Heiligenberg was born in Berlin (Germany) in 1938. At the age of only 15, he met Konrad Lorenz (see Box 2.1), at the time head of the Max Planck research group in Buldern (Westphalia). Under the guidance of Lorenz, and while still in school, Heiligenberg performed his first behavioral observations on fish and birds. During that time, Lorenz, together with Erich von Holst (see Box 2.3), established the Max Planck Institute for Behavioral Physiology in Seewiesen. Heiligenberg followed Lorenz to do his Ph.D. work, an analysis of the influence of motivational factors on the occurrence of behavioral patterns in a cichlid fish. This study, completed in 1963 under the guidance of Konrad Lorenz and the well-known sensory physiologist

Hansjochem Autrum of the University of Munich, already revealed features that became characteristic of Heiligenberg's work: exploration of a biological phenomenon of general interest through the application of a strict analytical and quantitative methodology. In using this approach, Heiligenberg was years ahead of his time. The approach had its roots in Heiligenberg's keen interest in mathematics and physics, subjects, which in addition to zoology and botany, he also formally studied at the university.

During his postdoctoral years in the laboratory of Horst Mittelstaedt at the Max Planck Institute for Behavioral Physiology, Heiligenberg continued to analyze motivational processes. He was the first to succeed in a quantitative demonstration of the law of heterogenous summation (see Chapter 3). In 1972, upon invitation by Theodore Bullock, he joined the Neurobiology Unit of the Scripps Institution of Oceanography of the University of California, San Diego. At the same institution, he established his own laboratory in 1973, and he remained there as a member of the faculty until his death in 1994. It was during this period of roughly 20 years that he performed his work on the jamming avoidance response.

Walter Heiligenberg was not only an extraordinarily gifted researcher of enormous devotion to, and enthusiasm for, science. He was also an outstanding craftsman who liked to design and assemble his own mechanical and electronic devices. After his first wife died of cancer in 1991, he married a musician who lived in Munich. Following their marriage, he traveled back and forth between San Diego and Munich. On one of these trips, on his way to San Diego, he arrived in Chicago earlier than expected. He, therefore, decided to take a flight different from the one for which he was scheduled. A few minutes before landing at Pittsburgh, this plane went into a nose dive and crashed. All 132 people aboard the aircraft, including Walter Heiligenberg, perished.

Together with a German postdoctoral fellow, Henning Scheich, Ted Bullock carried out a thorough characterization of the jamming avoidance response and even made a computer model that accurately predicted the dynamics of the input–output relationship for a range of different stimuli. Their model included several key parameters of the stimulus but lacked a measure of the assumed function, namely the accuracy of detecting objects. Such a test was devised by another German scientist, Walter Heiligenberg (see Box 7.1), who came to the Scripps Institution of Oceanography at the beginning of the 1970s. Over a period of 20 years, he, together with his group and collaborators (notably Theodore Bullock, Carl Hopkins, Joseph Bastian, Leonard Maler, Catherine Carr, John Dye, Gary Rose, Masashi Kawasaki, Clifford Keller, and Walter Metzner), studied the jamming avoidance response in great detail. They succeeded in revealing both the principal behavioral rules and the major neural components underlying this behavior. The results of this research provide an excellent example to illustrate how **sensory information** and **motor programs** are **integrated** to generate a biologically important behavior pattern in response to a stimulus.

The system and its components

Electric organs and electroreceptors

Eigenmannia sp. (in most studies the species *E. lineata* or *E. virescens* are used) is a teleost fish belonging to the order Gymnotiformes. Figure 7.1 shows a photograph of *Eigenmannia lineata*. The genus *Eigenmannia* lives in freshwater habitats in South and Central America. Like all other members of its order, it is distinguished by its ability to produce **electric fields**. This is achieved by discharging an **electric organ** in the tail. The electric organ is composed of modified muscle cells called **electrocytes**, each of which produces a discharge of some tens of a millivolt. As shown in Fig. 7.2, many of these cells are arranged in series. By this arrangement, and the synchronous depolarization of the electrocytes, the rather low voltages of the individual discharges add up to a sum potential of the order of 1 V.

The electric signals can be monitored by placing two stainless steel electrodes close to the head and tail of the fish, respectively. The electric organ discharges approximately 200–500 times per second. The corresponding frequency range of 200–500 Hz is specific to the species. Although different individuals discharge at different frequencies within this range, the frequency of each individual fish is extremely regular. Moreover and as demonstrated by Fig. 7.3, these so-called **electric organ discharges** are, in terms of their waveform, rather simple, resembling a train of sine waves.

The regularity of the electric organ discharges is determined by the action of an endogenous oscillator in the medulla oblongata, an area located in the brainstem of these fish. This oscillator, called **pacemaker**

Figure 7.1 The knifefish *Eigenmannia lineata*. This weakly electric gymnotiform fish, as well as other species within its genus, have become one of the premier model systems in neuroethology. (Courtesy: G.K.H. Zupanc.)

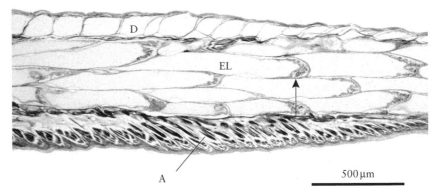

Figure 7.2 Parasagital section through the electric organ of *Eigenmannia* near the caudal end of the anal fin (orientation: dorsal side, up; ventral side, down; rostral end, left; caudal end, right). The electric cells (electrocytes; EL) are distinguished by their long, slender shape. They are arranged in series along the longitudinal axis of the body. At their caudal end, where the membrane exhibits pronounced evaginations (arrow), the electrocytes are innervated by electromotoneurons. A, muscle fibers of anal fin; D, dermis with scales. (Courtesy: G.K.H. Zupanc and F. Kirschbaum.)

nucleus, shows a similar high degree of regularity as the electric organ. It sends one volley of spikes down the spinal cord to trigger one discharge cycle of the electric organ discharges.

The fish are able to sense their own electric discharges, as well as electric signals of other fish, or of abiotic sources, through various types of **electroreceptors**. Two of the major receptor types are shown in Fig. 7.4. This illustration is based on a microphotograph taken from a

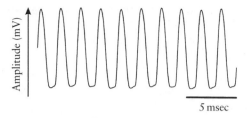

Figure 7.3 Electric organ discharge of *Eigenmannia*. The recording shows 10 discharges produced within approximately 20 msec, thus corresponding to a frequency of roughly 500 Hz. Note an enormous degree of regularity in frequency. (Courtesy: G.K.H. Zupanc.)

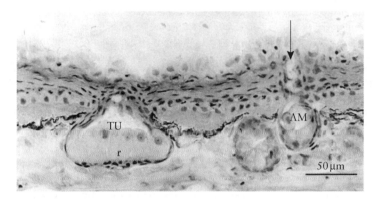

Figure 7.4 Two types of electroreceptor organs in the skin of *Eigenmannia*. In the left half of the microphotograph, a tuberous (TU) electroreceptor organ, composed of individual receptor cells (r), is shown. This type of electroreceptor is tuned to ac signals with frequencies in the range of the fish's own electric organ discharge. In the right half of the microphotograph, an ampullary electroreceptor organ (AM) is visible. This receptor organ type is characterized by the long, large-diameter canal (arrow). Ampullary receptors respond maximally to dc and low-frequency ac signals. (Courtesy: G.K.H. Zupanc and F. Kirschbaum.)

histological section through the skin of *Eigenmannia*, and thus reveals the morphological structure of these receptors. Both types of electroreceptors are distributed all over the body, but are most abundant in the head region. Electroreceptors are derived from mechanoreceptors of the lateral line system, but specialized in the detection of electric currents of low amplitude.

Major components of the electric system: electroreceptors; electric organ; and central structures devoted to the processing of electrosensory information and to the motor control of the electric organ discharge.

Electrolocation

Since the fish is continuously 'surrounded' by its own electric field, electric signals of neighbors are perceived as perturbations in the feedback from their own discharges, rather than as discrete electric events. Similarly, an object in the fish's vicinity distorts the electric field and alters the current

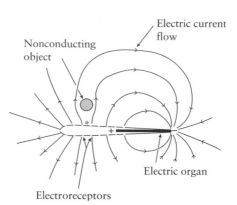

Figure 7.5 Schematic drawing of an electrolocating electric fish. The black bar indicates the location of the electric organ. Electroreceptors are found in pores of the body surface. Their density is highest in the rostral part of the body. While the interior of the body is of relatively low resistance, the resistance of the skin is high, forcing the current to flow through the pores occupied by the electroreceptors. An object with an impedance different from that of the surrounding water (in the drawing, a non-conducting object is shown) causes a distortion of the electric field, as represented by the lines of current flow. This leads to an alteration in the pattern of transepidermal voltage in the area of the skin nearest the object (indicated by *). This alteration, which represents the electric image of the object, is monitored by the electroreceptors. (After **Heiligenberg, W.** (1977).)

pattern perceived by the array of electroreceptors closest to the object. This is schematically shown in Fig. 7.5 by demonstrating the effect of a non-conducting object on the current pattern. The alteration in the current pattern on the fish's body surface represents the **electric image** of the object. The analysis of such images enables *Eigenmannia*, like other electric fish, to **electrolocate** objects in their own environment. Electrolocation can, therefore, be considered a form of 'seeing' with the body surface.

The electrolocation capability was first demonstrated by Hans Lissmann (1909–1995) of the University of Cambridge (UK) in the 1950s. Lissmann trained the weakly electric fish *Gymnarchus niloticus* to distinguish between objects of different conductivity, for example two cylindrical porous pots filled with solutions differing in salinity. In the course of training, approach of one porous pot was rewarded with food, while approach of the other one was followed by punishment consisting of chasing away the fish. The fish's ability to discriminate the objects based on differences in conductivity was tested by analyzing the differences in approach when the porous pots were presented, without rewarding the fish for a 'correct' response or punishing it for an 'incorrect' response. The positive results of these and other experiments demonstrated that the fish are, indeed, capable of electrolocation, and that they can distinguish objects solely based on their electric properties—even when the differences in conductivity are minor. As key experiments in the discovery of a 'sixth sense' of electric fish, Lissmann's findings paved the path for numerous subsequent investigations on this animal group.

The jamming avoidance response

Problems in the ability to electrolocate arise if a neighboring fish with a discharge frequency similar to the fish's own electric organ discharge frequency comes close. Then, **the two electric fields interfere with each other,** resulting in phase and amplitude modulations of each of the two electric signals, as will be shown in detail below. It has been demonstrated that these modulations severely impair the fish's ability to electrolocate. This effect is most detrimental, if the difference between the fish's frequency and the frequency of the neighbor is approximately 20 Hz.

To avoid the detrimental interference of the two electric fields, the fish shifts its own frequency away from the neighbor's frequency such that it maximizes the frequency difference. This behavior is called **jamming avoidance response.** It leads to the following behavioral pattern:

1. If the neighbor's frequency is higher than the fish's own frequency, the fish lowers its discharge frequency.
2. If the neighbor's frequency is lower than the fish's own frequency, the fish raises its own discharge frequency.

These observations demonstrate that the fish is able to determine the **sign of the frequency difference.** This difference, commonly termed 'Df', is defined as the neighbor's frequency minus the fish's frequency.

Example: (1) The frequency of the fish is 400 Hz, that of the neighbor 410 Hz. Then, the frequency difference is +10 Hz, with the sign of Df being positive. This will cause the fish to lower its frequency. (2) The frequency of the fish is, again, 400 Hz, but the frequency of the neighboring fish is now 397 Hz. Then, the frequency difference is −3 Hz, with the sign of Df being negative. As a result, the fish will raise its frequency.

How does the fish determine the sign of the frequency difference, and how does it implement these behavioral rules at the neural level?

Behavioral experiments

Determination of the sign of the frequency difference without internal reference

If a human engineer were confronted with the task of determining the sign of the frequency difference, the likely solution would be as follows. Since

---

Electric fish obtain information about objects in their closer vicinity through analysis of distortions of the self-generated field surrounding them.

The jamming avoidance response involves a shifting of the fish's own frequency away from the frequency of the interfering signal.

Frequency difference (Df): frequency of the neighbor's signal minus frequency of the fish's signal.

the fish's own discharge frequency is determined by the frequency of the pacemaker nucleus, the fish could use the frequency of this oscillator as an internal reference. Thus, by comparing this frequency with the frequency of the interfering signal produced by the neighbor, the fish would be able to determine the sign of the frequency difference between the two electric organ discharges.

Very much to the disappointment of the biologist, however, this is not what *Eigenmannia* does. The following two experiments have ruled out this possibility and revealed some details of the behavioral mechanism underlying the jamming avoidance response.

Curare: generic name for various types of unstandardized extracts derived mainly from the bark of the tropical plants Strychnos and Chondrodendron. Prepared for use as an extremely potent arrow poison by Indians in South America. The physiological active ingredient of curare is the alkaloid tubocurarine, which is employed as a relaxant of skeletal muscles during surgery to control convulsions. The muscle-relaxant effect is caused by interference of the alkaloid with the action of acetylcholine at the neuromuscular junction.

In the first experiment, the fish's electric organ discharge was silenced by application of **curare**. (This is possible because the electric organ of *Eigenmannia* is derived from muscle cells, and curare acts as a relaxant of skeletal muscles. On the other hand, curare does not affect neurons; thus, the frequency of the neuronal pacemaker nucleus continues to oscillate at a normal rate.) Then, the electric organ discharge of the fish was replaced by a mimic of similar amplitude and frequency. This can be done by placing one electrode into the fish's mouth and another electrode at the tip of its tail (Fig. 7.6). Such an arrangement of electrodes results in an electric field geometry similar to that produced by the natural electric organ discharge. It is sufficient to use an **electric sine wave**, as it completely mimics the fish's discharges. Similarly, the electric field of a neighboring fish can be perfectly mimicked by an electric sine wave applied through a pair of external electrodes straddling the fish.

When the fish is exposed to this experimental condition, it will execute a correct jamming avoidance response. It will lower its pacemaker frequency, if the mimic of the neighbor's signal has a frequency slightly higher than the mimic of its own discharges. On the other hand, it will raise its pacemaker frequency, if the frequency difference between the two signals is of opposite sign. As Fig. 7.6 demonstrates, this shifting of the pacemaker frequency in the correct direction can be elicited repeatedly and in a highly reliable manner.

These results correspond to the observations made under natural conditions when two fish of similar frequency meet. However, they do not tell much about the behavioral mechanism involved in determining the sign of the frequency difference. Therefore, the frequency of the fish's discharge mimic was changed, for example to a frequency 50 Hz below the frequency of its pacemaker nucleus. Then, when confronted with a mimic of a neighbor, the fish responded, as if the new, 50 Hz lower frequency would be its own frequency. It obviously uses the electric field frequency as a reference, rather than the 'true' internal frequency of the pacemaker nucleus.

In the second experiment, the fish was placed into a two-compartment chamber. Around its pectoral region, an electronically tight seal was fitted

(a)

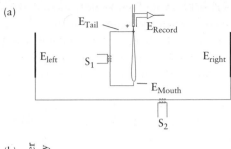

(b)

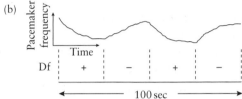

Figure 7.6 (a) Jamming avoidance response evoked in a curarized fish after its silenced electric organ discharge has been replaced by a sine wave substitute (S$_1$) applied through electrodes between the fish's mouth and tail (E$_{Mouth}$ and E$_{Tail}$, respectively). The fish is stimulated with a second sinusoidal electric stimulus (S$_2$) simulating the electric field of a neighbor. This stimulus is delivered transversely through one electrode placed on the left side (E$_{left}$) and another electrode placed on the right side (E$_{right}$). During the experiment, the pacemaker frequency is monitored by recording the spinal volley traveling from the pacemaker nucleus to the electric organ through a suction electrode placed over the tip of the tail (E$_{record}$). (b) The sign of the frequency difference of the fish' discharge mimic and the neighbor's mimic (defined as Df) is positive, if the frequency of the neighbor is higher than that of the fish. It is negative, if the frequency of the fish's mimic is higher than that of the neighbor's mimic. In the experiment shown, the sign of Df is switched every 25 sec. The results demonstrate that the fish executes a correct jamming avoidance response. It lowers its pacemaker frequency, if the neighbor's frequency is higher (positive Df), and it raises its pacemaker frequency, if the neighbor's frequency is lower (negative Df) than the frequency of the fish's mimic. A similar result can be obtained even if the frequency of the fish's substitute differs from the frequency of the pacemaker. This demonstrates that the fish uses the external electric field frequency, rather than the 'true' internal frequency of the pacemaker nucleus, as a reference. (After **Heiligenberg, W., Baker, C., and Matsubara, J.** (1978).)

so that practically no signal of its own electric organ discharge could any longer be detected at its head surface. In this region, the density of the electroreceptors is higher than in any other part of the body surface.

When a jamming stimulus, that is, a stimulus within a frequency range of 20 Hz above or below the fish's own discharge frequency, was presented to the head alone, the fish did not perform a jamming avoidance response. Yet, when the jamming signal was allowed to leak into the chamber containing the trunk, a proper jamming avoidance response was evoked. Similarly, a response was elicited by allowing the fish's electric organ discharge to leak into the chamber with the head.

Indirectly, this experiment supports the notion that the fish does not make internal reference to the frequency of the pacemaker. More important yet, this result demonstrates that the fish needs to be exposed to a mixture of its own signal and the interfering signal in some parts of its body surface in order to execute a jamming avoidance response.

Behavioral rules governing the jamming avoidance response

The first of the two experiments described above demonstrates that a potential interference of the electric fields of two fish with similar frequency does not require application of the natural discharge signals. Rather, it can be adequately simulated by replacing the fish's discharges by a sine wave applied through electrodes placed into the mouth and at the tail, and by applying the neighbor's mimic through a separate pair of electrodes. Due to this arrangement, the two electric fields have different geometries. This causes the electric current originating from the neighboring fish to affect the electroreceptors at different parts of the body to different degrees. As can be seen from Fig. 7.7, this is mainly due to differences in the angle by which the current of the neighbor hits the electroreceptors, and to differences in the distance from the neighbor to the different parts of the fish. This ensuing difference in the current paths of the two fields also causes the mixing ratio of the two signals to vary across the fish's body surface.

The variation in the mixing ratio is important for the execution of the jamming avoidance response. This could be shown by a simple modification of the above experiment. The mimics of the two discharge stimuli are added electronically, and their sum is presented through the pair of electrodes in the fish's mouth and at the tail. Then, a jamming avoidance response could no longer be elicited. Under this **identical geometry** condition, the two electric fields differ in frequency, but are spatially identical. This is in contrast to the natural situation, under which the two electric fields exhibit **separate geometry**, that is, they are *not* spatially identical. As a consequence of the identical geometry condition, the mixing ratio of the two currents no longer varies across the fish's body surface.

Various experiments suggest that *Eigenmannia* uses the following mechanism, solely based on afferent information, to determine the sign of the frequency difference. To illustrate this mechanism, it is sufficient to represent the fish's electric organ discharge by a sine wave stimulus. We will call this stimulus S_{Fish}, and the electric organ discharge of the neighbor, which is also represented by a sine wave stimulus, $S_{Neighbor}$.

Since S_{Fish} originates from an internal current source—the electric organ in the tail—it will recruit all electroreceptors on the fish's body surface evenly and at nearly the same phase of the stimulus cycle. By

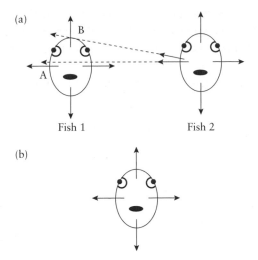

Figure 7.7 Separate (a) and identical (b) geometry conditions. Under separate conditions, the two electric fields are produced by two separate electric current sources, for example two separate technical dipoles or, as shown here, two individual electric fish. Since the current of the fish's discharge originates from an internal source, it penetrates the body surface perpendicularly and at similar intensities at all points. This also holds for substitutes of the electric organ discharge supplied through electrodes placed in the fish's mouth and at its tail. On the other hand, electroreceptors only sense the current component oriented perpendicularly to the local body surface. As a consequence, electroreceptors located at points A and B of Fish 1 are hit by electric current (indicated by dotted lines) from Fish 2 at different angles. Point A is more strongly affected by the neighbor's current than point b. Therefore, the combined signal is more strongly modulated in A than in B. By contrast, under identical geometry conditions, the mimics of the two discharge stimuli are added electronically, and their sum is presented through a pair of electrodes placed in the fish's mouth and at the tail. As a consequence of this stimulation situation, electroreceptors at each part of the body are affected in a similar way. (After **Heiligenberg, W.** (1991).)

contrast, the effect of $S_{Neighbor}$ is quite different. Since this stimulus originates from an external source—the electric organ of the neighboring fish—it will affect only some areas on the body surface of the fish, namely those where the current flow has a component perpendicular to the skin surface (Fig. 7.7). Moreover, since a neighbor is at some distance, and the intensity of an electric field in water decreases dramatically with distance, the amplitude of $S_{Neighbor}$ is normally smaller than that of S_{Fish} at every point of the fish's surface.

Due to these features of the two stimulus signals, the mixing of S_{Fish} and $S_{Neighbor}$ causes **modulation of amplitude and phase** of the mixed signal in reference to the pure S_{Fish} signal. The degree of modulation depends on the amplitude ratio of the two interfering signals. If the ratio is not equal to one, the phase shift will cycle periodically within the 'beat' cycle. The frequency of this beat cycle is equal to the frequency difference between the two signals.

Does the pattern of the amplitude and phase modulation contain information about the sign of the frequency difference between S_{Fish} and $S_{Neighbor}$? In other words, using this information, can we determine which of the two signals has the higher frequency?

The answer is 'yes'. We can demonstrate this by using an approach which requires a certain understanding of physical principles underlying the interference of two waves. Before discussing the rather complex biological situation, we will first demonstrate the principle using a very simple theoretical example (Fig. 7.8).

Suppose two sine waves, S_1 and S_2, interfere with each other. The maximum amplitude of the first sine wave is termed A_{1max} and the maximum amplitude of the second sine wave A_{2max}. The two waves have the same frequency and are exactly in phase (i.e. when the first wave reaches its minimum or maximum, the second wave does so as well).

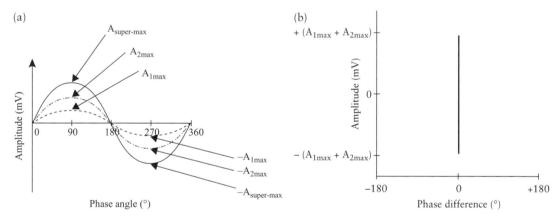

Figure 7.8 (a) Superposition of two sinusoidal waves, S_1 (dotted line) and S_2 (dashed line). For sake of simplicity, the two sine waves have been chosen such that they exhibit identical frequencies and no phase difference. Their superposition results in another sine wave, S_{super} (closed line), of identical frequency. The instantaneous amplitudes of the new sine wave are obtained by addition of the instantaneous amplitudes of S_1 and S_2. Thus, the maximum amplitudes in positive and negative direction, $A_{Super-max}$ and $-A_{Super-max}$, are equivalent to $A_{1max} + A_{2max}$ and $(-A_{1max}) + (-A_{2max})$. (b) A plot of the instantaneous amplitudes over a complete cycle of the mixed signal as a function of the phase difference results in a straight line parallel to the y axis. The two endpoints of this line are represented by the sum of the maximum amplitudes of the two original waves in both negative and positive direction. (Courtesy: G.K.H. Zupanc.)

We can determine the amplitude of the mixed signal, S_{Super}, by adding the instantaneous amplitudes of the two waves. Further, we know that the phase difference between the two waves is, at any given point, zero, since they were chosen such that they are exactly in phase. When we now plot the phase difference on the x axis, and the amplitude values obtained at any time point within one complete cycle of the mixed signal on the y axis, the resultant function represents a straight line parallel to the y axis at the x value of zero. This line extends from $-(|A_{1max}| + |A_{2max}|)$ to $+(|A_{1max}| + |A_{2max}|)$. These two extreme values are reached at phase angles of 270° and 90°, respectively. An amplitude of zero results at phase angles of 0°, 180°, and 360° (the latter value is equivalent to 0° of the subsequent cycle). Over the time of one complete cycle of the mixed signal (i.e. between 0° and 360°), the resultant amplitude travels from 0 to $+(|A_{1max}| + |A_{2max}|)$, then back to 0, further to $-(|A_{1max}| + |A_{2max}|)$, and finally back to 0.

Now back to the more complex mixed signal occurring during the jamming avoidance response. Here, typically, not only the amplitude of the mixed signal is different from either of the two original signals, but also the phase angle between the mixed signal and the fish's discharge is unequal to zero. Similar to the very simple example discussed above, the amplitude and phase values of the mixed signal are recorded for successive S_{Fish} cycles. These values are, then, plotted in a two-dimensional, amplitude-versus-phase plane (Fig. 7.9). As a result, a circular trajectory is obtained, which repeats itself at a rate equal to the beat frequency of the mixed signal. The sense of rotation reflects the sign of Df: clockwise rotation indicates that the frequency of $S_{Neighbor}$ is lower than the frequency of S_{Fish}; counterclockwise rotation reflects an opposite sign of the frequency difference, that is, the frequency of S_{Fish} is lower than the frequency of $S_{Neighbor}$.

Example: (2) If S_{Fish} displays a frequency of 400 Hz and $S_{Neighbor}$ a frequency of 402 Hz, then the amplitude–phase plot defines a circular trajectory that rotates counterclockwise. This trajectory completes two full cycles per second, since it has a beat frequency of 2 Hz.

The amplitude-*versus*-phase plot shows that, if *Eigenmannia* could determine the direction of the rotation of the resulting graph, it could immediately gauge the direction in which it has to shift its frequency to perform a correct jamming avoidance response. It should raise its frequency for clockwise rotation and lower its frequency for counterclockwise rotation.

One crucial aspect of these considerations is the question of how the fish could extract the phase information out of the mixed signal. This would be relatively simple, if the fish could sample a pure S_{Fish} signal (i.e. a signal without interference of a $S_{Neighbor}$ signal) somewhere from its body surface. It could, then, use this pure S_{Fish} signal as a reference to

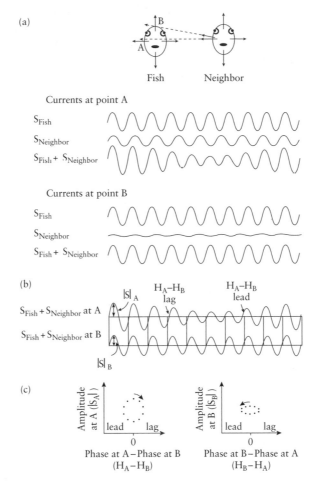

Figure 7.9 Schematic representation of the interference pattern resulting from the addition of two sine waves of slightly different frequencies. (a) The two sine waves, S_{Fish} and $S_{Neighbor}$, are used as mimics of a fish's electric organ discharge and the discharge of a neighboring fish, respectively. The amplitude of the sine wave S_{Fish} has been chosen such that it is dominant, compared to the amplitude of $S_{Neighbor}$. This resembles the situation where the fish's own signal interferes with the neighbor's signal near the fish's body. The resulting signal, $S_{Fish} + S_{Neighbor}$, exhibits cyclic modulations in both amplitude and phase compared to the fish's signal. However, the combined signal, $S_{Fish} + S_{Neighbor}$, is more strongly modulated in point A than in point B, due to differences in the angle at which the neighbor's signal hits the fish's signal. The length of one modulation cycle of the mixed signal's envelope (e.g. the length between two minima of the amplitude) is referred to as the length of the beat cycle. (b) Comparison of the combined signals perceived in point A and point B. The diagrams show that the differences in the timing of their zero-crossings ('phase', H) are modulated over the course of the beat cycle. Compared to the signal in point B, the signal in point A lags in phase, while its amplitude falls, and it leads, while its amplitude rises. This particular order in amplitude and phase modulation is caused by the fact that, in the example chosen, the neighbor's discharge frequency is lower than the fish's own frequency. If, on the other hand, the neighbor's frequency were higher, then a fall in the amplitude of the combined signal in point A would be paired with a phase lead, and a rise in amplitude would be paired with a phase lag. (c) The joint modulations in amplitude

determine the relative phase shifts associated with the mixed signal over one beat cycle. It has been shown, however, that the fish is able to determine the relative phase, even if no pure representation of S_{Fish} is available anywhere on its body surface.

The solution for the fish is to sample amplitude and phase information from many points on its body surface, compare the phase inputs from different pairs of points, and let the amplitude-*versus*-phase plot that shows the larger amplitude modulation win. These pair-wise comparisons indicate the correct rotation, as long as the fish's signal dominates. This is the case in most parts of the body. Due to the rapid decrease in intensity of the electric field of the neighboring fish with increasing distance from the fish, the amplitude of $S_{Neighbor}$ is, typically, much lower than the amplitude of S_{Fish}. As a result, in the majority of the pair-wise comparisons, the correct sign of the frequency difference is indicated. Thus, a 'democratic' decision based on the votes of such individual comparisons will, finally, lead to a correct determination of the sign of the frequency difference, and, thus, to a correct jamming avoidance response.

Overall, this detailed behavioral analysis shows that the jamming avoidance response of *Eigenmannia* is driven by a **distributed system** of contributions arising from the evaluation of inputs from a large number of pairs of points. Importantly, **no evidence of a central decision maker** has been found.

Neuronal implementation

Electrosensory processing I: electroreceptors

How are the behavioral rules for execution of a correct jamming avoidance response implemented at the neural level? Obviously, performance of this behavioral response affords accomplishment of two major

---

and phase can be recorded in a two-dimensional state plane, with the amplitude plotted on the ordinate and the differential phase (i.e. the phase in one point relative to the phase in the other point) plotted on the abscissa. Then a circular trajectory results, which repeats itself at a rate equal to the beat frequency of the mixed signal. If the more strongly modulated signal in point A is analyzed in reference to the less strongly modulated signal in point B, a clockwise rotation of the circular trajectory results. In contrast, if the signal in point B is analyzed in reference to point A, then an opposite sense of rotation is obtained. However, the latter graph is characterized by a smaller amplitude modulation. Application of the rule that the graph with the larger amplitude modulation wins in this pair-wise comparison, the conclusion drawn by the fish would be that the frequency of the neighbor discharges its electric organ at a lower frequency than the fish itself does. (After **Heiligenberg, W.** (1991).)

tasks: First, extraction of the sign of the frequency difference between the fish's own discharge frequency and the frequency of the neighboring fish by **electrosensory processing** of phase and amplitude information. Second, translation of the determination of the sign of the frequency difference into a **change of the motor output,** that is, of the pacemaker frequency. This will, then, lead to a correct shift in the frequency of the electric organ discharge.

In *Eigenmannia,* electrosensory information is perceived by two types of electroreceptors: ampullary receptors and tuberous receptors. The morphological differences between the two types have already been shown in Fig. 7.4. These receptors also differ in their physiological properties. The **ampullary receptors** are tuned to dc and low-frequency ac signals. Their possible role in the jamming avoidance response is, at present, unclear. The **tuberous receptors** are tuned to ac signals with frequencies in the range of the fish's own electric organ discharge. They can be further divided into two subtypes referred to as **P-type receptors** and **T-type receptors**. Afferents of the P-type receptors ('probability coders') fire intermittently and increase their rate of firing, in a probabilistic manner, with a rise in stimulus amplitude. Afferents of T-type receptors ('time coders') fire one spike on each cycle of the stimulus. These spikes are phase-locked, with little jitter, to the zero-crossing of the signal.

Figure 7.10 schematically illustrates how the T-unit and P-unit afferents respond to phase and amplitude modulations caused by interference of the fish's discharges and the discharges of the neighboring fish. The T-unit afferents fire one spike, with a fixed latency, at each positive zero-crossing of the mixed signal. The sampling of such zero-crossings of the (somewhat different) mixed signal at different parts of the body and a comparison of the differences in firing of action potentials of the T-units reflect the **differential phase** between the respective mixed signals on two parts of the body surface. The firing rate of the P-unit afferents reflects the **amplitude** of the mixed signal. They generate more spikes at large amplitudes than they do at low amplitudes.

When the responses of the P-units and the T-units to the mixed signal are plotted over one period of the beat cycle in a similar way as the amplitude–phase values have been plotted in the two-dimensional plane (see Fig. 7.9), a similar circular trajectory is obtained. This shows that the pattern of amplitude and phase modulations are reflected by the joint activation of P-units and T-units.

Electrosensory processing II: electrosensory lateral line lobe

The primary afferents of the tuberous electroreceptors project to three maps of the **electrosensory lateral line lobe** in the hindbrain. According to their

Tuberous electroreceptors can be divided into two types: P-type receptors encode the amplitude, whereas T-type receptors encode the phase of the perceived electric signal.

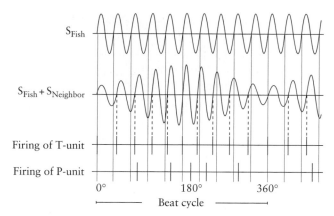

Figure 7.10 Schematic representation of the coding of phase and amplitude modulations by T-type and P-type tuberous electroreceptors. The top trace represents the fish's electric organ discharges, the second trace represents the mixed signal resulting from the addition of the fish's signal and the neighbor's signal. This interference pattern is characterized by modulations of both the instantaneous amplitude and the instantaneous phase. The latter type of modulation can be best appreciated by comparing the timing of the zero-crossings (indicated by broken lines) of the mixed signal compared to that of the pure fish's signal. The T-units code the phase by firing a single action potential within each cycle of the signal and, at a fixed latency, with reference to the timing of the positive zero-crossings. The P-units code the amplitude of the signal's envelope by modulating the probability of firing of action potentials with the instantaneous amplitude modulations. (After **Heiligenberg, W.** (1991).)

anatomical locations, these maps are referred to as lateral, centrolateral, and centromedial segments. The primary afferents of the ampullary electroreceptors project to a fourth map called medial segment. Each of these **maps** is **somatotopically ordered**, that is, they preserve the spatial order of electroreceptors on the body surface. In these maps, the head region of the fish is by far over-represented, thus reflecting the high density of electroreceptors in this region, compared to other areas of the body surface. Each primary afferent of the tuberous electroreceptors provides, via collaterals, input to each of the three maps. This suggests that electrosensory information received via tuberous electroreceptors is present in triplicate form within the electrosensory lateral line lobe.

Within each of the three maps, information encoded by T-type receptors and P-type receptors is processed separately (Fig. 7.11). This mode of action is referred to as **parallel processing**. Input from the T-type receptors is received by **spherical cells**. One spherical cell is connected, via **electrotonic synapses**, with several T-units within its receptive field. To initiate the generation of spikes in spherical cells, several afferent action potentials have to arrive within a narrow time window. This ensures that individual afferent signals that arrive out of synchrony with the majority

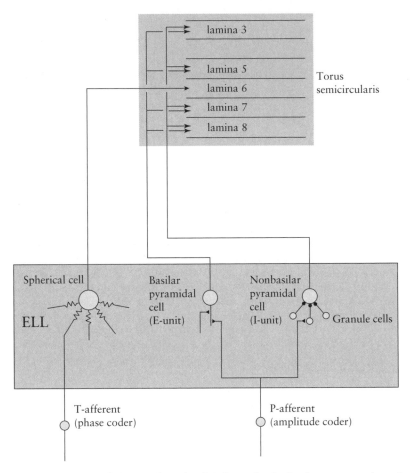

Figure 7.11 Encoding of phase and amplitude information in the electrosensory lateral line lobe (ELL). Information encoded by the T-type receptors and P-type receptors is processed separately. Primary afferents from the T-type receptors (phase coders) form electrotonic synapses (indicated by zig-zag lines) with spherical cells in the ELL. A single spherical cell may receive input from several T-afferents. Primary afferents from P-type receptors (amplitude coders) form excitatory synapses (indicated by triangles) on the basilar dendrites of basilar pyramidal cells (called E-units). Via collaterals, they also excite granule cells which, in turn, inhibit (indicated by circles) non-basilar pyramidal cells (called I-units). As a result, E-units fire in response to a rise in stimulus amplitude, whereas I-units fire in response to a fall in stimulus amplitude. Spherical cells project exclusively to lamina 6 of the torus semicircularis. Pyramidal cells project to various laminae above and below lamina 6. (After **Heiligenberg, W.** (1991).)

In the electrosensory lateral line lobe, amplitude and phase information are processed in parallel.

of the signals from a receptive field do not contribute to the production of action potentials. Consequently, spherical cells code the phase of the stimulus with less jitter than do individual T-unit afferents.

P-unit afferents provide input to two cell types of the electrosensory lateral line lobe. First, they form excitatory synapses on the basal dendrites of basilar pyramidal cells. Second, they inhibit **non-basilar**

pyramidal cells through local interneurons. As a result of this connectivity, a rise in stimulus amplitude, causing an increase in P-unit firing, will excite basilar pyramidal cells directly and inhibit non-basilar pyramidal cells indirectly. By contrast, a fall in stimulus amplitude, which causes a decrease in P-unit firing, will release non-basilar pyramidal cells from inhibition and lead to an increase in their rate of firing. Thus, excitation of basilar pyramidal cells reflects a rise in the amplitude of the stimulus signal, whereas excitation of non-basilar pyramidal cells indicates a fall in stimulus amplitude.

Lesioning experiments have shown that the centromedial segment of the electrosensory lateral line lobe is both necessary and sufficient for the jamming avoidance response. The 'ampullary' medial segment appears also to have certain effects on this behavior. However, the exact role of ampullary information for the jamming avoidance response is currently unclear, especially since current theories can sufficiently explain this behavior based only on tuberous information.

Electrosensory processing III: torus semicircularis

The next station in processing electrosensory information is the **torus semicircularis** in the midbrain. The major connections between this structure and the electrosensory lateral line lobe are shown in Fig. 7.11. As this wiring diagram indicates, the torus is divided into various **laminae**. The phase-coding spherical cells of the electrosensory lateral line lobe project exclusively to **lamina 6** of the torus. By contrast, both types of amplitude-coding pyramidal cells send axons to various laminae below and above this layer.

A network of neurons within lamina 6 computes phase differences between any two points on the body surface, and thus the differential phase. Neurons in the layers below and above lamina 6 show a variety of response characteristics: some cells respond only to amplitude modulations, other only to phase information, and still others to both forms of modulation. The **convergence of amplitude and phase information** appears to be achieved by vertical connections between the different layers. While some cells in laminae 5 and 7 have dendritic fields limited to their respective layer, other neurons extend their dendrites into lamina 6. These cells respond to modulations in differential phase. Neurons within laminae 5 and 7 then project to deeper laminae, notably lamina 8c. Cells in these deeper layers of the torus are the first in the neuronal hierarchy that are able to recognize the sign of the frequency difference between the fish signal and the signal of the neighboring fish. However, the **sign selectivity** of these neurons depends on the orientation of the jamming signal, which is determined by the orientation of the neighbor relative to the fish. Thus, the sign selectivity information is still ambiguous.

In the torus semicircularis, amplitude and phase information converge.

Electrosensory processing IV: nucleus electrosensorius

Cells that encode the sign of the frequency difference between the fish's signal and the discharges of the neighbor unambiguously (i.e. independently of the orientation of the jamming stimulus) are found in the **nucleus electrosensorius**. This area in the diencephalon receives input from various layers of the torus semicircularis. However, the somatotopic order associated with the various toral layers is lost in the nucleus electrosensorius, suggesting pronounced spatial convergence of the projection from the torus to the nucleus electrosensorius. Moreover, the sign-selective neurons in the latter nuclear region are more sensitive to interfering signals than the 'sign-selective' neurons of the torus. While the most sensitive neurons found in the torus can discriminate phase modulations as small as approximately 10 μsec, the sensitivity of neurons in the nucleus electrosensorius is increased by roughly one order of magnitude. They can discriminate phase modulations as small as 1 μsec. These, as well as other response properties make the sign-selective neurons of the nucleus electrosensorius resemble the jamming avoidance response itself. This suggests that the nucleus electrosensorius is essential for the execution of the jamming avoidance response.

Stimulation experiments combined with morphological analysis have revealed several distinct clusters of neurons within the nucleus electrosensorius. Two of these clusters are involved in the control of the jamming avoidance response (Fig. 7.12). A small dorsal area can be stimulated by iontophoretic application of *L*-glutamate to cause smooth rises in the frequency of the electric organ discharges. Since the nucleus electrosensorius is commonly abbreviated 'nE', this area is referred to as nE↑ (pronounced 'en-ee-up'). Conversely, *L*-glutamate stimulation of a more ventrally located area, called nE↓ (pronounced 'en-ee-down') causes a smooth fall in discharge frequency. The time course of both of these frequency alterations is similar to those observed during the jamming avoidance response. Bilateral lesions of these two areas eliminate frequency shifts from the jamming avoidance response, thus providing further support for the above hypothesis that the nE is a crucial link in the control of the jamming avoidance response.

In contrast to the torus semicircularis, neurons of the nucleus electrosensorius encode the sign of the frequency difference between the fish's signal and the discharges of the neighbor's discharges unambiguously.

Motor control

The above dissection of the individual steps involved in sensory processing of information relevant to the jamming avoidance response has shown that two separate pathways lead to the recognition of positive and negative frequency differences, respectively. Similarly, motor control of the resulting frequency changes is also mediated by two distinct pathways. Neurons of the nE↑ (excitation of which, finally, leads to rises

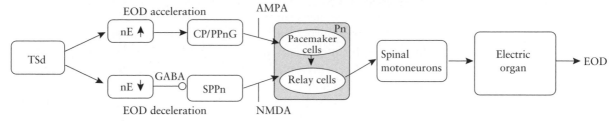

Figure 7.12 Flow diagram of the motor control of the jamming avoidance response in *Eigenmannia*. Arrowheads indicate excitatory synapses, open circles inhibitory connections. Various layers of the torus semicircularis (TSd) project to the nucleus electrosensorius (nE). Stimulation of one area, the nE↑, causes accelerations of the electric organ discharge, while stimulation of a different area, the nE↓, results in decelerations. The nE↑ innervates, via excitatory synapses (indicated by arrow), the central posterior/prepacemaker nucleus (CP/PPn) in the dorsal thalamus. The nE↓, on the other hand, provides, via inhibitory (GABAergic) synapses (indicated by circle), input to the sublemniscal prepacemaker nucleus (SPPn). Final motor control is achieved in the pacemaker nucleus (Pn) of the medulla oblongata. Neurons of the CP/PPn innervate pacemaker cells in the pacemaker nucleus. This input is mediated by AMPA-type of glutamate receptors. On the other hand, neurons of the SPPn innervate relay cells, which are also situated within the pacemaker nucleus. This input is mediated by NMDA receptors. As a final step, the relay cells project to spinal motoneurons that innervate the electric organ. Synchronous depolarization of the electrocytes comprising the electric organ generates the electric organ discharge (EOD). (After **Metzner, W.** (1999).)

in discharge frequency) project to a small region in the dorsal thalamus called **CP/PPn-G**. 'PPn' stands for '**prepacemaker nucleus**' and 'G' for 'gradual frequency rises.' The PPn-G is the dorsolateral portion of a larger cell assembly, the so-called **central posterior nucleus** (CP). The CP/PPn-G is comprised of small ovoidal neurons which innervate one particular cell type within the pacemaker nucleus, the pacemaker cells. This input is mediated by α-amino-3-hydroxy-5-methyloxazole-propionic acid (**AMPA**)-**type of glutamate receptors**. Iontophoretic application of the excitatory transmitter *L*-glutamate to the CP/PPn-G elicits gradual accelerations of the electric organ discharges. Conversely, bilateral lesions of this cell group abolish frequency rises of the jamming avoidance response. Injection of 6-cyano-7-nitroquinoxaline-2,3-dione(CNQX), which selectively blocks AMPA-type receptors, into the pacemaker nucleus has a similar effect as lesions applied to the CP/PPn-G; it results in elimination of a jamming avoidance response to negative frequency differences.

Neurons of the nE↓ (excitation of which, finally, leads to decreases in discharge frequency) project to a different region, the **sublemniscal prepacemaker nucleus** in the mesencephalon (Fig. 7.12). Cells of this nuclear assembly, in turn, innervate the relay cells within the pacemaker nucleus. This input is mediated by *N*-methyl-*D*-aspartate (**NMDA**)-**type** of **glutamate receptors**. Interestingly, bilateral lesions of the sublemniscal prepacemaker nucleus obliterate not only decelerations of the electric organ discharge during the jamming avoidance response, but also reduce the resting frequency of the discharges (i.e. the frequency when no jamming signal is presented). This indicates that neurons of the sublemniscal prepacemaker nucleus provide a **tonic input** to the pacemaker nucleus. Injection of 2-amino-5-phosphonovaleric acid (APV), an antagonist of NMDA-type of glutamate receptors, into the pacemaker nucleus has shown similar results as lesions of the sublemniscal prepacemaker nucleus.

The hypothesis that neurons of the sublemniscal prepacemaker nucleus provide a tonic input to the pacemaker nucleus could, indeed, be confirmed by injection of the inhibitory transmitter γ-aminobutyric acid, commonly known as **GABA**, into the sublemniscal prepacemaker nucleus. Such experiments result in decreases of the discharge frequency, thus resembling the effect observed after L-glutamate stimulation of the nE↓. Furthermore, injection of the GABA$_A$-receptor antagonist bicuculline blocks any of such effects of L-glutamate stimulation of the nE↓. Based on these results, it has been hypothesized that the nE↓ provides an inhibitory input, mediated by GABA, to the sublemniscal nucleus. Upon activation, this input diminishes the tonic excitatory input of the sublemniscal nucleus onto the relay cells in the pacemaker nucleus, causing a deceleration of the electric organ discharge.

While fibers of the CP/PPn-G appear to terminate mainly on the dendrites of the pacemaker cells, fibers of the sublemniscal prepacemaker nucleus synapse upon the somata of the relay cells (Fig. 7.12). Strong electrotonic coupling between the relay cells and the pacemaker cells ensures that the input arising from the sublemniscal nucleus affects not only the relay cells, but also the pacemaker cells.

In the final stage of the motor control, the pacemaker cells drive the relay cells, which, in turn, transmit the command pulses generated by the pacemaker cells to spinal motoneurons of the electric organ. This is done in a one-to-one fashion, resulting in one discharge of the electric organ at each command pulse produced by the pacemaker nucleus.

The final motor control of the jamming avoidance response is mediated through input to the pacemaker nucleus, originating from the 'G' portion of the central posterior/prepacemaker nucleus and the sublemniscal prepacemaker nucleus.

Reflections on the evolution of the jamming avoidance response

Both the behavioral and the neural mechanisms governing the jamming avoidance response are far from perfection. As we have seen, an engineer would design a neural network for determination of the sign of the frequency difference between the fish's own discharge and that of a neighbor very differently by making use of the pacemaker frequency as an internal reference. However, evolution does not exhibit a goal-oriented, long-term planning to produce simple and elegant solutions. When, in the course of the phylogenetic development of *Eigenmannia*, the need for a jamming avoidance response arose, neural implementation of this behavior was only possible by modification of pre-existing structures. Obviously, none of these pre-existing structures allowed the fish to simply incorporate internal information about the discharge frequency.

On the other hand, other neural elements, used for the control of different behaviors, could readily be modified to subserve specific functions in the context of the jamming avoidance response. Such considerations are supported by comparative investigations. The genus *Sternopygus* within the order Gymnotiformes does not perform a jamming avoidance response. However, these fish already have brain neurons that encode amplitude and phase information. These neurons are used for electrosensory analysis of object movements in the context of electrolocation. It appears reasonable to think that, through a few modifications in the associated neural network, these 'ancestral' neurons could easily be employed for the control of the jamming avoidance response. This is what might have happened in the course of the evolution of *Eigenmannia*.

> The imperfection of the jamming avoidance response can be explained by its likely evolutionary development.

Summary

▨ The weakly electric fish *Eigenmannia* continuously generates, by means of an electric organ, wave-type electric discharges. The discharge rate is determined by the frequency of the pacemaker nucleus, an endogenous oscillator in the medulla oblongata. The fish sense their own electric currents, as well as those of neighbors, through electroreceptors.

▨ Electrolocation involves the analysis of perturbations of the fish's own electric field by objects in the closer vicinity.

▨ Object-induced changes in the electric current patterns of the fish's own electric field may be masked by modulations caused by a neighbor's electric organ discharges, especially if the frequency of the neighbor is close to the fish's own frequency. The fish avoids the detrimental effects of such a signal interference, or 'jamming', by shifting the frequency of its own electric organ discharge away from the frequency of the interfering signal. This behavior is called 'jamming avoidance response.' Consequently, if the neighbor's frequency is higher than the fish's own discharge frequency, the fish lowers its frequency. If the neighbor's frequency is lower than the fish's own discharge frequency, the fish raises its frequency.

■ The fish determines the sign of the frequency difference between its own signal and that of the neighbor not by making reference to the pacemaker frequency. Rather, it evaluates afferent information contained in the interference, or 'beat' pattern, which results from the mixing of its own discharge and that of the neighbor.

■ Both amplitude and phase of the mixed signal are different from the fish's own signal. This difference is analyzed by comparing the differential phase in different parts of the body surface. Thus, the jamming avoidance response is driven by a distributed system of contributions resulting from evaluations of inputs from pairs of points. There is no evidence of a central controller or decision maker.

■ Extraction of differential phase and amplitude information, essential to determine the sign of the frequency difference, is mediated by two types of tuberous electroreceptors, T-type receptors ('time coders') and P-type receptors ('amplitude coders').

■ Primary afferents of the T-type receptors terminate on spherical cells in the electrosensory lateral line lobe of the hindbrain. In the same brain region, primary afferents of P-type receptors form excitatory connections with basilar pyramidal cells, and inhibitory connections, via local interneurons, with non-basilar pyramidal cells.

■ Spherical cells project exclusively to lamina 6 of the torus semicircularis in the midbrain. Both types of pyramidal cells project to various laminae above and below lamina 6. Within this layer, differential phase is computed. Convergence of differential phase and amplitude information is achieved through vertical connections between the laminae. Although neurons within the torus exist that fire at higher rate for one sign of the frequency difference than for the opposite sign, their response is still ambiguous, as it depends on the orientation of the jamming signal.

■ Unambiguous sign-selective neurons occur in the next station of electrosensory processing, the nucleus electrosensorius in the diencephalon. Excitation of a subdivision of the nucleus electrosensorius, the nE↑, raises the discharge frequency, whereas excitation of a second subdivision, the nE↓, leads to frequency decreases.

■ Frequency increases and frequency decreases are mediated by two separate motor pathways. Frequency increases are controlled by a subnucleus of the diencephalic central posterior/prepacemaker nucleus, the CP/PPn-G, whose neurons innervate the pacemaker neurons of the pacemaker nucleus. Frequency decreases are controlled by the mesencephalic sublemniscal prepacemaker nucleus, which in turn projects to the relay cells of the pacemaker nucleus.

■ Command pulses generated by the pacemaker cells drive the discharges of the electric organ in a one-to-one fashion.

Recommended reading

Heiligenberg, W. (1991). *Neural nets in electric fish*. MIT Press, Cambridge/Massachusetts.

A comprehensive synopsis of the work of Walter Heiligenberg and associates on the jamming avoidance response. Written for an informed audience. Therefore, suitable only for advanced graduate students or instructors, who would like to gain a deeper understanding of the subject.

Metzner, W (1999). Neural circuitry for communication and jamming avoidance in gymnotiform fish. *Journal of Experimental Biology*, **202**, 1365–1375.

An update of the research on the jamming avoidance response, which can also be used as a good starting point.

Questions

7.1 How do weakly electric fish electrolocate? Suggest a behavioral experiment to verify the hypothesis that, based on their electrolocation ability, the fish can analyze various features of objects, including size and direction of movement.

7.2 What is the jamming avoidance response of gymnotiform fish? Describe the computational rules and the neural correlates underlying this behavior.

Neuromodulation: the accommodation of motivational changes in behavior

■ Introduction

■ Neuronal plasticity as the basis of motivational changes in behavior

■ Structural reorganization

■ Biochemical switching

■ Summary

■ Recommended reading

■ Questions

Introduction

As has been demonstrated in previous chapters, major factors that determine the behavior of an animal are stimuli originating from the environment. Different stimuli typically elicit different behaviors, as different intensities of a stimulus may elicit different frequencies, durations, or intensities of a behavioral pattern. In the case of an 'ideal' reflex (which, probably, does not exist in the real world, but is a helpful construct for theoretical considerations), these environmental stimuli are the only factors that determine the behavioral response. This situation is schematically summarized in Fig. 8.1(a). In this diagram, the animal is represented by a 'black box.' Input to the black box indicates the stimulus originating from the environment, whereas the output represents the animal's response. Very often, however, this input/output relationship is not fixed.

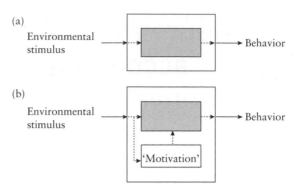

Figure 8.1 Input–output relationship in 'ideal' reflexes and motivational systems. (a) Block diagram of a hypothetical 'ideal' reflex. The behavioral response of an animal is solely determined by a given stimulus. This stimulus evokes, whenever applied, an identical response in the animal. (b) Block diagram of a motivational system. The behavioral response of an animal is not only determined by a stimulus, but also by the motivational state of the animal. Identical stimuli applied at different times can, therefore, evoke different responses. Note that the stimulus may also affect the motivation of the animal. (Courtesy: G.K.H. Zupanc.)

Example: *Seasonally breeding animals are characterized by dramatic changes in many behaviors, particularly those associated with courtship and aggression. During the breeding season, males may respond to a proper 'female' stimulus with courtship behavior, whereas outside the breeding season the identical stimulus rarely, or even never, elicits courtship.*

Motivation: the physiological state of an animal which defines the frequency and intensity of occurrence of a behavior when elicited by a given endogenous or exogenous stimulus.

The variation in the animal's response, even under identical stimulation regimes, has led to the assumption that variable internal causal factors exist which mediate the relationship between external stimuli and resulting behaviors. The influence of such internal factors on the neural control of the respective behavioral pattern is referred to as **motivation**. Structures within the central nervous system from which a modulatory influence originates may be regarded as 'motivational' components of the neural network involved in the generation of the behavior. This situation is sketched in Fig. 8.1(b).

The proposal of the existence of motivational components does not make any predictions where the modulatory influence exerts its effect within the neural chain connecting the sites of sensory perception with those that initiate motor action. In particular, the existence of a single central structure devoted to the exclusive control of the level of motivation (we could call such a structure a 'motivational center') is not required. In a given instance, the only effect of motivational factors could, for example, be to modulate the perception and processing of sensory stimuli. Central structures controlling the motor output may then modify their response solely on the basis of altered sensory information relayed to them, rather than due to a direct influence of motivational factors.

In the following, we will discuss neural mechanisms that mediate motivational changes in behavior. The underlying principles will be illustrated by referring to some well-studied model systems.

Neural plasticity as the basis of motivational changes in behavior

Motivational changes in behavior are possible only if the neural network underlying this behavior exhibits the potential for **neural plasticity**. This is not the case in the 'ideal' reflex. Therefore, and as mentioned above, the response in this type of behavior is solely determined by the stimulus originating from the environment. In contrast, plastic neural networks can produce different behavioral outputs, even under identical stimulation regimes.

Research on a variety of different behavioral systems suggests that the neural mechanisms mediating motivational changes in behavior can be grouped into two major categories. These categories are subsumed under the terms **structural reorganization of neural networks** and **biochemical switching of neural networks**.

Motivational changes in behavior are possible only in networks exhibiting the potential for neural plasticity.

Two major neural mechanisms that mediate motivational changes in behavior: structural reorganization of networks and biochemical switching of neural networks.

Structural reorganization

The principle of structural reorganization involves a variety of structural modifications in neural networks underlying behavior. They include the following:

* generation of new neurons and glial cells
* elimination of older cells
* changes in the dendritic structure of neurons
* retraction and outgrowth of axons.

While until a few decades ago it was believed that the adult brain of mammals, and particularly of primates, is incapable of producing new neurons, in recent years the potential for the generation of new neurons (**neurogenesis**) has been demonstrated in the brain of all vertebrates classes, including that of primates and even of humans. In 'lower' vertebrates, this capability of adult neurogenesis is especially pronounced, both in terms of the overall number of neurons produced and the number of brain regions exhibiting proliferative activity.

Although the functional significance of adult neurogenesis still affords detailed analysis, this phenomenon obviously provides excellent means for the restructuring of neural networks, and, thus, for the production of dramatic changes in their properties, finally leading to alterations in the behavioral output. An example of the behavioral significance of this mechanism will be presented at the end of this book, when we discuss the involvement of adult neurogenesis in the improvement of learning and the formation of new memories in Chapter 11.

Dendritic plasticity: seasonal changes in chirping behavior of weakly electric knifefish

Another important mechanism leading to dramatic changes in the properties of a neural network involves alterations in the dendritic structure of neurons. This mechanism and its relationship to seasonal behavioral changes has been examined in detail in the knifefish *Eigenmannia* sp. Normally, this weakly electric fish produces, by means of a specialized electric organ, electric discharges distinguished by their enormous regularity, in both frequency and amplitude. Short-term modulations consist of complex changes in frequency and amplitude, which may be followed by a complete cessation of the electric organ discharge, as shown in Fig. 8.2. When transformed into an acoustic signal, these modulations resemble the sound produced by crickets and are, hence, called **chirps**.

Further behavioral observations have shown that chirps are produced almost exclusively during the breeding season, when these seasonally breeding fish reproduce. The functional significance of chirping behavior is illustrated by play-back experiments. Stimulation of isolated gravid females with pre-recorded male chirps can even induce egg laying, thus suggesting that this behavior functions as a powerful communicatory signal.

In the natural habitat in South America, the breeding season coincides with the tropical rainy season. In the laboratory, maturation of gonads

> Chirps: transient frequency and amplitude modulations of the electric organs discharge.

200 msec

Figure 8.2 Electric organ discharge of the knifefish *Eigenmannia* sp. during courtship. The oscilloscope trace reveals two types of discharge sequences: one with constant frequency, and the other with brief periods of frequency modulations (arrows). The first modulatory event is followed by a complete cessation of the fish's discharge. (Courtesy: G.K.H. Zupanc.)

and development of the associated behaviors can be induced by simulating the rainy season. This is achieved by daily addition of deionized water ('rain water') to the aquarium, which leads to a rise in the water level and to a decrease in the conductivity of the aquarium water. Furthermore, circulated water is periodically sprinkled onto the surface of the aquarium water, thus imitating rain drops. Taken together, these individual steps, applied over several weeks, lead to changes in the environment of the fish similar as those caused by the heavy rain falls during the tropical rainy season. The possibility to induce gonadal recrudescence through subjecting the fish to such rainy regimes has enabled researchers to study the neural basis of the seasonal changes in chirping behavior.

The electric organ discharge is controlled by an endogenous oscillator in the medulla oblongata, the so-called **pacemaker nucleus**. The pacemaker nucleus triggers each discharge of the electric organ by one spike. The regularity of the oscillations is an intrinsic property of the pacemaker nucleus, as this structure continues to fire with a similar frequency as in the living animal, after being isolated and kept in a petri dish containing oxygenated artificial cerebrospinal fluid. Therefore, it cannot be the pacemaker nucleus itself that controls chirping behavior. Rather, input to this nucleus must be responsible for the control of this behavioral pattern.

Morphologically, input to a brain nucleus can be revealed by **neuronal tract tracing**, a technique described in detail in Chapter 3. Using this approach, the input to the pacemaker nucleus was identified by Walter Heiligenberg and his group at the Scripps Institution of Oceanography of the University of California, San Diego at the beginning of the 1980s. Figure 8.3 summarizes the experimental procedure. First, Heiligenberg and his group injected horseradish peroxidase into this nucleus of anesthetized fish. Then, following a post-injection survival time of several days (during which the enzyme is transported in retrograde direction within the axon) and histochemical processing of the brain tissue, they screened sections of the brain for labeled cells. The neurons traced through this approach form a small bilateral cluster in the dorsal thalamus, which is part of the diencephalon. Since this cellular assembly provides input to the pacemaker nucleus, it is referred to as the **prepacemaker nucleus**. Later studies showed that this nucleus is part of a larger complex called central posterior nucleus. That portion of the latter complex which provides input to the pacemaker nucleus is, therefore, now referred to as the **central posterior/prepacemaker nucleus**.

Detailed morphological and physiological analysis of the central posterior/prepacemaker nucleus has revealed that only approximately 100 neurons within this nucleus are involved in the control of chirping behavior. These neurons are rather large and give rise to three or four dendrites, which show a moderate degree of branching. Based on this morphological appearance, these neurons are categorized as **multipolar**.

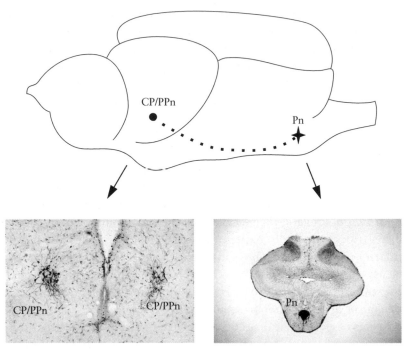

Figure 8.3 Identification of the structural correlate of input to the pacemaker nucleus by neuronal tract tracing. The drawing on top is a side view of the brain of *Eigenmannia* sp. The pacemaker nucleus (Pn) is located in the brainstem area, the central posterior/prepacemaker nucleus in the diencephalon. The two microphotographs below show cross-sections through the brain at the level of the pacemaker nucleus (right) and the central posterior/prepacemaker nucleus (left). In the experiment, the enzyme horseradish peroxidase was applied to the pacemaker nucleus in the anesthetized animal. After a survival period of three days, during which axons innervating the pacemaker nucleus took up the enzyme and transported it in retrograde direction toward the cell bodies, the fish was killed and the brain cut into thin sections. After histochemical processing of these sections, the tracer became visible as a black precipitate. Screening of the brain sections revealed that the input to the pacemaker nucleus arises from a bilateral cluster of neurons in the diencephalon, called central posterior/prepacemaker nucleus. On each side of the brain, a number of labeled cell bodies comprising this brain nucleus, together with part of their dendritic arbors, are visible. (Courtesy: G.K.H. Zupanc.)

Chirps are controlled by a subnucleus of the central posterior/prepacemaker nucleus in the dorsal thalamus.

Application of the excitatory transmitter *L*-glutamate to the subnucleus comprised of the large multipolar neurons elicits modulations of the electric organ discharge resembling the chirps in the intact animal. A similar effect has electrical stimulation of these neurons. These physiological experiments strongly support the hypothesis that the large multipolar neurons of the central posterior/prepacemaker nucleus play a crucial role in the control of chirping behavior.

After having identified the central structure controlling chirping behavior, it was possible to approach another exciting aspect of this behavior, namely, the question of what neuronal mechanism is responsible for the enormous increase in the motivation to chirp when the fish sexually mature. That the increase is not just caused by changes in the intrinsic properties of the central posterior/prepacemaker nucleus can be shown by a simple experiment. When *L*-glutamate is injected into this nucleus, chirps can be elicited in any post-juvenile fish—regardless of its state of maturity. So what, then, causes the seasonally induced motivational changes in chirping behavior?

This question was tackled by Günther Zupanc in the laboratory of Walter Heiligenberg in the late 1980s. They first established the following two groups of fish: a first group which was kept under artificial rainy season conditions, until the fish developed ripe gonads; and a second group which was kept under non-rainy ('dry') season conditions, so that the fish did not sexually mature. Then, the large multipolar neurons of the central posterior/prepacemaker nucleus were labeled by retrograde tracing from the pacemaker nucleus. Finally, individually labeled neurons were reconstructed to obtain a complete picture of their morphology.

Comparison of such reconstructed neurons in the two groups of fish revealed a close correlation between the dendritic morphology of the chirp-controlling neurons of the central posterior/prepacemaker nucleus and the state of sexual maturity of the fish. As Fig. 8.4 shows, neurons of sexually mature fish exhibit a well-developed dendritic arbor with

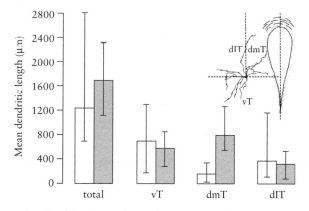

Figure 8.4 Mean length of dendritic arbors of individually reconstructed chirp-controlling neurons of the central posterior/prepacemaker nucleus of immature (open bars) and mature (closed bars) females in the knifefish *Eigenmannia* sp. The neurons were retrogradely traced by application of horseradish peroxidase to their projection site, the pacemaker nucleus, and histochemical processing of the brain sections. A reconstructed neuron of a mature female, based on such retrograde tracing, is shown in the inset. The dendrites extending from the labeled cell bodies define three different dendritic territories, called ventral territory (vT), dorsomedial territory (dmT), and dorsolateral territory (dlT). The increase in overall length (in the diagram referred to as 'total') by approximately 40% during sexual maturation is mainly due to a significant proliferation in the dorsomedial territory. As shown in the inset, in mature fish some of the dendrites of the dorsomedial territory travel close to the wall of the third ventricle. The vertical lines at each bar indicate the range of individual data. (After **Zupanc, G. K. H., and Heiligenberg, W. (1989).**)

dendrites extending in three directions: dorsolaterally, dorsomedially, and ventrally. In contrast, neurons of sexually immature fish almost completely lack dendrites in the dorsomedial field, while the dendrites in the dorsolateral and ventral territories of the dendritic arbor are similarly developed as in mature fish.

Investigations at the electron microscopic level have shown that the dendrites of the large multipolar neurons of the central posterior/prepacemaker nucleus receive input mainly from excitatory synapses. By contrast, the cell body region is, to a large extent, contacted by inhibitory synapses. The excitatory input originates from neurons of another diencephalic region, the nucleus electrosensorius. Neurons within this sensory processing station respond specifically to various electric stimuli, including chirp-like modulations of the electric organ discharge.

As dendrites of the dorsomedial territory comprise roughly half of the total dendritic arbor of neurons of the central posterior/prepacemaker nucleus, their retraction after the end of the breeding season leads to a dramatic reduction in the input received from the nucleus elecrosensorius. On the other hand, the sexual maturation during the breeding season re-establishes the interface between the sensory part and the motor part of this neural system. Figure 8.5 summarizes the structural changes taking place in the course of gonadal maturation and regression.

At the synaptic level, the interruption of the connection with axons of the nucleus electrosensorius, and thus the loss of predominantly excitatory synapses, leads to a shift of the balance between excitatory input and inhibitory input toward a higher degree of inhibition. The 'veto' exerted in the somatic regions of the large multipolar neurons of the central posterior/prepacemaker nucleus appears to cause the suppression of electrical activity in these chirp-controlling neurons outside the breeding season, thus resulting in the cessation of chirping behavior.

> Seasonally induced changes in chirping behavior are accompanied by alterations in dendritic morphology of central posterior/prepacemaker nucleus neurons.

Seasonal variation in dendritic morphology of motoneurons in white-footed mice

Similar changes in dendritic morphology, as they occur in the central posterior/prepacemaker nucleus, have been observed in white-footed mice (*Peromyscus leucopus*), a species common in temperate habitats in wide areas of North America. This seasonal breeder exhibits marked fluctuations in reproductive behavior. Feral mice breed in the spring and summer and are reproductively quiescent during the winter. As laboratory experiments have shown, gonadal recrudescence is stimulated by long day lengths, and gonadal involution takes place as the day lengths shorten. Nancy Forger and Marc Breedlove, then at the University of California, Berkeley, found that, in concert with these seasonal endocrine and behavioral changes, the motoneurons of a specific neuronal assembly, the

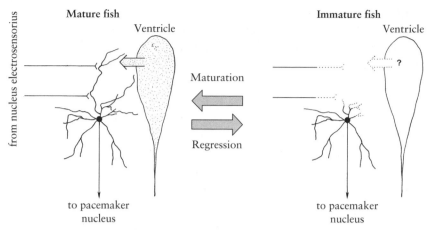

Figure 8.5 Model of the structural changes in the neural network underlying seasonal alterations in chirping behavior in the weakly electric knifefish, *Eigenmannia* sp. In the course of the rainy season, which triggers gonadal recrudescence, dendrites in the dorsomedial territory of the chirp-controlling neurons of the central posterior/prepacemaker nucleus grow out and make, predominantly excitatory, synaptic contact with axons originating from the nucleus electrosensorius. These structural changes may be induced by, yet unknown, molecular factors diffusing from the ventricle into the brain parenchyma. After the onset of the dry season, the dendrites of the dorsomedial territory retract. As a result, the link between the sensory side (nucleus electrosensorius) and the motor side (chirp-controlling neurons of the central posterior/prepacemaker nucleus) is interrupted. This reduces dramatically the excitatory synaptic input received by the chirp-controlling neurons, thus leading to a cessation of chirping behavior. (Courtesy: G.K.H. Zupanc.)

spinal nucleus of the bulbocavernosus, also undergo pronounced alterations in morphology.

This nucleus is located in the lumbar region of the spinal cord. It consists of approximately 200 motoneurons in adult males, but only of 60 motoneurons in adult females. In males, the majority of these motoneurons innervate the muscles bulbocavernosus and levator ani. These muscles attach to the base of the penis, and are involved in producing erection and ejaculation. Adult females lack the bulbocavernosus and levator ani. Instead, the motoneurons of their spinal nucleus of the bulbocavernosus innervate a sexually non-dimorphic muscle—the anal sphincter. This is also the case with the remaining spinal nucleus of the bulbocavernosus motoneurons in males.

The morphological changes that occur in the spinal nucleus of the bulbocavernosus after exposure of males to short day lengths include a shrinkage of the somata and nuclei of the motoneurons, as well as a reduction of their dendritic trees. These effects are reversed by reinstatement of long day lengths.

Further studies by Elizabeth Kurz, Dale Sengelaub, and Arthur Arnold of the University of California at Los Angeles have shown that sex steroid

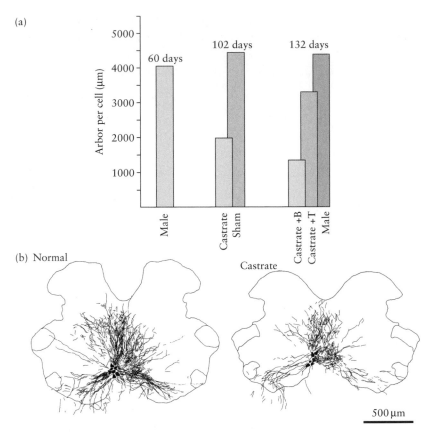

Figure 8.6 (a) Average length of dendritic arbor per motoneuron of the spinal nucleus of the bulbocavernosus in white-footed mice of different experimental groups: normal males of 60 days of age (Male, 60 days); males castrated (or sham-castrated) at 60 days of age and killed at 102 days (Castrate/Sham, 102 days); normal males at 132 days of age (Male, 132 days), as well as males castrated at 60 days of age, which received blank implants (Castrate + B, 132 days) or testosterone implants (Castrate + T, 132 days) at 102 days and were killed at 132 days. (b) Reconstructed dendritic arbors of spinal nucleus of the bulbocavernosus motoneurons of a normal male (left) and a castrated male (right). The overall dendritic length is reduced in the castrated males. (After **Kurz, E. M., Sengelaub, D. R., and Arnold, A. P. (1986).**)

hormones play a crucial role in these structural alterations of the spinal motoneurons. The principal result of their investigation is summarized in Fig. 8.6(a). In this diagram, the average length of the dendritic arbor per cell of the motoneurons has been plotted for the following experimental groups:

- males at the age of 60 days
- males castrated (or sham-castrated, which served as controls) at the age of 60 days, and killed six weeks later, when they were 102 days old

- males also castrated at 60 days of age, but then implanted at 102 days with Silastic tubes that either contained testosterone or were empty (the latter served as controls); mice of this group were killed at 132 days of age, and compared with normal males of the same age.

The motoneurons of the spinal nucleus of the bulbocavernosus were visualized by **retrograde tracing**. Cholera toxin conjugated to horseradish peroxidase was injected into the bulbocavernosus muscle (for a detailed discussion of neuronal tract-tracing techniques, see Chapter 3). Forty-eight hours later, the mice were killed. During this post-administration survival time, the tracer substance was taken up by the axonal terminals making synaptic contact with the muscle, transported in retrograde direction towards the motoneuron cell body, from where it finally filled the dendrites. Sections through the spinal cord were processed histochemically so that the motoneurons labeled with tracer substance became visible.

It is evident from the diagram that the dendritic arbor was clearly reduced in castrated males killed six weeks after the castration, relative to that of sham-castrated males. The extent of this reduction is illustrated in Fig. 8.6(b), which shows reconstructed dendritic arbors of a normal male and a castrated male.

The decrease in dendritic length could be reversed by treatment with testosterone, and after four weeks of testosterone treatment the dendritic arbor of the cell was largely restored to normal levels. These changes in dendritic morphology are likely paralleled by a modulation of the number or the organization of synaptic inputs, and thus by profound alterations of the properties of the neural network involved in control of copulatory behavior.

A similar **endocrine-controlled mechanism** is thought to be at work in feral white-footed mice. According to this hypothesis, exposure to short day lengths causes regression of the testes, which reduces the titer of circulating androgens. This results in a number of morphological changes in androgen-sensitive structures, including changes in dendritic morphology and the size of the motoneurons of the spinal nucleus of the bulbocavernosus, as well as the size of the bulbocavernosus muscles. These changes lead then to the season-related changes in reproductive behavior.

Structural reorganization mediated by glial cells

Changes in neuronal excitation may not only be caused by changes in the number or the morphology of neurons, but also by alterations in the organization of the network provided by **glial cells**. An example of such a mechanisms has been found in the so-called **magnocellular hypothalamo-neurohypophysial system** of mammals, which forms part of the anterior hypothalamus. As electron microscopic examination has shown, the magnocellular neuronal somata are almost completely separated from their neighbors by fine processes of astrocytic glial cells, so that only approximately 1% of the total neural membrane is in direct apposition without intervening glial processes. In response to changes in the animal's physiological state, for example during periods of water deprivation or lactation, the astrocytes rapidly retract, thus resulting in a tremendous increase in the number of neuronal elements directly juxtaposed. This glial withdrawal leads to an enhancement of neuronal excitability which is believed to be directly linked to the observed behavioral changes.

When to use 'structural reorganization'?

The strategy to produce motivational changes by modifying the structure of either neurons or glial cells, or both, and thus the properties of the associated neural network, appears to be especially used in the following cases:

• changes in the motivation underlying a behavior are rather slow
• alterations at the behavioral level are dramatic
• a new motivational state becomes manifest for rather long periods of time.

All three features are characteristic of seasonally breeding animals. It appears, therefore, possible that structural reorganization of neural networks is a mechanism commonly employed in such animals to control seasonal changes in behavior, although this does not exclude other mechanisms.

Biochemical switching

Modulation of the stomatogastric ganglion

A second set of mechanisms employed to accommodate motivational changes in behavior is subsumed under the term **biochemical switching**. This mechanism has been particularly well studied in the **stomatogastric ganglion** of decapod crustaceans.

As shown in Fig. 8.7(a), the complete **stomatogastric nervous system** of lobsters and crabs consists of the following elements:

• a single **stomatogastric ganglion**
• a single **esophageal ganglion**

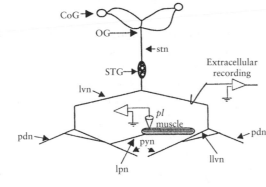

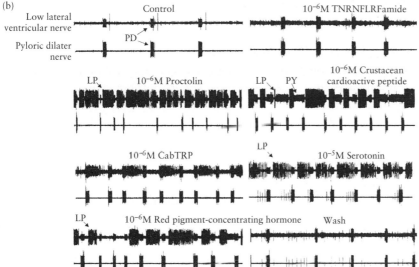

Figure 8.7 (a) Schematic organization of the stomatogastric nervous system of the American lobster (*Homarus americanus*). The paired commissural ganglia (*CoG*) and the single esophageal ganglion (*OG*) provide, via the stomatogastric nerve (*stn*), input to the stomatogastric ganglion (*STG*). Muscles of the foregut are innervated by neurons of the stomatogastric ganglion via several motor nerves, including the lateral ventricular nerve (*lvn*), the pyloric dilator nerve (*pdn*), the pyloric nerve (*pyn*), the lateral pyloric nerve (*lpn*), and the low lateral ventricular nerve (*llvn*). (b) The experiment shows the effects of various modulators on the motor patterns generated by the isolated stomatogastric ganglion. The ganglion is physiologically isolated by blocking the stomatogastric nerve. This results in rhythmic activity only of the pyloric dilator (*PD*) neuron of the stomatogastric nerve, as revealed by extracellular recordings from the low lateral ventricular nerve and the pyloric dilator nerve after bathing the ganglion in saline as a control. Bath application with the peptide TNRNFLRFamide leads to a modest increase in frequency and intensity of the bursts produced by the pyloric dilator neurons. In contrast, application of proctolin, crustacean cardioactive peptide, *Cancer borealis* tachykinin-related peptide (CabTRP), serotonin, and red-pigment-concentrating hormone all strongly activate the lateral pyloric neurons, as seen on the trace of the recordings from the low lateral ventricular nerve. They also increase the frequency of the pyloric dilator burst, as seen on the trace of the recordings from the pyloric dilator nerve. Between each application, the preparation is washed with saline. The final application of saline recovers the original pattern of bursts generated by pyloric dilator neurons. (After **Marder, E., and Richards, K. S.** (1999).)

- bilaterally paired **commissural ganglia**
- the **stomatogastric nerve**.

The stomatogastric ganglion contains 26–30 neurons. This rather low number of neurons has enabled researchers to achieve, through **simultaneous physiological recordings**, quite a comprehensive understanding of the properties of this neural network. Further information on the fundamental significance of the individual network components has been accumulated by selectively eliminating single neurons within the ganglion through **photoinactivation** after intracellular injection of fluorescent dyes—a technique described in detail in Chapter 3.

Most of the neurons of the stomatogastric ganglion are motoneurons that make excitatory connections with the muscles of the stomach. There are no intervening premotor interneurons, a feature that distinguishes the stomatogastric ganglion from most other motor systems. This makes it possible to record the motor pattern directly from the neurons of the stomatogastric ganglion.

The muscles innervated by the stomatogastric ganglion neurons move different regions of the foregut.

- The **gastric mill** consists of three teeth. The rhythmic movements of the gastric mill muscles determine the **gastric mill rhythm**.
- The **pylorus** abuts the gastric mill region. Its filtering and sorting movements are produced by a set of muscles that control both the pylorus and its valves. The corresponding rhythm is called the **pyloric rhythm**.

The stomatogastric ganglion of decapod crustaceans consists of individually identified neurons that control rhythmic movements of the foregut.

The stomatogastric nerve contains approximately 250 fibers. Relatively few of these fibers originate from stomatogastric ganglion neurons. Most of them arise from somata in the commissural ganglia and the esophageal ganglion. Additional input to the neuropil of the stomatogastric ganglion originates from the gastro-pyloric receptor neurons. The latter are sensory neurons that respond to stretch of several of the stomach muscles. All types of input to the neuropil of the stomatogastric ganglion contain a large number of modulators. The exact nature of these substances may, to a certain extent, vary between species. In the lobsters *Homarus americanus* and *Homarus gammarus*, the following **transmitters** and **neuromodulators** have been identified: histamine; serotonin (also called 5-hydroxytryptamine); dopamine; γ-amino-butyric acid, GABA; proctolin; red-pigment-concentrating hormone; tachykinin-related peptide; allostatin-related peptides; β-pigment-dispersing hormone-like peptide; cholecystokinin; nitric oxide; SDRNFLRFamide; and TNRNFLRFamide. (The names of the latter two peptides indicate their sequences, using the one-letter code of amino acids.)

Input to the stomatogastric ganglion, mediated by the stomatogastric nerve, contains a large number of neuromodulators.

Originally, it was thought that separate sets of neurons within the stomatogastric ganglion control the gastric mill rhythm and the pyloric rhythm. However, more recent research, mainly performed in the laboratory of Eve Marder at Brandeis University in Waltham, Massachusetts, has shown that this is not the case. Rather, the work of her group has demonstrated that individual neurons may participate in the production of either of the two rhythms. The exact firing pattern of the neurons is defined by the actual modulatory environment—the anatomical network of the stomatogastric ganglion provides only a physical backbone upon which the modulatory inputs can operate. This makes it possible that a single network can produce multiple variations in the behavioral output under different conditions. For this concept, Peter Getting and Michael Dekin of the University of Iowa coined the term **polymorphic network**.

Polymorphic networks: anatomically defined networks whose modulation results in multiple functional modes of operation.

An example of the neuromodulatory action of six of the substances present in the stomatogastric ganglion of *Homarus* on its motor pattern is shown in Fig. 8.7(b). In this experiment, input from the anterior ganglia was removed by placing a vaseline well filled with sucrose on the stomatogastric nerve. This treatment blocks impulse traffic down the nerve. The major results are as follows:

- When applying saline as a control, only the so-called pyloric dilator neurons are rhythmically active.
- Bath application of TNRNFLRFamide produces modest increase in frequency and intensity of the motor pattern.
- Application of proctolin, crustacean cardioactive peptide, red-pigment-concentrating hormone, and serotonin all strongly activate lateral pyloric neurons and increase the frequency of the pyloric dilator burst.

Modulation of crayfish aggressive behavior

Another example demonstrating the importance of neuromodulators in the setting of the motivational state of an animal also stems from work on crustaceans. Such investigations, mainly conducted by Robert Huber of the University of Graz, Austria, and Edward Kravitz of Harvard University in Boston, Massachusetts, make use of the fact that crayfish readily engage in aggressive encounters when placed together in an aquarium. The fights, which can be easily quantified, escalate in a probabilistic manner until one of the opponents retreats.

While under normal circumstances animals faced with much larger opponents quickly withdraw from the encounter, administration of **serotonin** into the hemolymph of freely moving individuals leads to an alteration of their behavioral strategy to retreat and act as subordinates. As a consequence, fights last considerably longer compared to control animals. Thus, serotonin injected into subordinate animals appears to

Administration of seratonin to the hemolymph of subordinate crayfish can change the aggressive motivation toward higher levels.

change the aggressive motivation of the crayfish toward higher levels. At present, the central sites responsible for this effect are unknown.

Modulation of the modulators

Although the examples given above suggest a rather unidirectional relationship between neuromodulators and behavior, the actual situation is certainly more complex. This has been indicated in a study conducted, again on crayfish, by Shih-Rung Yeh, Barbara Musolf, and Donald Edwards of the Georgia State University in Atlanta, Georgia. Their investigations have shown that the modulatory effect of serotonin on the **lateral giant interneuron**, a command neuron controlling escape, is itself modulated by the social status and the social history of the animal. In dominant crayfish, the response of the lateral giant interneurons triggered by serotonin is transiently increased, whereas in subordinates it is transiently inhibited. These slow, but reversible, modulatory alterations appear to result from changes in the population of serotonin receptors.

What makes modulators suitable for neuromodulation?

A number of properties make catecholamines and neuropeptides well suited to mediate motivational influences within the central nervous system. Among them are, in particular, the following two features:

- Neuropeptides and catecholamines are frequently **released in a non-synaptic fashion**. This contrasts with 'classical' transmitters, such as acetylcholine, glutamate, or GABA, which are typically released at synaptic specializations. Such specialized synaptic zones are visible under the electron microscope as electron-dense (black) thickenings at the presynaptic membrane. These thickenings represent specialized protein structures at which the synaptic vesicles dock to initiate the fusion with the presynaptic membrane and the release of the synaptic content.

- In the transmission process involving catecholamines and neuropeptides, there is often no focal relationship between the site of release and the location of the corresponding receptors. This, again, contrasts with the situation known from classical transmitters. At the neuromuscular junction, where acetylcholine is used as a transmitter, cholinergic receptors are located at high densities immediately opposed to the site of transmitter release on the postsynaptic membrane. The lack of this focal relationship between the site of release and the site of ligand–receptor interaction is referred to as **ligand–ligand receptor mismatch**.

Both properties, the non-synaptic release and the ligand-ligand receptor mismatch, lead to a 'diffuse' effect of catecholamines and neuropeptides. As a consequence, an endogenous ligand may interact with receptors at more than just one site in the central nervous system, and some of these sites may be fairly distant to the site of release. Not surprisingly, these sites may be involved in the control of more than just one behavior. Such a notion is in agreement with behavioral observations: motivational factors which affect the occurrence of one type of behavior tend to influence the probability of other behavioral patterns as well.

When to use 'biochemical switching'?

The strategy to produce motivational changes through biochemical switching appears to be especially suitable in the following cases:

- the changes in the motivation underlying a behavior are rather fast
- the alterations in the propensity to execute a behavior do not occur in an all-or-none function; rather, fine gradual differences within a broad range of possibilities are to be accommodated.

Such fast, and gradual changes in motivation are typical, for example, of behaviors associated with feeding and ingestion. Indeed, biochemical switching has frequently been found in neural circuitries involved in control of these behaviors.

Summary

- In an 'ideal' reflex, the relationship between a stimulus and the resulting behavior is fixed. Modulation of this input-output relationship is caused by internal—motivational—factors.

- Motivational changes in behavior can occur only if the underlying neural network exhibits the potential for neural plasticity.

- Two major mechanisms that mediate motivational changes in behavior involve structural reorganization of the neural network underlying the control of the respective behavior; and chemical modulation of this network through neuromodulators, especially serotonin, catecholamines, and neuropeptides ('biochemical switching').

- Alterations in dendritic morphology of neurons, as one mechanism to structurally reorganize neural networks, has been implied to play a role in the dramatic behavioral changes, such as those that occur in seasonally breeding animals.

■ Structural reorganization may further be mediated by non-neuronal cells, for example, by the degree to which glial cells intervene between neurons. This leads to alterations in the physiological properties of the neural network, and thus, of the behavioral outcome.

■ Biochemical switching as a mechanism to modulate behavior has, among other systems, been demonstrated in the stomatogastric ganglion of decapod crustaceans. The modulators, mainly neuropeptides, are contained in axons innervating this ganglion. These modulators operate upon the 'polymorphic' network to produce different behavioral outputs.

■ Modulators themselves may be modulated by the social status and the social history of an animal.

■ Among the cellular properties that make neuromodulators well suited to mediate motivational changes in behavior are the frequently observed release in a non-synaptic fashion and the lack of a focal relationship between the site of release and the site of ligand–receptor interaction ('ligand–ligand receptor mismatch').

Recommended reading

Harris-Warrick, R. M., and Marder, E. (1991). Modulation of neural networks for behavior. *Annual Review of Neurosciences*, **14**, 39–57.

An excellent review summarizing the concept of polymorphic networks. Most principles associated with this concept are illustrated using the stomatogastric ganglion of decapod crustaceans.

Marder, E., and Calabrese, R. L. (1996). Principles of rhythmic motor pattern generation. *Physiological Reviews*, **76**, 687–717.

An exhaustive review of how rhythmic movements are controlled by central pattern-generating neural networks, and how the behavioral output is changed by modulatory input.

Zupanc, G. K. H. (1996). Peptidergic transmission: from morphological correlates to functional implications. *Micron*, **27**, 35–91.

A comprehensive review of morphological features characterizing synaptic and non-synaptic modes of neuropeptide transmission. An understanding of these features is essential to appreciate the biochemical switching mechanism.

Zupanc, G. K. H., and Maler, L. (1997). Neuronal control of behavioral plasticity: the prepacemaker nucleus of weakly electric gymnotiform fish. *Journal of Comparative Physiology A*, **180**, 99–111.

This review article provides an overview of chirping behavior and its neural substrate, the central posterior/prepacemaker nucleus, in weakly electric gymnotiform fish. Special emphasis is placed on the discussion of neural mechanisms accommodating motivational changes in this behavioral pattern.

Questions

8.1 What is motivation? How can animals, at the neuronal level, accommodate short-term and long-term motivational changes in behavior?

8.2 Why are neuropeptides much better suited to mediate motivational changes in behavior than 'classical' transmitter substances?

8.3 Adult neurogenesis has been demonstrated in the central posterior/prepacemaker nucleus of weakly electric knifefish. Design experiments to verify the hypothesis that this phenomenon is causally linked to changes in chirping behavior.

Large-scale navigation: migration and homing

9

- Introduction
- Modes of migration
- Genetic control of migratory behavior
- Homing
- Approaches to study animal migration and homing
- Mechanisms of long-distance orientation in birds
- Homing in Salmon
- Orientation in sea turtles
- Summary
- Recommended reading
- Questions

Introduction

The migration and homing of animals, some of which travel in the course of their journeys over thousands of kilometers, have fascinated Man for centuries. The best-known example are the annual journeys of many migratory birds. They travel in the fall from the place of their birth to the wintering grounds in warmer regions. In the spring, they make a similar migratory journey, but this time in reverse direction.

Focusing on such long-distance migrations, we will summarize, in the first part of this chapter, some of the basic facts of these phenomena, such as the modes of migration and the specificity of homing. Then, we will take a closer look at the general methods used to study animal migration. The last, and the largest, portion of this chapter will be devoted to a detailed discussion of the behavioral and neural mechanisms involved in orientation during migration and homing by using representatives of three particularly well-studied animal groups—birds, fish, and sea turtles.

Figure 9.1 Differential distribution and migratory routes of the two major European populations of the white stork (*Ciconia ciconia*). While birds of the western population take a southwesterly route to their wintering grounds in western parts of Africa, members of the eastern population fly around the eastern end of the Mediterranean to spend the winter in eastern parts of Africa. (After **Schüz, E.** (1971).)

Modes of migration

During the large-scale migrations, the entire species, or at least subpopulations of the respective species, use **specific routes** from their point of origin to the final target site.

Examples: Based on their distribution and migratory behavior, the European white storks (Ciconia ciconia) can be divided into the following two populations (Fig. 9.1):

1. *Members of the **western population** migrate to their wintering grounds in western parts of Africa through Spain and across the Straits of Gibraltar.*

2. *Members of the **eastern population** nest farther east in Europe and spend the winter months in East and South Africa; they take an eastern route around the eastern end of the Mediterranean Sea.*

A similar phenomenon is observed in the American golden plover (*Pluviatis dominica*) (Fig. 9.2). The adults breed in arctic regions of Canada.

- Each fall, the members of the **Pacific population** fly over open ocean to their winter range in Hawaii, the Marquesas Islands, and the Low Archipelago. In the spring, the birds take the same route back.

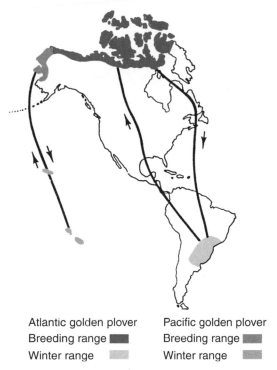

Atlantic golden plover   Pacific golden plover
Breeding range ▇▇   Breeding range ▇▇
Winter range ▢   Winter range ▢

Figure 9.2 Distribution and migration of the American golden plover (*Pluviatis dominica*). Members of the Pacific population breed in Alaska and fly over open ocean to their winter range in Hawaii, the Marquesas Islands, and the Lower Archipelago. By contrast, the Atlantic population breeds more easterly and spends the winter in South America. Members of the latter population use different routes for the southbound and northbound migration. (After **Keeton, W. T.** (1980).

- At the same time, members of the **Atlantic population** cross, in the course of their southbound journey, the Atlantic from Labrador and Nova Scotia to the Lesser Antilles and northeastern South America to continue to their winter range in temperate latitudes in northern Argentina, Uruguay, and Southern Brazil. In the spring, they take a different route, flying northwest through Central America and north over the Great Plains back to the Arctic.

Interestingly, in contrast to the adults, almost all of the young golden plovers of the Atlantic population take the inland route via Central America also during their first fall journey.

In a very few animal species, the **annual migratory cycle is completed by several generations**, rather than by one individual. Such animals represents rare cases of a biological rhythm that is longer than the life of

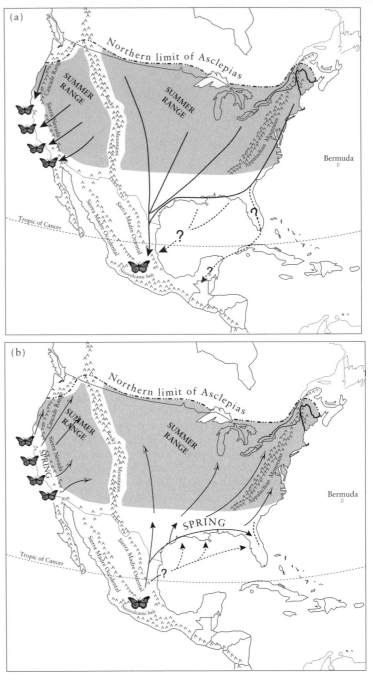

Figure 9.3 Migratory routes of the eastern population of the monarch butterfly. This population breeds, over several generations, east of the Rocky Mountains. (a) In the fall, the butterflies migrate southwards to overwinter in Central Mexico. (b) In the spring, they re-enter the USA along the Gulf Coast states, where the females lay eggs and die. The adults of the resulting first spring generation migrate further and produce the next generation. One large subpopulation of this second generation, mainly located in the Midwest, continues the migration eastwards over the Appalachians, where two or more summer generations follow. Also indicated is the northern limit of the milkweed plants (genus *Asclepias*). (After **Brower, L. P.** (1996).)

the organisms exhibiting this rhythm. Due to this complexity, details of such a cycle have been revealed in only a very few instances. One of the better-understood examples is the monarch (*Danaus plexippus*), a North American butterfly species of the family Nymphalidae. One large population of the monarch breeds east of the Rocky Mountains. In the fall, the butterflies migrate through Central Texas into Mexico where they follow the Sierra Madre Oriental to finally overwinter in high-altitude forests of the Transverse Neovolcanic Belt in Central Mexico (Fig. 9.3). During this journey, they migrate up to 3 600 km over a period of about 75 days. This corresponds to an average per-day travel distance of approximately 50 km.

In early spring, a rapid northward journey takes place so that by early April both sexes, now approximately half-a-year-old, reach the Gulf Coast states of the USA. The females lay their eggs on the resurgent spring milkweeds (genus *Asclepias*) and die. The offspring of this first spring generation continue the migration northeastward to the Great Lakes region and southern Canada, where they arrive by early June. Along their way, they lay eggs, which give rise to the second spring generation of adults. This new generation migrates due east, laying again eggs along their way, thus resulting in a third generation. While one or two more short-lived breeding generations are produced, the butterflies spread eastward across the Appalachian Mountains. From late August to early September, the fall migration is again under way.

An unusual example: in monarch butterflies, the annual migratory cycle is completed not by one, but several generations.

Genetic control of migratory behavior

It has been suspected for a long time that the **urge to migrate** and the **sense of migratory direction** are under **direct genetic control**. Such an assumption is not trivial, as, theoretically, the first departure of obligate migrants could be triggered by experienced conspecifics. However, in more recent years convincing evidence in support of the inheritance hypothesis has been obtained. This is mainly due to the pioneering work of Peter Berthold and Andreas Helbig of the Ornithological Station Radolfzell, which is run as a department of the Max Planck Institute for Behavioral Physiology in Seewiesen, Germany.

These investigations have taken advantage of the existence of various populations of blackcaps (*Sylvia atricapilla*), a European warbler. These populations differ in terms of their migratory behavior. For example, in Central Europe blackcaps are fully migratory. By contrast, on the Cape Verde Islands (islands in the Atlantic, off Africa) a resident population exists. When these two populations were crossbred, 40% of the F_1 hybrids were migratory. This result demonstrates that the urge to migrate can be bred into a non-migratory birds population. On the other hand,

Figure 9.4 Breeding areas (light gray) and main wintering grounds (dark gray) of the blackcap. The major routes taken during the fall migration are indicated by arrows. Based on the differences in the main migratory directions, this European warbler species can be divided into two populations. One population (1) uses a southwest route to migrate to the wintering grounds in the western Mediterranean region and Northwest Africa. Birds of the other population (2) take, in the fall, a southeastern route to reach the wintering ground in East Africa. (After **Helbig, A. J.** (1991).)

the fact that not all hybrids become migratory indicates that it is not a single genetic locus that determines the urge to migrate. Rather, a multi-locus system has been proposed.

In a second experiment, blackcaps from two Central European populations which differ considerably in the route they take during the fall migration were crossbred:

Crossbreeding experiments using birds of different populations provide information about genetic factors controlling migratory behavior.

- One population breeding in Germany uses mainly a western route to the wintering areas in western Mediterranean regions and Northwest Africa (Fig. 9.4).

- A second population of blackcaps have their breeding areas in Austria. They take an eastern route to their wintering range in East Africa (Fig. 9.4).

Crossbreeding produced phenotypically intermediate offspring. These birds were oriented toward mean directions intermediate between those of the two parental populations (Fig. 9.5).

Based on these results, it is thought that migratory birds inherit information on the direction to travel and how long to fly in this direction. They, then, complete their first migration using a 'clock-and-compass' mechanism (see below).

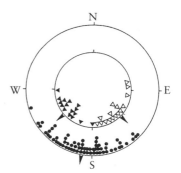

Figure 9.5 Directional choices of hand-raised blackcaps during the early part of the fall migration season. The inner circle represents the choices of the parental generation. The solid triangles indicate the behavior of birds from western Germany, the open triangles that of birds from eastern Austria. The outer circle, with the filled small circles, represents the mean choices of individual birds of the F_1 generation. The mean directions exhibited by the three groups are indicated by the arrowheads. While the directional preferences are clearly distinct between the German and the Austrian populations, the hybrids exhibit a mean directional choice intermediate between those of the two parental populations. (After **Helbig, A. J.** (1991).)

Homing

While **migration** merely describes the directed locomotory activity of an animal during long-distance journeys, **homing** refers to the ability of an animal to specifically return to its 'home'. The home can be the nest with the offspring to which a parental bird returns after foraging. In such instances, the time spent between leaving the home and returning to it is, typically, short —often in the order of minutes. In other cases, a considerable amount of time, sometimes years, may elapse until the animal returns to its home. Sea turtles, for example, lay eggs into underground nests on the beach. Upon emerging from these nests, the hatchlings quickly move to the sea, until they reach the ocean. After migration across thousands of kilometers in the open sea, they return as adult to their natal beach to nest.

The **precision of homing** has been well studied in birds. Some avian species are even able to return to their nest after being transported in a closed box over considerable distances and released at a location unfamiliar to them. This may resemble the situation when a wild bird, after being displaced by strong winds from its nesting area, has to find its way back home. In the experiment, marking, often with a colored band, makes it possible to unambiguously identify the bird upon arrival and time of its return.

One of the most spectacular examples of homing in wild animals was reported by Geoffrey Matthews of the University of Cambridge, England, in 1953. Matthews performed a large number of displacement experiments on Manx shearwater (*Puffinus puffinus*), ocean birds feeding on small fish and crustaceans, which they catch with their bills in the water. They can rest on the surface of the ocean, but most of the time they are seen flying near the water. Manx shearwaters come ashore only during the time of reproduction, when they breed in colonies in burrows on islands and the top of cliffs. Between intervals of incubation or care of the young, they may make foraging excursions of a hundred or more kilometers in length. Further, incubation is shared between the parents, and frequently, one bird takes a spell of several days without feeding. This allows the

Manx shearwater, one of the 'champions' among birds: homing over thousands of kilometers has been documented.

investigator to remove one of the parents from the nest to conduct experiments without endangering the breeding success of the birds.

To test the homing ability of Manx shearwaters, Matthews released individuals at various distances from the nest. The most extreme experiment was conducted on a bird that was transported by airplane from its burrow off the coast of Wales to Boston, Massachusetts. Just $12\frac{1}{2}$ days later, this bird reappeared at its nest. During that time, it must have flown over more than 4500 km across the Atlantic, with an average speed of 360 km per day!

While homing experiments on wild birds are rather cumbersome to perform, the **homing of domestic pigeons** is more readily accessible. Experiments can be carried out throughout the year, rather than being restricted to the breeding season. Moreover, pigeons are highly tolerant to experimental manipulation. Therefore, numerous laboratory experiments have been performed, including many on their sensory capabilities. Some selected examples will be discussed below.

The enormous homing ability of pigeons is likely related to the breeding behavior of their ancestors. Domestic pigeons presumably originate from wild Mediterranean rock doves (*Columba livia*), which nest on cliffs and forage for food on nearby fields. For centuries, these birds have been bred and selected for fast and reliable homing, especially in their function as carrier pigeons transporting messages in small capsules attached to their legs. They are raised in small sheds called **lofts**, which remain their permanent homes throughout life. Starting a few months after birth, the pigeons are systematically trained by transporting them to increasingly distant points from the home loft where they are released to fly back home. Upon completion of this training, the most capable pigeons are able to home from more than 2000 km in two or three days.

Approaches to study animal migration and homing

To study migratory behavior in the field, research was initially restricted to the collection of **distribution data** about breeding, stopover, and winter ranges of the various species. Although these observations were, in the history of migration research, of enormous importance to establish that many animals are migrating, numerous questions remained to be answered. For example, this approach could not answer the question whether or not populations of a given species from different areas migrate to different target regions.

Therefore, the introduction of **tags** attached to the animal was a milestone in the historical development of migration research. For birds, cylindrical plastic or metal bands, loosely fitted around one leg, are employed. Each tag has inscribed a unique identification number and the name of an addressee that can be informed if the animal is found. Butterflies are typically tagged by glueing round paper discs to one wing. For fish, metal or plastic tags applied to jaws, opercula, or fins are often used.

Since the introduction of this method, millions of animals have been tagged. An enormous amount of information has been accumulated through the banding of birds. The corresponding data show that the **degree of recovery** is relatively high in large birds, such as ducks and geese, where it reaches sometimes 20% and higher. This is due to the fact that these birds are readily visible and many of them heavily hunted. In small songbirds, on the other hand, the percentage of recovery is very low, with values typically far below 1%. This low recovery rate requires large-scale banding programs to yield meaningful data on the migrational pattern of a given species. Despite this limitation, the specificity of the tagging method has provided extremely valuable information on the migration and homing of animals.

Two further technologies successfully applied to navigation research are **radar** and **microtransmitters**. The radar echos produced by aggregations of animals can be used either to screen for migratory activity within a certain area (**surveillance radar**) or to detect and follow a single target (**tracking radar**). Since the resolving power of radar is relatively low, it typically allows the researcher only to follow aggregations of animals, but not single individuals. By contrast, radio tracking through microtransmitters attached to animals provides the possibility to track individuals. Signals emitted by the microtransmitter are received by antenna that are stationary, carried on a vehicle or boat, or attached to an aircraft. More recently, this method has been further enhanced by **satellite-based radiotelemetry**. The major limitations of this extremely powerful method are caused by the costs associated with the satellite time, the lifetime of the batteries required to operate the microtransmitter, and the weight of the transmitter/battery, which currently restricts the use to animals with relatively large body mass.

Most information available on animal migration and homing is based on tagging experiments.

Mechanisms of long-distance orientation in birds

Among migratory animals, more studies on mechanisms involved in orientation have been conducted in birds than in any other animal group. These investigations have resulted in an impressive amount of behavioral data; yet, due to the complexity of the mechanisms involved, the interpretation of the data has often proven difficult. In the last couple of years, these behavioral data have increasingly been supplemented by the results of sensory, physiological, and neurobiological experiments.

Sun compass

Experimental evidence that the **sun** may provide an important visual cue for diurnal migrants was first reported by Gustav Kramer from the Max Planck Institute for Marine Biology in Wilhelmshaven, Germany. In 1950, he published, partially with his associate Ursula von St Paul, a series of papers on this topic. Earlier observations have shown that caged migratory birds exhibit a pronounced **migratory restlessness** at the time when they normally start their journeys. To further investigate this phenomenon under controlled conditions, Kramer and von St Paul placed the birds in circular cages which were designed in such a way that they displayed a perfect symmetry. Any visual landmark was excluded so that only the sky and the sun were visible. Thus, the birds had only celestial cues, but no landmarks, available for orientation.

When starlings (*Sturnus vulgaris*), diurnal migrants, were used in the experiments, the birds attempted to fly predominantly in the direction of their normal migratory path, as long as they could see the sun. If the sun was completely obscured by clouds, the headings of the birds became rather random. On the other hand, if the sunlight was deflected by a mirror at a certain angle, the birds altered the direction of the attempted flight accordingly (Fig. 9.6).

How do the starlings determine the direction they head to during the migration restlessness? To answer this question, Kramer and von St Paul

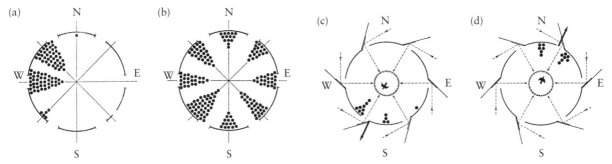

Figure 9.6 Spontaneous directional headings of a starling during the time of migratory restlessness, as examined in the circular cage used by Gustav Kramer. Each dot represents the mean direction of the bird's body exhibited during 15 sec of migratory restlessness. Broken arrows show the incidence of light from the sky through the six windows. Thick arrows indicate the mean direction of activity. (a) When the sun is directly visible through one of the six windows, the bird heads in northwest direction. (b) Under heavily overcast sky, the directional activity of the starling is rather random. In a modified set-up, each window has an opaque screen, and a mirror deflects the light so that it appears to enter the cage at an angle approximately 90° counterclockwise (c) and clockwise (d) from its normal direction. Now the bird attempts to fly in southwest and northeast direction, respectively. (After **Kramer, G.** (1950b).)

conducted a second set of experiments. To avoid the restrictions imposed upon the experimenter by the rather short seasonal period of migratory restlessness, a **food conditioning approach**, which could be employed at any time of the year, was used.

The birds were placed in a circular cage with an array of food chambers arranged at equal spacing around the periphery. The starlings were, then, trained to look for food in a certain chamber lying in a particular compass direction. Each food chamber was covered with a slotted rubber membrane. It was, therefore, impossible for the birds to know whether the feeder actually contained food, before they had thrust their bills through the slot and picked up any grain provided. Despite these difficulties, the birds maintained the training direction—even if the cage was rotated or the experiments were conducted at different times of the day.

Kramer and von St Paul concluded from these results that the birds must use the sun for orientation and possess an **internal clock** to compensate for the apparent movement of the sun across the sky. In other words, the birds can extract compass information from the position of the sun, independent of whether the bird sees the sun low in the east in the morning or low in the west in the evening. This mechanism is referred to as **time-compensated sun compass**.

The time-compensation mechanism of the sun compass by an internal clock was confirmed in subsequent experiments carried out by Klaus Hoffmann. He trained starlings in a circular cage to obtain food from a feeder in a certain direction using the sun as a cue, similarly as was the case in the experiments by Gustav Kramer and Ursula von St Paul. Then, he reset or shifted the bird's internal clock experimentally. This can be done by confining the bird to a light-proof room and exposing it to artificial photoperiods. This resulted in the bird misreading the sun compass in a predictable manner: a clockwise shift of the internal clock by x hours leads to a counterclockwise misinterpretation of the sun compass by $(360/24)x$ degrees, and vice versa.

> *Example: A starling is trained to look for food in a feeder located in northwest direction. Then, the bird is exposed for 4–6 days to an artificial day beginning and ending 6 h later than the natural day. When it is returned to the experimental cage and exposed to natural sunlight, it preferentially orients toward the northeast instead of the northwest (Fig. 9.7). The shift of its internal clock by 6 h in counterclockwise direction has resulted in a 90° clockwise misreading of the sun compass.*

Similar experiments in which the internal clock was shifted behind or ahead of the local photoperiod were conducted by Klaus Schmidt-Koenig from the University of Tübingen, Germany, on homing pigeons. After the clock shift, the birds were taken to a place far from the home loft, and released singly. The direction in which the pigeon vanished from sight was

A time-compensated sun compass allows an animal to extract compass information from the position of the sun, independent of the time of the day.

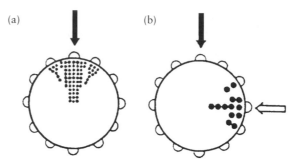

Figure 9.7 Change of orientation of a food-conditioned starling after experimental shifting of its internal clock. (a) Test during the training phase under natural-day conditions. (b) Test after exposing the starling for several days to an artificial day, which is six hours behind local time. Each dot represents one critical choice without food reward. The black arrows indicate the original training direction, the open arrow indicates the direction expected if the clock is successfully shifted. (After **Hoffmann, K.** (1954).)

recorded and the data entered in a compass diagram. While control pigeons (i.e. pigeons that were not subjected to a clock shift) vanished roughly in the direction pointing to the home loft, time-shifted pigeons vanished with an error angle reflecting the direction and degree of the time shift as described above.

It is likely that birds are able to determine the position of the sun and make use of this information for sun compass orientation, even if the sun is obscured by clouds. However, this is possible only if at least some patches of blue sky are still visible. It is believed that under such circumstances the birds use the polarization pattern of the blue sky for determining the position of the sun. This polarization vision is strongest in the ultraviolet range of light, where polarization is especially strong. Indeed, as revealed by conditioning experiments, pigeons are highly sensitive in the ultraviolet range (Fig. 9.8), and they can perceive the plane of polarization of light. This sensitivity to ultraviolet light, the ability to perceive the plane of polarized light, and the possible use of these mechanism for sun compass orientation resemble, in a striking way, the sensory capabilities and orientation mechanisms of honey bees.

Taken together, these behavioral experiments firmly established the existence of a time-compensated sun compass used by diurnal migratory birds and homing pigeons for orientation.

Star compass

The experiments conducted by Gustav Kramer and confirmed by many others have demonstrated that birds can use the position of the sun for compass orientation. This can explain the directed restlessness of diurnal migrants. But how do **nocturnal migrants** orient?

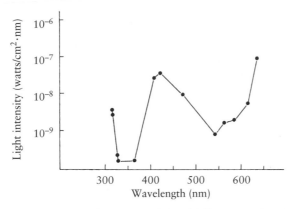

Figure 9.8 Behavioral sensitivity of a pigeon to light of different wavelengths. The pigeon shows high sensitivity to light in the ultraviolet range (i.e. at wavelengths between 325 and 360 nm), as indicated by the low intensity of light necessary to elicit a behavioral response. (After **Kreithen, M. L.** (1979).)

This question is especially intriguing, since nocturnal migratory birds too exhibit a directed, rather than a random, migratory restlessness. To identify the cues directing this activity, Franz Sauer from the University of Freiburg, Germany, performed experiments on warblers (genus *Sylvia*)— a nocturnal migratory bird—in circular cages with a glass top, similar to the set-up used by Kramer. When these birds were kept in the cage indoors, their activity appeared disoriented. However, when they were kept in the same cage outdoors so that they could see the starry sky, their activity became oriented during the time when they normally started their seasonal migration. In the fall, they showed preference for southern directions, while in the spring they oriented predominantly northward. This indicated that cues provided by the night sky may be used for orientation.

To analyze this phenomenon quantitatively, Stephen Emlen, then a graduate student at the University of Michigan, Ann Arbor, and his father, John Emlen, from the University of Wisconsin, Madison, developed the **funnel technique** (Fig. 9.9). The bird is kept in a circular cage with an ink pad covering the bottom. Blotting paper lines the sides in a funnel-like fashion. When the bird attempts to leave the cage, ink footprints are produced on the blotting paper. These footprint data can be transcribed into vector diagrams by a variety of methods, for example by measuring the density of the blackening employing computer-aided image analysis (Fig. 9.10).

Using this method with indigo buntings (*Passerina cyanea*), Stephen Emlen demonstrated that the orientation of the birds was the same under the natural night sky and the stationary sky of a planetarium. When

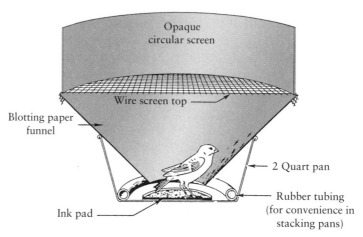

Figure 9.9 Schematic drawing of the 'funnel' experimental set-up designed by Stephen and John Emlen. Single birds are kept in a closed, funnel-like cage. The bottom of this cage is covered with an ink pad. When the bird attempts to leave the cage, it produces footprints on the blotting paper lining the sides. (After **Emlen, S. T., and Emlen, J. T.** (1966).)

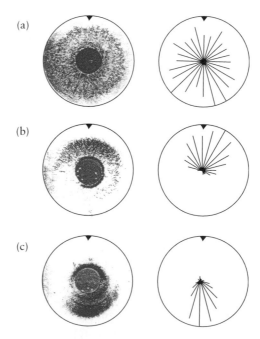

Figure 9.10 Footprint data produced by indigo buntings in the 'funnel' set-up introduced by the Emlens. Left column: Footprint raw data. Right column: Corresponding transcription into vector diagrams. (a) Rather random orientation. (b) Orientation directed predominantly north-northeast. (c) Orientation directed predominantly south-southwest. (After **Emlen, S. T., and Emlen, J. T.** (1966).)

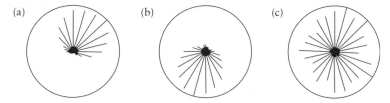

Figure 9.11 Preferential orientation of indigo buntings (*Passerina cyanea*) in planetarium experiments, as revealed by the funnel technique. The original footprint data have been transcribed into vector diagrams. (a) Orientation of indigo buntings in the spring under a spring planetarium sky. (b) Orientation in the spring under a spring planetarium sky horizontally rotated by 180°. The rotation of the starry sky results in the migratory restlessness exhibited roughly opposite to the one shown before rotation. (c) Control experiment with the stars shut off. (After **Emlen, S. T.** (1967a).)

horizontally rotating the planetarium sky by 180° in the spring, the indigo buntings exhibited their migratory restlessness predominantly to the south, rather than to the north (Fig. 9.11). This demonstrated the existence of a **star compass**. Further experiments have shown that time compensation is **not** used in order to take into account the movement of the stars during the night. It is thought that the birds learn, individually, the different star constellations near the pole star so that they can draw from them direction. It has been shown that indigo buntings do this by watching that part of the nocturnal sky that rotates least. In the northern hemisphere, this is the area around **Polaris**. How exactly the birds obtain compass information from the constellation of the stars around Polaris is unknown. In principle, several star patterns are available to achieve this.

Planetarium experiments have provided compelling evidence of the use of star-compass information by nocturnal migratory birds.

Emlen also demonstrated that indigo buntings learn to read the star compass in a **sensitive period** between the time when the birds leave the nest and the beginning of the fall migration. If they have not seen the nocturnal sky during the time before their first migration, they are unable to use the star compass—even if they are exposed to starry skies thereafter.

Magnetic compasses and maps

Although the demonstrations of the existence of a sun and star compass can be considered major discoveries, these mechanisms cannot, by far, explain all observations. For example, these two mechanisms can be used to determine direction of orientation (**compass function**), but there is no evidence that they enable the animal to determine its position relative to its goal. The latter task can be achieved only by mechanisms incorporating a **map function**. Theoretical consideration predicted that this requires a system in which two physical parameters are combined to a bicoordinate navigation system. Another observation that cannot sufficiently be explained by the existence of a sun and star compass is the fact that frequently migratory birds maintain correct orientation, even if the sky is completely obscured by clouds. Such observations, as well as the proposal of a bicoordinate map mechanism, sparked early speculations that birds may use **geomagnetic cues** for fulfilling these tasks.

The first indication that sensory and central structures for the detection of geomagnetic cues may, indeed, exist came from experiments on European robins (*Erithacus rubecula*), a passerine species that migrates at night. Captive individuals of this species become restless at the times of the year when their free-living conspecifics migrate. Then, they prefer to stay at that side of the cage that points toward their migratory direction. To record this activity, the birds were kept in an octagonal cage with radial perches (Fig. 9.12). In this cage, the individuals could move freely around the central structure, which provided in its inner part food and shelter. Each hop is recorded by one of the radial perches through an electromechanical system. The 'hop scores' of one night are, then, processed vectorially to determine the mean direction of the restlessness shown by the bird.

In the first set of experiments it was confirmed that the migratory restlessness of European robins exhibited the correct directionality, even if no visual cues were available. Evidence that the birds use the earth's magnetic field for orientation was obtained by Wolfgang Wiltschko from the University of Frankfurt, Germany, in the late 1960s. When the magnetic north was artificially rotated by Helmholtz coils arranged

Geomagnetic cues could, in theory, be used by magnetoreceptive animals to establish a bicoordinate map.

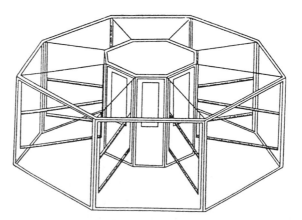

Figure 9.12 Schematic drawing of the octagonal cage used for testing migratory restlessness in European robins. The birds can move freely around the central structure, whose inner part they can enter through holes, to obtain food. Their activity is electromechanically recorded when they hop in the peripheral part of the cage onto one of the radial perches. (After **Merkel, F. W., and Fromme, H. G. (1958).**)

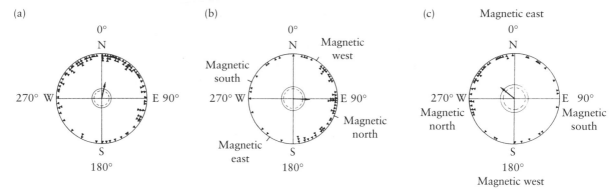

Figure 9.13 Orientation of European robins in various magnetic fields during the migratory spring season. Each triangle indicates the mean bearings of a single experimental night. The arrows represent the average vector of all individual responses within one test group. When the normal earth magnetic field with the magnetic north at 0° (a) is altered so that magnetic north is located at 120° (b) or 270° (c), the birds change the predominant direction of their activity accordingly. (After **Wiltschko, W., and Wiltschko, R. (1996).**)

around the octagonal test cage, the birds altered their directional preference accordingly (Fig. 9.13).

Analysis showed that the avian magnetic compass is notably different from the technical magnetic compass used by humans. To appreciate this difference, it is necessary to have a closer look at the **earth's magnetic field**. At first approximation, the earth's geomagnetic field can be described as a **dipole field** whose poles lie near the geographical poles. The field lines leave the ground at the southern magnetic pole and, then, curve around the earth to re-enter its surface at the northern pole (Fig. 9.14). In other words, the magnetic field vector points perpendicularly away from ground at the southern pole, perpendicularly into ground at the northern pole, and is parallel to the earth's surface at the magnetic equator. At locations different from the poles and the equator, the angle between the magnetic vector and the horizon varies between these two extremes. This angle is called **inclination** or **dip**.

Several experiments have shown that, in contrast to the technical compass of humans, which points to magnetic 'north' and 'south' (therefore called **polarity compass**), the magnetic compass of birds is an **inclination compass**. It defines 'poleward' as the direction along the earth's surface where the angle between the magnetic field vector and the gravity vector becomes minimal. By contrast, 'equatorward' is the direction along which the angle formed between the magnetic field vector and the gravity vector becomes maximal. However, the inclination compass ignores the polarity of the magnetic field. This property has an important advantage over a polarity compass. Since the polarity of the

In contrast to the technical compass, the magnetic compass of birds is an inclination compass.

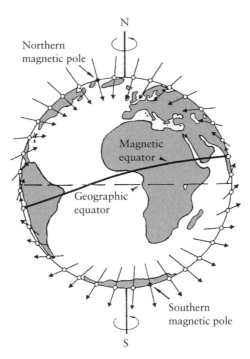

Figure 9.14 Schematic view of the geomagnetic field of the earth. In the southern hemisphere, the magnetic field lines leave the earth to re-enter it in the northern hemisphere. The arrows are a vector representation of the magnetic field lines. Their lengths indicate the intensities, their orientation the direction of the field lines. (After **Wiltschko, W., and Wiltschko, R.** (1996).)

earth's magnetic field has reversed repeatedly in the past, a polarity compass might have led to devastating misguidance of the birds over evolutionary times.

The finding of Wolfgang Wiltschko that the restlessness behavior of migratory birds can be altered by artificial magnetic fields inspired others to conduct similar experiments in homing pigeons. After failed attempts by several scientists, William Keeton from Cornell University in Ithaca, New York, reported in 1971 that bar magnets glued to the backs of pigeons resulted in disorientation, as long as either inexperienced birds were used, or experimental birds were released under total overcast, that is, when the sun compass was not available. These findings also explained why so many previous studies had failed to produce positive results. As long as the sun is visible, the birds appear to prefer using the sun compass. Only when the sun compass cannot be used at all, pigeons rely entirely upon magnetic cues.

The results of Keeton were further substantiated by Charles Walcott, then at the State University of New York at Stony Brook, now at Cornell University. Walcott attached battery-operated coils around the head of

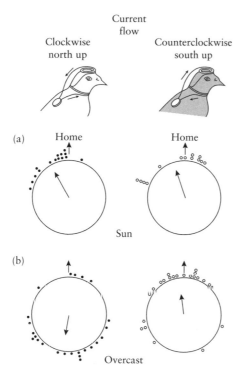

Current
flow

Clockwise
north up

Counterclockwise
south up

(a) Home

Home

Sun

(b)

Overcast

Figure 9.15 Orientation of homing pigeons wearing Helmholtz coils in the sun (a) and under overcast (b). In one group of pigeons, the current flowed clockwise, resulting in a north-up orientation of the magnetic field (closed circles, left column). In the second group of pigeons, the current flew counterclockwise, resulting in a south-up orientation of the magnetic field (open circles, right column). The circles in the periphery indicate the vanishing bearings of individual birds. The arrows in the center represent the average vector of the oriented response. Homeward direction is upward. The results demonstrate that inversion of the vertical component of the natural magnetic field alters the orientation behavior of the pigeons under overcast, but not when the sun is visible. (After **Walcott, C., and Green, R. P.** (1974).)

pigeons. Current flowing through the coils changed the magnetic field around the bird's head. Similar to Keeton's finding, the coils had little effect on the behavior of the pigeons when the sun was visible. Under overcast, the effect depended upon the direction of the current. When the magnetic north of the induced magnetic field pointed upward ('north-up arrangement'), the test birds showed a tendency to fly in the opposite direction to their home loft (Fig. 9.15). Under this experimental conditions, the resultant magnetic field shows roughly an inversion of the vertical component of the natural field. Correspondingly, if the resultant magnetic field pointed downward ('south-up arrangement'), the pigeons tested flew in the correct direction toward their home loft.

Evidence that magnetic cues may not only be involved in determining direction, but also in gathering information for navigation, was obtained

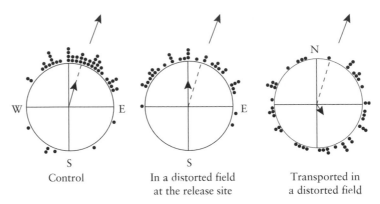

Figure 9.16 Effect upon orientation of young, inexperienced pigeons following exposure to distorted magnetic fields. The closed circles in the periphery indicate the vanishing bearings of individual birds. The arrows in the center represent the average vector of the oriented response. Open arrows point in the direction of the home loft. While pigeons that were subjected to a distorted field only at the release site show a similar response as pigeons not exposed to any artificial magnetic field, the orientation of pigeons transported in a distorted field is severely impeded. (After **Wiltschko, R., and Wiltschko, W.** (1978).)

by accident. To transport their pigeons to the release sites, Wolfgang and Roswitha Wiltschko used a Volkswagen (VW) squareback. Sometimes, they observed that the orientation of the pigeons was disturbed. This was regularly the case when the pigeons' crate was placed on top of the engine. As it turned out, it was the generator attached that produced a magnetic field and, thus, caused this 'VW effect'.

This finding was confirmed in subsequent, systematically conducted experiments. When young, inexperienced pigeons were transported to the release site in a distorted magnetic field, they were disoriented. If, however, a second group of inexperienced pigeons were exposed to a distorted magnetic field for an equal amount of time at the release site, their homing behavior did not appear to be affected (Fig. 9.16). These findings show that it is not the exposure to a distorted magnetic field itself which causes disorientation, but **transport in the distorted field**. It appears as if the pigeons store information in the course of their outward journey. In the simplest case, they may integrate the different directions recorded by their magnetic compass over time during the outbound journey and define the reversed route as the correct way back home.

It is interesting that older and more experienced pigeons are hardly affected by magnetic distortion during their outward journey. Apparently, young inexperienced pigeons and older experienced pigeons use different navigational strategies.

The large body of behavioral evidence has led scientists to propose the existence of specific **magnetoreceptors** in migratory birds. The main questions arising from this proposal are as follows:

- What is the structure of the proposed magnetoreceptors?
- Where are they located?
- How do they function?
- How do they interact with other sensory and central systems?

Although none of these questions has been satisfactorily answered to date, significant progress has been made toward a neurobiological understanding of this phenomenon over the last few years. This is especially due to the work done by Robert Beason from the State University of New York at Geneseo, now at the National Wildlife Research Center of the Animal and Plant Health Inspection Service/United States Department of Agriculture at Sandusky, Ohio.

In his studies, Beason has used bobolinks (*Dolichonyx oryzivorus*), a bird that breeds in the northern part of the USA and in southern Canada, across North America. In the fall, bobolinks migrate, primarily during the night, to their winter ranges in southern Brazil and northern Argentina. Due to the enormous distance between the summer ranges and the winter ranges, bobolinks have the longest migratory pathway of any New World landbird.

A number of experiments have demonstrated that, for orientation during their journey, bobolinks use both **visual cues originating from the stellar constellation** and **magnetic cues derived from the earth's magnetic field**. These experiments have further indicated that two mechanisms exist for detection of magnetic fields. One appears to be associated with photoreceptors, is moderately sensitive to changes in the earth's magnetic field, and may feed information into an inclination-compass system. The second mechanism is photoreceptor independent, shows high sensitivity to alterations in the earth's magnetic field, and could possibly be used to extract information to generate the proposed bicoordinate magnetic map (see Box 9.1).

While little is known about the nature of the photoreceptor-associated magnetoreceptors in bobolinks, several lines of evidence have suggested that iron deposits, probably **magnetite**, in the area of the upper beak, are involved in transducing information from the earth's magnetic field. This region appears to be innervated by the **ophthalmic branch of the trigeminal nerve**. Electrophysiological recordings from the ophthalmic nerve and the trigeminal ganglion have revealed the presence of units that are highly sensitive to small changes in the magnetic field. The most sensitive units respond to changes as small as 200 Nanotesla (nT), which corresponds to less than 0.5% of the earth's total field. Most commonly, these units increase the rate of firing in response to alterations in the surrounding magnetic field (Fig. 9.17).

Magnetoreceptors in birds appear to fall into two categories: ones associated with photoreceptors, others being photoreceptor independent.

BOX 9.1   Mechanisms of magnetoreception

A wealth of behavioral data have provided compelling evidence for the existence of a **magnetic sense** in a variety of animals, including vertebrate and invertebrate species. Much less is known about the structure and function of the proposed **magnetoreceptors**, and many of the corresponding theoretical considerations are still awaiting empirical proof. Three major hypotheses on how animals could detect the earth's magnetic field have been brought forward.

The first mechanism, proposed by Adrianus Kalmjin from the Scripps Institution of Oceanography of the University of California, San Diego, involves electromagnetic induction. It may be used by elasmobranchs, a systematic subclass comprised of sharks and rays. These predominantly marine fishes possess electroreceptive organs known as ampullae of Lorenzini. When the fish, surrounded by seawater—which acts as a highly conductive medium—move in any direction (except parallel) to the field lines defining the earth's magnetic field, current flow is induced (**electromagnetic induction**). This physical phenomenon is observed whenever a closed conductive circuit is moved within a magnetic field. It has been hypothesized that the highly sensitive electroreceptors within the ampullae of Lorenzini can detect the tiny alterations in voltage associated with the induced current.

The second mechanism possibly used for detection of magnetic fields has been proposed by Klaus Schulten and his group from the Beckman Institute of the University of Illinois at Urbana-Champaign. His hypothesis can be subsumed under the term **chemical magnetoreception**. This mechanism is based on the influence of external magnetic fields on the spin orientation of an electron during its transfer from an excited donor molecule to an acceptor molecule. The spin is caused by the electron rotating about its own axis, which results in a weak magnetic field. Electron-transfer processes leave both the donor molecule and the acceptor molecule with an unpaired electron. The two electrons can have either parallel spins or opposite spins. When after a brief period the transferred electron returns to the donor molecule, the new relationship of the two spins may, under the influence of external magnetic fields, differ from the one before the transfer. This may be the case even if these fields are as weak as those of the earth's magnetic field. However, a reversal of the original spin relationship occurs only if a number of requirements are met. This is frequently the case in electron-transfer processes induced by photo-excitation, that is, by the absorption of light. Alterations in spin orientation affect the chemical properties of the donor and the acceptor molecules, and thus those of subsequent reactions involving these two molecules.

While the proposal of the existence of chemical magnetoreception in animals is mainly based on theoretical considerations, several studies have provided empirical evidence for a link between magnetoreception and photoreception. This includes electrophysiological responses to magnetic fields found in parts of the central nervous system that receive input from the visual system. Moreover, the response of some of the units from which recordings were taken has been shown to be modulated, through illumination of the associated photoreceptors, with light of different wavelengths.

The third mechanism proposed to mediate the reception of magnetic fields is based on **magnetite** (Fe_3O_4). Attempts to isolate crystals of this mineral from tissue were successful in various animals. Such experiments were sparked after Richard Blakemore, then a graduate student at the University of Massachusetts at Amherst and working at the Woods Hole Oceanographic Institution, reported in 1975 that magnetotactic bacteria, which are able to swim along magnetic field lines, contain magnetite. The tiny crystals of this molecule act like permanently magnetized bar magnets. In animals, it is thought that geomagnetic field information is transduced to the nervous system when the magnetite crystals tries to twist into alignment with the earth's magnetic field, thereby exerting pressure or torque on (secondary) mechanoreceptors. An alternative possibility to achieving transduction is that the mechanical force originating from the movement of the magnetite crystal opens ion channels directly.

The strongest evidence to date for the existence of a **magnetite-based magnetoreception** mechanism has been obtained in bobolinks and trouts (see text). In the trout, magnetite appears to be contained in the olfactory lamellae. This region within the nose is innervated by a branch of the fifth cranial (the trigeminal) nerve. Interestingly, in the bobolink, a passerine transequatorial migrant, the region that contains magnetic material (presumably magnetite) is also innervated by a branch of the trigeminal nerve (see text). This similarity in the innervation pattern may point to a common mechanism shared by animals as diverse as fish and birds.

It is possible that different sets of magnetoreceptors exist in the same animal. They may subserve different functions. It has been hypothesized that magnetoreceptors associated with the visual system could provide directional information (**compass sense**). This information is, however, insufficient for an animal to find its destination, as long as it does not know its position relative to its destination. Positional information could, in principle, be inferred from local properties of the earth's magnetic field, as the field intensity and the inclination of the field lines vary in a systematic way across the earth's surface. Such information could be extracted from the local geomagnetic parameters by magnetoreceptors based on magnetite (**map sense**).

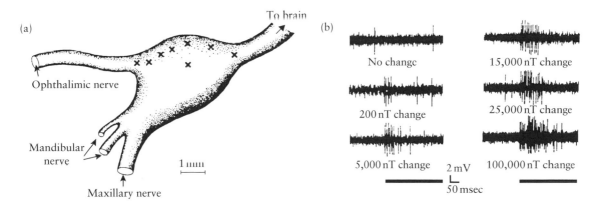

Figure 9.17 Electrophysiological recordings from the trigeminal ganglion of the bobolink during alterations in the ambient magnetic field. (a) Schematic structure of the trigeminal ganglion with the ophthalmic branch. The crosses indicate the locations from which recordings were taken. (b) Responses from one unit in the ganglion during changes in vertical magnetic-field intensity. The duration of the stimulus is indicated by the horizontal bars. The upper trace in the left column shows the spontaneous activity of the unit. The other traces show the responses of the same unit to changes in the magnetic field as indicated. (After **Semm, P., and Beason, R. C.** (1990).)

Additional support for a magnetite-based magnetoreception system has come from magnetization experiments. If bobolinks were subjected to brief magnetic pulses of high intensity and different polarities, they were oriented in directions different from one another and from controls. It is thought that exposure of the birds to these magnetic pulses alters or destroys the magnetite-based magnetoreceptors, and thus leads to the disorientation observed.

Olfactory navigation in homing pigeons

In 1971, the Italian scientist Floriano Papi and his group of the University of Pisa first proposed what has become known as the **olfactory hypothesis**. The essential idea of this hypothesis is best summarized in Papi's own words: "Pigeons acquire their homing ability at the loft by learning to recognize the odor of the loft area, as well as foreign odors carried by the winds. Associating these different 'foreign' odors with the direction from which they come, pigeons gain information about odors prevailing in the surrounding areas and build up their 'olfactory map'. This olfactory experience at the loft enables pigeons to use odors perceived during the outward journey and over the release area to establish the home direction. The sun or the magnetic compass will be employed to select the bearings deduced thereby."

Initial experimental evidence for this proposal came from experiments in which pigeons were subjected to olfactory deprivation. This was achieved by various methods, including sectioning of the olfactory nerve, plugging of the nostrils with cotton, insertion of tubes into the nasal passages, and application of local anesthetics to the nasal epithelium. These treatments resulted in a poor homing performance, as well as in rather random vanishing bearings. However, such negative effects were observed only when the pigeons were released at unfamiliar sites, but not at sites of which they had learned certain features through prior experience.

While Papi and his group interpreted the behavior of anosmic pigeons released at unfamiliar sites as evidence for the hypothesis that olfactory cues are necessary for navigation, others (including William Keeton) raised doubts about the **specificity** of the approaches employed. According to Keeton's interpretation, it may also be possible that the treatments affected the birds' motivation to fly or their ability to process non-olfactory navigational information.

Strong support for the olfactory hypothesis came from experiments in which the direction of the natural air flow was deflected within the loft. Pigeons that were, for example, exposed to wind blowing in direction opposite to the natural one showed, on average, vanishing bearings directly away from home, whereas controls flew toward home when released at the same site.

Despite positive results, as the ones produced by the wind reversal experiments, the olfactory hypothesis has remained controversial. This is due to several problems that, regardless of considerable effort to resolve, have remained elusive. For example, it is unknown how gradients of odors, which are predicted by the olfactory hypothesis to exist, are maintained over hundreds of kilometers in the atmosphere. Further, nothing is known about the exact chemical nature of the proposed olfactory cues in the atmosphere. Moreover, there are some indications

Despite many experiments, the olfactory hypothesis of pigeon homing remains controversial.

that the answer to the question whether or not a pigeon uses olfactory cues for navigation crucially depends on the environment to which the bird is exposed during the first few months of its life. Wolfgang Wiltschko and his group, for example, obtained evidence that pigeons reared in a loft sheltered from winds were largely unaffected by anosmia and were homeward-oriented, even when released at unfamiliar sites. Thus, birds raised under such conditions may use cues different from olfactory ones for navigation. If this is the case, it will add considerable complexity to the phenomenon of homing in pigeons.

Homing in Salmon

Salmon are famous for their ability to return to their natal stream for reproduction after lengthy foraging migrations in the sea. A large number of extensive studies have suggested that the precision with which salmon find their way home is achieved by using different navigational mechanisms during the journey in the ocean and in the in-river migration. These mechanisms will be discussed in detail below.

Life cycle of Salmon

One of the best-studied salmonid species is the coho salmon (*Oncorhynchus kisutch*). Its life cycle is summarized in Fig. 9.18. Coho salmon spawn in late October through December in fast-flowing streams along the Pacific coast of North America. The female digs nests (the so-called **redds**) by lying on her side and rapidly beating with her tail, until a nest pocket results. In each of the several nests, the female lays 500–1000 eggs, which are fertilized by the male. Using her mouth and tail, the female covers the eggs with gravel from the bed of the river. In this phase, which lasts a few days, the female guards the nests. Both partners, by then exhausted, drift down the river and die.

During the time that follows, the gravel provides protection for the eggs. One to two months after egg deposition, the larvae hatch. The hatchlings are called **alevins**. They remain buried in the gravel for another four months. During that time, they are nourished by the yolk sac.

After the yolk sac is absorbed, the alevins emerge from the gravel. However, only approximately 1% of the coho salmon survive past the larval stage. The survivors, called **parrs**, establish territories and feed on plankton and small insects. The young salmon remain in the natal stream until they are 18 months old. During that time, they reach a total length of approximately 10 cm.

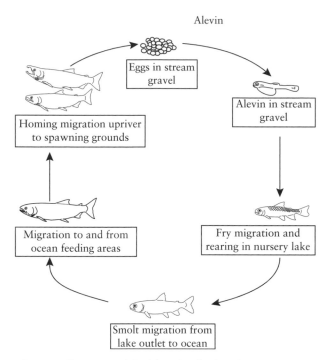

Figure 9.18 Schematic illustration of the life cycle of coho salmon (*Oncorhynchus kisutch*), indicating the main developmental processes. (Courtesy: G.K.H. Zupanc, based on **Childerhose, R. J., and Trim, M.** (1979) and **Quinn, T. P., and Dittman, A. H.** (1990).)

At the end of this phase, the young salmon (now referred to as **smolts**) give up their territories, form large schools consisting of thousands of individuals, and swim down the river toward the sea. Concurrently, they undergo a major metamorphosis, most commonly called **smolt transformation**. This transformation is induced by thyroid hormones and encompasses, among other features, disappearance of the parr marks and change of their coloration into silver, as well as an adjustment of their osmoregulation to life in seawater.

In the sea, the salmon grow rapidly. After only 18 months in the ocean, they reach a length of 1 m and weigh 5 kg. During the **sea phase**, they may travel over enormous distances. Some salmon have been found in the sea more than 5000 km away from the mouth of their home river. To illustrate the size of such feeding ranges, Fig. 9.19 shows the marine distribution of another well-examined salmonid species, the sockey salmon (*Oncorhynchus nerka*).

The spawning migration begins in midsummer of the third year and involves an extensive journey from the open water into shore areas near the home-river system. Upon arrival near the coast, and for the next

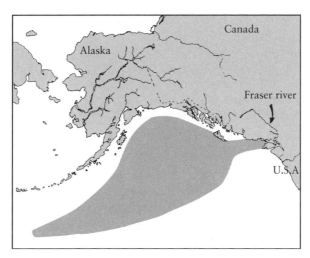

Figure 9.19 Marine distribution of sockey salmon (*Oncorhynchus nerka*) population originating from the Fraser River in British Columbia, Canada. The distribution has been inferred from results of tagging experiments and analysis of scale markings, which differ among fish from different watersheds. (After **Quinn, T. P., and Dittman, A. H. (1990)**.)

month or so, both re-adaptation of the salmon's osmoregulation to freshwater and full sexual maturation take place. While in early June the gonadal mass is still very small, typically comprising only 1–2% of the total body weight, by the end of August the gonads have grown tremendously, so that they contribute roughly 50% to the fish's total body weight. At that time, the fish also stop feeding and mobilize fat reserves.

As the fish migrate upstream, they develop secondary sexual characteristics, such as brightly-colored sides and hooked upper jaws. Moreover, their gastro-intestinal tract is now completely absorbed and the entire body cavity, except for heart and kidney, is filled with ripe eggs and testes. After spawning, the adults die and leave it up to the offspring to repeat the cycle.

The above life cycle is characteristic of coho salmon. In other salmonid fishes, the pattern of migration and reproduction may vary considerably. In some salmonids, for example the Atlantic salmon (*Salmo salar*), the males often spawn before going to sea, but then, after a phase of significant growth, return to the home river to reproduce once more. In masu salmon (*Oncorhynchus masou*), males remain throughout their life in freshwater, where they spawn. Only the females show the characteristic **anadromous** pattern, that is, the migration from seawater to freshwater for the purpose of reproduction. Again other species spend most of the time of their sea phase in coastal waters near the mouth of their natal river, rather than migrating thousands of kilometers to oceanic feeding grounds. Moreover, salmon can even complete their life cycle without ever going to sea. This was artificially forced in a population of coho

Salmon are anadromous migrants: born in freshwater, they grow up to adulthood in the sea and return to freshwater to reproduce.

salmon by introducing them into the Great Lakes in the late 1960s. These salmon have accepted the freshwater lakes as a substitute for marine environments during their adult life. However, similar as their Pacific conspecifics, they return to their natal stream to spawn.

Precision of homing

The precision of the salmon's homing ability has been studied by capturing the fish in their natural tributary before the seaward migration. The salmon are then marked with fin chips, external tags, or magnetic wire implants, and released back into the river.

The results of a large number of such studies are in remarkable agreement. They indicate that, because of high mortality in the ocean, only about 0.5–5% of the original downstream migrants survive to spawn. Of these survivors, roughly 95% return to their natal streams. The remaining 5% or so stray and spawn in non-natal rivers. It is thought that straying ensures colonization of new habitats. Straying can also occur as an avoidance response to degradation in water quality, as was observed when volcanic ash contaminated streams following the eruption of Mount St. Helens in the state of Washington, in 1980.

Transplantation experiments

To examine whether the salmon's homing is based on inherited or learned components, a series of transplantation experiments have been performed. In these studies, the fish were transplanted from their natal tributary to a different river. The results of these experiments can be summarized as follows:

- salmon transplanted before undergoing smolt transformation return to the river of release;
- salmon transplanted after smolt transformation return to their natal tributary;
- the apparent learning of certain features of the home river takes place rapidly and irreversibly and, thus, resembles the process of **filial imprinting** known especially from birds (see below).

The olfactory imprinting hypothesis

What are the cues of the home river that salmon become imprinted to? A breakthrough in answering this question was made in 1951 when Arthur Davis Hasler, a pioneer in the research on homing of salmon (see Box 9.2), and his student Warren Wisby of the University of Wisconsin in Madison presented their **olfactory imprinting hypothesis**. The core element of this

BOX 9.2   Arthur Davis Hasler

Arthur Davis Hasler (right) and his student Peter Hirsch (left), performing a physiological experiment on a salmon. (Courtesy: F. Albert.)

Arthur Davis Hasler, albeit best known for his research on the homing of salmon, has made major scientific contributions in a variety of biological disciplines. Hasler was born in Lehi, Utah, in 1908. After completing his undergraduate degree from Brigham Young University, he attended graduate school at the University of Wisconsin, Madison. Although determined to become a limnologist, he decided to first obtain training in a more rigorous subject, physiology, since limnology at that time was largely confined to descriptive studies. In his thesis, he investigated the physiology of digestive enzymes of copepods and cladocerans. In 1937, he received his

Ph.D., and in the same year he joined the faculty of his *alma mater*, as an instructor. By 1948, he had reached the rank of full professor. During his tenure, he established the Laboratory of Limnology in Madison, of which he was the director until his retirement in 1978.

Based on his own roots—his grandfather had immigrated to the USA from Switzerland—Hasler maintained close ties to Europe, especially to scientists in Germany. Among them, he was mostly inspired by the work of Karl von Frisch, whom he met for the first time in 1945, immediately after the end of the Second World War, in his capacity as an officer of the United States Air Force Strategic Bombing Survey, in Germany. These interactions led to the establishment of the study of sensory physiology of fish as one major area of his own research. The second area where he gained international reputation was limnology. He was instrumental in the shift from the classical, descriptive approach toward a new experimental orientation, in which the rigor of the experiments of the laboratory was applied to research in the field. However, most of his effort was devoted to the elucidation of the mechanisms underlying homing in salmon (see text). Besides his academic achievements—from his school more than one hundred doctoral and master's degree students graduated—Hasler also became widely known for his initiatives in conservation ecology, and international peace programs.

Arthur Davis Hasler died in 2001 at the age of 93.

hypothesis is the proposal that, in a sensitive period during the smolt phase, salmon become imprinted to the odor of their home river and use, during adulthood, this characteristic odor to find the way back home.

In formulating this hypothesis, Hasler was inspired by the work of two Austrian scientists working in Germany, Karl von Frisch and Konrad Lorenz. Von Frisch had discovered that the skin of schooling minnows, if injured by a predator, releases tiny amounts of a specific chemical substance called *Schreckstoff* (alarm substance). This substance is contained in alarm cells in the skin of the fish (Fig. 9.20). Release of *Schreckstoff*, which occurs only if the skin is broken, causes the other members of the school to disperse and hide. Since this substance is present in the water after the injury only in minute amounts, von Frisch's finding suggests that fish must be able to smell with high sensitivity.

Olfactory imprinting hypothesis: salmon become imprinted to the odor of their home river during the smolt phase; after having reached sexual maturity, they use this information to find the way back home.

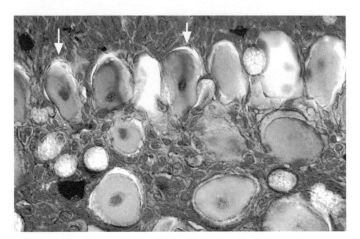

Figure 9.20 Cell containing *Schreckstoff* in the skin of tench (*Tinca tinca*). Two club-like cells are indicated by arrows. This alarm substance is released only if the skin is broken after injury, for example, caused by the bite of a predator. (Courtesy: G.K.H. Zupanc and H. Altner.)

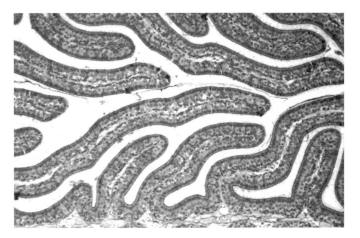

Figure 9.21 Section through the nasal cavity of a black-mouthed dogfish (*Galeus melanostomus*). Accommodation of the numerous olfactory cells is achieved by an enormous folding of the olfactory epithelium. (Courtesy: G.K.H. Zupanc and H. Altner.)

This enormous sensitivity was subsequently confirmed by the German zoologist Harald Teichmann. He showed that European eels (*Anguilla anguilla*) can detect β-phenylethyl-alcohol in concentrations as low as 1 part of the alcohol in $3 \cdot 10^{18}$ parts of water! This corresponds to only three molecules of β-phenylethyl-alcohol in the nostril of an eel. The enormous sensitivity is partly achieved by great folding of the olfactory epithelium, which leads to the accommodation of a large number of olfactory cells in the nasal cavity of the fish (Fig. 9.21).

The second prominent scientist who had a profound impact on the work of Arthur Hasler and his group was Konrad Lorenz. By investigating geese, he found a process of rapid and irreversible learning during a critical period of development. Lorenz called this learning process **imprinting**. In geese, the critical period occurs shortly after birth. The gosling forms then a permanent attachment to the first moving object seen—normally its mother, but, in an experimental situation, it can be almost any object, including the human observer.

Hasler's merit was to have the idea that imprinting, usually involving visual or acoustic sensory channels, could also take place by making use of olfactory cues. A prerequisite of this phenomenon is that

(1) every stream has a characteristic and persistent odor that the fish can perceive;

(2) the salmon are able to discriminate between the odors of different streams; and

(3) the salmon remember the typical odor of the home stream when they return to freshwater after the sea phase.

Laboratory experiments

As a first step to verify the olfactory imprinting hypothesis, and by using a reward/punishment paradigm, Hasler and Wisby trained bluntnose minnows and coho salmons to discriminate between waters collected from two streams. These **conditioning experiments** showed that the fish could distinguish between different water samples, even if the water was collected in different seasons. The latter is an important point, as it suggests that the factor detected by the fish is long-lasting, and remains present throughout the year. On the other hand, when the nasal sacs were cauterized, the fish were no longer able to discriminate between different water samples. Also, when the organic fraction of the water was removed, the fish failed to discriminate different waters.

Electrophysiological studies have, in general, confirmed the conclusion that salmon can distinguish their home water from other waters. In these experiments, migrating adult salmon are captured after arriving in their natal tributary. **Electroencephalograms** are taken by inserting a recording electrode into the olfactory bulb. Then, the nasal sacs are flushed with water from the stream in which the fish are captured and with water from other streams. Typically, high-intensity electroencephalograms are elicited by presentation of home-stream water, but not by water from other streams.

Imprinting to artificial substances

Strong evidence that supports the hypothesis formulated by Hasler and Wisby comes from experiments with artificially imprinted salmon. In the

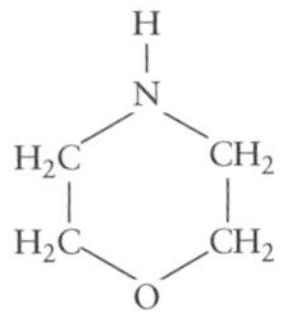

Morpholine

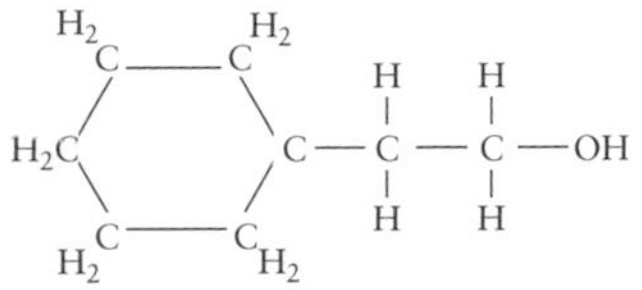

β–Phenylethyl-alcohol

Figure 9.22 Chemical formulae of morpholine and β-phenylethyl-alcohol, two chemical substances widely used for artificial olfactory imprinting of salmon.

Strong evidence supporting the olfactory hypothesis is provided by experiments in which salmon are imprinted to artificial substances.

first set of such experiments, Hasler's group used morpholine (Fig. 9.22). This heterocyclic amine is not found in natural waters and acts neither as a deterrent nor as an attractant. Furthermore, it is cheap, highly stable in the natural environment, extremely soluble in water, and can be detected by coho salmon even at very low concentrations.

For the experiments, salmon from the population artificially introduced to Lake Michigan were used. Smolts of identical genetic stocks, hatched and raised under uniform conditions, were divided into two groups. One group was exposed, for 30 days, to water piped from Lake Michigan containing morpholine at a very low, but perceptible, concentration. The second group was also kept in Lake Michigan water for the same length of time, but no morpholine was added to this water. This latter group served as a control. Neither of the two groups had been exposed to water from any tributary of Lake Michigan during their early life history. The fish from each group were marked with different fin clips and released directly into Lake Michigan.

Eighteen months later, during the fall spawning season, an artificial homing stream was created by continuous addition of morpholine into Oak Creek, one of the tributaries of Lake Michigan. Indeed, a large number of fish imprinted to morpholine entered Oak Creek as adults, whereas the number of control fish captured in the same river was very low (Table 9.1).

In an additional control experiment, smolts were exposed to morpholine, and an equal number was left unexposed. In contrast to the previous

Table 9.1 Census record of coho salmon at Oak Creek

Experiment (census year)	Treatment and number of salmon released	Morpholine added to Oak Creek?	Number of salmon recovered?	Percentage of fish stock
1 (1972)	Imprinted: 8000	Yes	218	2.58
	Not imprinted: 8000		28	0.35
2 (1973)	Imprinted: 5000	Yes	437	8.74
	Not imprinted: 5000		49	0.95
3 (1973)	Imprinted: 5000	Yes	439	8.78
	Not imprinted: 5000		55	1.10
4 (1973)	Imprinted: 8000	Yes	647	7.89
	Not imprinted: 10 000		65	0.65
5 (1974)	Imprinted: 5000	No	51	1.00
	Not imprinted: 5000		55	1.10

(After **Hasler, A. D.**, and **Scholz, A. T.** (1983).)

series of experiments, no morpholine was added to Oak Creek or any other tributary of Lake Michigan during the fall of the following year. This time, only relatively few salmon were captured in Oak Creek, and the number was very similar in the two groups (Table 9.1). This underlines that it is morpholine, and not any other feature of Oak Creek, what attracts the imprinted salmon during adulthood.

In a second, more refined artificial imprinting experiment, coho smolts were divided into three equally large groups. The first group held at a hatchery was exposed to morpholine, the second to β-phenylethyl-alcohol, and the third was left unexposed. Fish of all three groups were released into Lake Michigan, midway between the mouths of two test streams, the little Manitowoc River and Two Rivers, located 9.4 km apart.

During the following year, morpholine was metered into one of the two test streams, and β-phenylethyl-alcohol into the other. Both streams and additional 17 other locations were surveyed for marked fish. This experiment was conducted twice, with a total of 45 000 fish used. The results are convincing: of the morpholine-exposed fish recovered, 95% were captured in the morpholine-scented stream; of the β-phenylethyl-alcohol exposed fish recovered, 92.5% were captured in the β-phenylethyl-alcohol-scented stream. By contrast, the number of untreated (control) fish was much lower than the number of correctly homing salmon in each of the two test streams.

Ultrasonic tracking

To monitor the behavior of morpholine-imprinted salmon, when during adulthood they encounter the morpholine scent of the artificial home river, the following experiment was conducted. Adult homing salmon, which had been exposed to morpholine as smolts, were captured in the course of the experiments described above. Before releasing them, an **ultrasonic transmitter** was inserted down the esophagus into their stomach. Then, each individual was tracked with a receiver on a boat that followed the signal emitted by the fish. Morpholine was introduced into an area located several hundred meters south of the release point. This area was located at the mouth of a small stream flowing into Lake Michigan. Through the water flow of the river, an 'odor barrier', extending from the mouth of the river to about 100 m offshore, was created. Thus, fish following the shoreline in south direction had to swim through this area.

When morpholine was present in the test area, morpholine-imprinted fish stopped their migration and remained in the area for roughly the time it took the water currents to dissipate the chemical (Fig. 9.23(a)). When morpholine was not introduced into the test area, morpholine-treated fish swam through without stopping (Fig. 9.23(b)). A similar behavior was

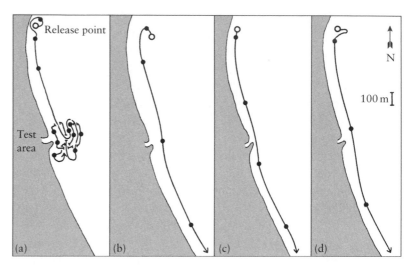

Figure 9.23 Tracking of movements of coho salmon in a test area when encountering the scent of morpholine previously used for imprinting. The movements of the salmon were tracked by ultrasonic transmitters inserted into their stomach. Recordings were taken in 15-min intervals. (a) Response of an imprinted salmon when morpholine was present in the test area. (b) Response of an imprinted salmon when morpholine was absent. (c) Response of an imprinted salmon when a chemical, other than morpholine, was present. (d) Response of a non-imprinted salmon when morpholine was present. (After **Hasler, A. D., and Scholz, A. T.** (1983).)

observed when a chemical other than morpholine was added to the water (Fig. 9.23(c)), or when non-imprinted salmon were tested in the morpholine-scented area (Fig. 9.23(d)).

Hormonal regulation of olfactory imprinting

What neuronal and endocrine factors mediate olfactory imprinting in salmon? Measurements of **thyroid hormones** in the blood serum by radioimmunoassay have shown that triiodothyronine and thyroxine (Fig. 9.24) increase roughly 5–10-fold at the beginning of the smolt transformation compared to the level during the pre-smolt and post-smolt stages. This, as well as other observations, suggests that the level of thyroid hormones may play a role in the process of olfactory imprinting.

This hypothesis has been confirmed by Allan Scholz, a student and co-worker of Arthur Hasler. Scholz artificially elevated the levels of triiodothyronine and thyroxine of pre-smolt coho salmon by injection of thyroid-stimulating hormone (TSH). Then, the fish were injected with gonadotropic hormone for 12 weeks to bring them into migratory disposition and to mimic the physiological state of naturally spawning salmon. The behavioral tests were conducted in the Ahnapee River at

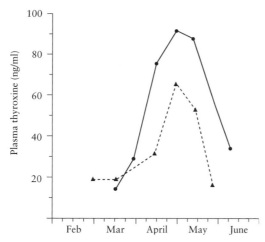

Figure 9.24 Plasma concentrations of thyroxine of coho salmon from two hatcheries (indicated by closed circles/solid lines and open triangles/dashed lines, respectively) during smolt transformation. Each symbol represents the mean of 10 samples. The brackets indicate the standard error. The thyroxine surge occurs just prior to smolt transformation. (After **Dickhoff, W. W., Folmar, L. C., and Gorbman, A.** (1978).)

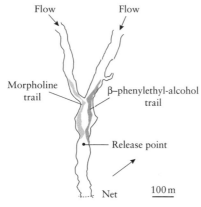

Figure 9.25 Study site on the Ahnapee River at Forestville, Wisconsin, used to test the response of salmon to imprinting substances. The experimental fish were released 150 m below the junction of the two tributaries. Addition of morpholine and β-phenylethyl-alcohol into either tributary produced odor trails traveling along opposite shores for some distance below the confluence. (After **Hasler, A. D., and Scholz, A. T.** (1983).)

Forestville, Wisconsin. The fish were released 150 m before the junction of two tributaries (Fig. 9.25). The water of the two tributaries arises from the same reservoir more upstream. The two tributaries are, thus, very similar in terms of their natural olfactory composition. This arrangement allowed Scholz to introduce morpholine or β-phenylethyl-alcohol into either arm of the river, while keeping all other factors constant.

The outcome of the experiments demonstrated that pre-smolts treated with TSH and simultaneously exposed to an artificial odor were able to select the tributary with the correct odor, when in migratory condition. In other words, they had become imprinted. Salmon injected with saline instead of TSH, or uninjected fish, swam downstream instead of upstream and did not select either of the two tributaries. The downstream behavior is typical of fish that do not become imprinted, or of imprinted fish, if the correct odor is not present in either tributary. Therefore, TSH injections appear to mimic the events that, under natural conditions, activate olfactory imprinting in smolts.

The model

In wild salmon, a complex interaction between thyroid hormones, migratory activity, and imprinting appears to take place. It has been shown that changes in thyroid hormone levels occur not only during certain stages of development, but are also caused by a number of environmental factors. For example, exposure to new environments, changes in water temperature or water flow rate, photoperiod, and lunar phase contribute to increases in thyroid hormone levels. On the other hand, artificially elevated thyroxine levels induce migration in salmon. It has, therefore, been hypothesized that developmentally and environmentally induced increases in thyroid hormone levels cause the salmon to migrate during the smolt transformation. The migratory activity, in turn, exposes the fish to new environments, and this leads to further rises in thyroid hormone levels. The resulting thyroid hormone surges increase the tendency to learn odors through the process of imprinting (Fig. 9.26).

The appealing feature of this model is that migration itself is involved in the control of imprinting. This could enable the salmon to learn a series of olfactory way points, rather than just a single odor (**sequential imprinting theory**). As adults, they could trace this odor sequence to find the correct natal stream.

The sequential imprinting hypothesis proposes that salmon learn a series of olfactory way points in the course of their downstream migration.

Possible neuronal mechanism of olfactory imprinting

How do thyroid hormones exert their influence on imprinting of salmon at the neuronal level? From work on other vertebrates, it is well established that thyroid hormones play a critical role during embryonic development. Both deficiency and excess of thyroid hormones are detrimental to normal brain development. These effects derive from interaction of thyroid hormones with a variety of cellular events, including cell proliferation, cell migration, outgrowth of neuronal processes, synaptogenesis, and myelin formation.

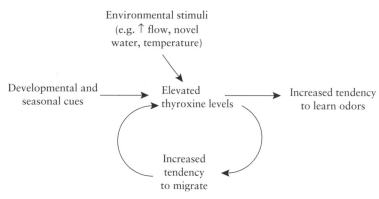

Figure 9.26 Hypothetical relationship between developmental factors, environmental stimuli, migratory behavior, thyroid hormones, and olfactory imprinting. Both developmental and environmental cues stimulate the production of thyroid hormones. This causes an increase in migratory activity, which leads to a further elevation of thyroid hormone levels. The resulting thyroid hormone surges exert, in turn, a positive effect upon the tendency to learn odors. (After **Dittman, A. W., and Quinn, T. P.** (1996).)

Some of these effects have also been demonstrated on the peripheral olfactory system of various vertebrate species. Moreover, it has been shown that the olfactory epithelium of smolting masu salmon has a higher density of thyroid hormone receptors than the one of parr. It is, therefore, possible that thyroid hormones prime olfactory imprinting through **regulation of postnatal neurogenesis in the olfactory epithelium**. The general implications of such a mechanism of behavioral plasticity are discussed in Chapters 8 and 11.

Open sea navigation: sun-compass orientation

While the olfactory imprinting hypothesis explains how adult salmon find the way from coastal areas to their natal stream, it cannot account for how they orient in the open sea. Hasler proposed **sun-compass orientation** as one possible mechanism. Together with Wolfgang Braemer and Horst Schwassmann, two German behavioral physiologists who had joined his laboratory, he demonstrated the use of the sun for orientation in several freshwater fishes. However, direct evidence for such a mechanism mediating open sea migration in a salmonid species is still lacking.

During the open-sea phase, salmon may use a sun compass and magnetic information for navigation.

Open sea navigation: magnetoreception

A second mechanism possibly used by salmon for long-distance navigation is the use of a **magnetic sense**. A key discovery to support this hypothesis was reported by the laboratory of Michael Walker of the University of

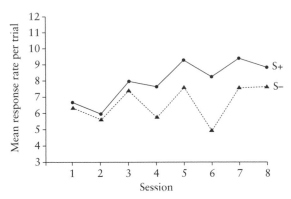

Figure 9.27 Results of magnetic discrimination experiments performed on rainbow trouts (*Oncorhynchus mykiss*). Employing a reinforced stimulus(S+) versus non-reinforced stimulus (S-) paradigm, the fish were trained to distinguish between the presence and the absence of a magnetic anomaly superimposed upon the earth's magnetic field. The response measured was the rate at which the fish struck a target in anticipation of a food reward, or of lack of reinforcement, at the end of each trial. Five S+ and S− trials were given in a balanced quasi-random order in each training session. The data shown are the mean response rate per trial exhibited by these fish. The vertical lines indicate the standard error of the mean. (After **Walker, M. M., Diebel, C. E., Haugh, C. V., Pankhurst, P. M., Montgomery, J. C., and Green, C. R.** (1997).)

Auckland, New Zealand, in 1997. For their study, Walker and associates examined rainbow trout (*Oncorhynchus mykiss*), a salmonid species. The fish were trained to discriminate between the presence and the absence of a magnetic anomaly induced by a Helmholtz coil and superimposed on the background of the earth's magnetic field. After a short training period, the fish responded consistently different to the reinforced stimulus (which could be either the presence or the absence of the magnetic anomaly) than to the non-reinforced stimulus (Fig. 9.27). This demonstrated that rainbow trouts can discriminate different magnetic fields.

As possible structural correlates of this magnetic sense, cells located near the basal lamina of the olfactory epithelium have been identified. These cells contain iron-rich crystals, of approximately 50 nm length, resembling the chains of magnetite present in magnetotactic bacteria. The presumptive **magnetoreceptive cells** are innervated by fine processes of the superficial ophthalmic ramus (also referrred to as **ramus ophthalmicus superficialis**) of the trigeminal nerve. Since the trigeminal nerve is the fifth cranial nerve and indicated by the roman number 'V', this ramus is called **ros V**. The cell bodies of this nerve, together with other somata, make up the anterior ganglion in the brain.

Stimulation of the fish with various magnetic fields (differing in direction, intensity, or both) and electrophysiological recordings from rosV has revealed units that respond to changes in the magnetic field

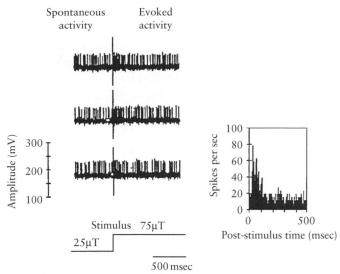

Figure 9.28 Electrophysiological responses to a step-like increase of magnetic field intensity from 25 to 75 μT, as indicated in the bottom trace on the left side. The top three traces show the peri-stimulus activity of a single unit in ros V of the rainbow trout. Before the onset of the stimulus, the unit is spontaneously active in the background magnetic field. After the onset of the stimulus, the traces show the activity of the unit for 1 sec. After the stimulus step, the firing rate increases for the first 100 msec. This is also evident from the post-stimulus time histogram of the responses exhibited by the same unit to 128 presentations of the stimulus (right). (After **Walker, M. M., Diebel, C. E., Haugh, C. V., Pankhurst, P. M., Montgomery, J. C., and Green, C. R.** (1997).)

presented. For example, addition of a magnetic field intensity of 50 microTesla (μT) to the background field of 25 μT, without changing the field direction, elicits excitatory responses immediately following the onset of the stimulus from certain units (Fig. 9.28).

Taken together, these results provide strong evidence that rainbow trouts, and possibly other salmonids as well, possess a magnetic sense. Whether salmon use this magnetic sense for long-distance orientation, such as during their migratory phase in the open ocean, remains to be elucidated.

Orientation in sea turtles

Life cycle of sea turtles

The third group of migratory animals that will be described in more detail are sea turtles. Several species migrate over enormous distances, often thousands of kilometers, across the open ocean to **nest on their natal beaches**. Sometimes, these sites are located on tiny, remote islands— circumstances that make navigation a particularly difficult task. The hatchlings, on the other hand, move immediately after emerging from their nests toward the open sea where they remain until the time when they return as juveniles and take up residence in feeding grounds along the North American coast. How do the hatchlings find the open sea, and what cues do the adults use for navigation back to their site of birth?

After an open-sea phase, females of sea turtles migrate back to their natal beaches to nest.

At the time when the baby turtles break through the surface of sand covering the underground nest, they are just a few centimeters long. By contrast, the adults of the largest sea turtle, the leatherback (*Dermochelys coriacea*), may exceed 2 m in length and weigh up to 600 kg. This makes the adult turtles rather unsuitable to carry out experiments in the laboratory. As a consequence, most research has focused on the navigational capabilities of the hatchlings.

Many of these investigations have been performed on loggerhead sea turtles (*Caretta caretta*). Hatchlings of this species emerge from their underground nests located beyond the tidal zone on the east coast of Florida. Immediately upon hatching, which nearly always takes place during the night, the turtles crawl quickly to the sea to avoid terrestrial predators, such as ghost crabs, foxes, and racoons. They then migrate into the Gulf Stream and the North Atlantic gyre, a current system that encircles the Sargasso Sea. In the floating sargassum seaweed, they find both shelter and food. Circumstantial evidence suggests that, in the course of the following years, the turtles swim around the Sargasso Sea, thereby crossing the entire Atlantic ocean (Fig. 9.29). This evidence includes an orderly progression of juvenile turtle sizes along the North Atlantic gyre. As juveniles, they return to the southeastern coast of the United States. The circle finally closes, when the females migrate back to their natal beaches to nest.

More direct evidence of the long-distance migrations of loggerhead turtles has recently been obtained for the Pacific population that is different from the one living in the Atlantic ocean. Using **genetic markers derived from mitochondrial DNA**, Brian Bowen from the University of Florida, Gainesville, together with several co-workers, has shown that approximately 95% of the pelagic feeding aggregates of this species found around Baja California originate from nesting areas in Japan. The remaining 5% are born on nesting ground in Australia. Thus, juvenile loggerhead turtles apparently traverse the entire Pacific Ocean in the course of their developmental migrations—a distance of at least 10 000 km!

Orientation on the beach

Research by several groups, notably that of Kenneth and Catherine Lohmann of the University of North Carolina, Chapel Hill, has shown that loggerhead sea turtles use at least three types of orientation cues to find, immediately upon hatching, their way to the ocean. These cues will be described in the following text.

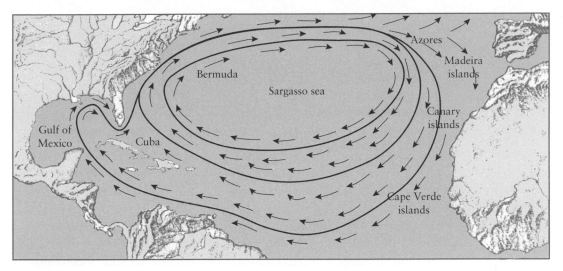

Figure 9.29 Migratory paths of Florida loggerhead sea turtles. Immediately after emerging from their nests on the beaches of Florida, the hatchlings enter the Gulf Stream and take an eastward route across the Atlantic by making use of the clockwise oceanic current encircling the Sargasso Sea. Then, they migrate south around the Sargasso Sea and westward back to the coast of Florida. The closed thick lines indicate the migratory paths of the sea turtles; the arrows indicate the flow of the ocean current. (After **Lohmann, K. J.** (1992).)

Hatchlings often cannot view the ocean directly, as irregularities in the beach surface and debris washed offshore may obstruct the view. Thus, they have to use indirect evidence to locate the ocean. Because water reflects more moonlight and starlight than does land, the horizon of the ocean is almost always brighter than the horizon toward land. Moreover, the level of the ocean is, typically, lower than that of most areas landward. Following these rules, the turtles normally move seaward by crawling **toward the lowest, brightest horizon.**

Magnetic orientation

A number of observations have shown that the hatchlings can orient to the earth's magnetic field. However, this is the case only if the hatchlings have been exposed to light. Turtles without prior exposure to light tested in darkness show no significant orientation in the magnetic field of the earth. Several lines of evidence make it likely that the phase of movement from the nest to the open ocean is important for this **acquisition of a magnetic compass.** As the hatchlings crawl across the beach, swim offshore, or both, they appear to transfer the initial seaward heading to a magnetic compass. This transfer of directional information may allow turtles to maintain offshore courses in deep water, where waves no longer move reliably toward land.

Experiments by Kenneth and Catherine Lohmann have shown that the preferred magnetic heading is acquired in three different ways: by crawling toward light; by swimming toward light; and/or by holding a course while swimming into waves.

Orientation on land

To elucidate details of the orientation mechanisms active on land, in one set of experiments the hatchlings were placed at one end of a 4.1m long runway with dim light placed at the opposite end. Upon reaching the other end, the light was extinguished and the turtles were transferred in darkness to a water-filled test arena where their orientation was monitored. The results of this experiment indicate that the orientation of the turtles depends on the location of the dim light. If the light is placed in the east, forcing the turtles to crawl eastward, the hatchlings subsequently swim eastward. If the light is in the west, establishing a westward course of movement in the runway, the hatchlings swim preferentially in westward direction in the test arena. Reversal of the magnetic field causes a reversal in the direction of orientation.

Waves as an orientation cue

After entering the ocean, the hatchlings establish a course toward the open sea. In this phase, the **wave direction** appears to be an important cue. This has been implied by a rather simple experiment conducted by the Lohmanns. Turtle hatchlings were released at sea near the coast of Florida on days when the waves moved in various directions. Then, the hatchlings swam preferentially directly into the waves (Fig. 9.30)—even when this behavior caused the turtles to adopt directions other than offshore.

The hypothesis that the direction of the waves plays an important role in establishing the course of orientation of the turtles is further substantiated by experiments conducted on days when no waves were present. Then, the hatchlings swam either in circles or exhibited a seemingly random orientation.

How do the hatchlings determine wave direction? One possible cue is the pattern of movement exerted by objects swimming under water near the surface of the ocean. Whenever waves pass above, such objects perform a **circular movement**. The direction of this orbital movement depends on the orientation of the object, say a turtle, relative to the waves (Fig. 9.31). When the turtle directly faces the approaching waves, one cycle of the circular movement consists of a directed acceleration of the turtle in different directions. These directions occur characteristically in the following sequence: upward–backward–downward–forward. When, on the other hand, the turtle is oriented in direction of the wave

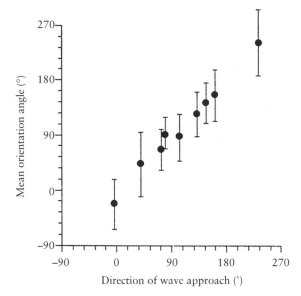

Figure 9.30 Effect of the direction of wave approach on the orientation in green turtle (*Chelonia mydas*) hatchlings. The turtles were released at sea, near the east coast of Florida, on days when the waves moved in various directions. Each data point is the mean orientation angle of a group of hatchlings shown with respect to the angle of the approaching wave at the time of release. The error bars indicate the angular deviation. The plot demonstrates that the hatchlings orient such that they swim directly into the waves—irrespective of the direction of the approaching wave. (After **Lohmann, K. J., and Lohmann, C. M. F.** (1996).)

movement, the sequence of accelerated movements is different, adopting the following order: upward–forward–downward–backward.

To examine whether the orbital movement provides cues to determine the direction from which the waves come, turtles were subjected to orbital movements in air using a wave motion simulator. Their attempted turning behavior was analyzed by monitoring the position of the rear flippers, which are used by swimming turtles as rudders for turning. To turn left, the hatchlings extend the left rear flipper until it is nearly perpendicular to the sagittal plane of the animal. Right turning, on the other hand, is accomplished by extending the right rear flipper. In the wave simulator, the turtles perform the same positional changes of the rear flippers to attempt turning.

The results of the experiments employing this behavioral paradigm are intriguing and can be summarized as follows (Fig. 9.32):

- orbital movements that simulate waves from the right evoke right turns in the hatchlings;
- orbital movements that simulate waves from the left elicit a left turning response;

After entering the ocean, sea turtle hatchling initially use the direction of waves to establish a course from the beach toward the open sea.

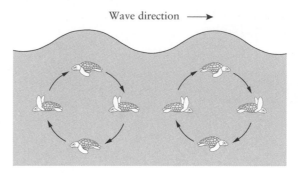

Figure 9.31 Pattern of orbital movement of a hatchling sea turtle swimming against (left) and with (right) the direction of wave propagation. The sequence of directional acceleration differs between the two orientations of the turtle. (After **Lohmann, K. J., Swartz, A. W., and Lohmann, C. M. F.** (1995).)

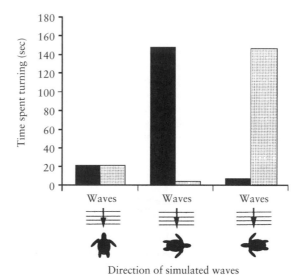

Figure 9.32 Effect of the direction of waves on the turning behavior of loggerhead hatchlings. The orbital movements associated with passing waves were simulated in air using a wave motion simulator. The waves imitated approach either from the front, left, or right. The black bars indicate the time spent by the hatchlings attempting to turn left. The gray bars represent the time during which the hatchlings tried to make a right turn. The behavior of the hatchlings was monitored over a total of 180 sec. (After **Lohmann, K. J., and Lohmann, C. M. F.** (1996).)

• orbital movements that simulate waves from directly in front result in only little turning behavior in either direction.

In addition, calculations have indicated that the accelerations associated with the orbital movements near the surface of the water exceed the detection threshold of the vertebrate inner ear. For waves 1 m in height, the peak acceleration has been estimated to be approximately 72 cm/sec^2. The minimum linear acceleration that can be detected by the vertebrate inner ear is, however, only 5 cm/sec^2.

These experiments and the theoretical considerations support the notion that sea turtle hatchlings can, indeed, monitor the **sequence of accelerations of the orbital movements** they experience underwater to determine the propagation direction of waves, and thus the direction of their movement in the ocean.

The use of the direction of waves for orientation is possible only within the wave refraction zone. Beyond this zone, probably magnetic compass orientation becomes the dominant mechanism guiding the turtles along an offshore course.

Summary

■ Many migratory animals travel, in annual cycles, over extremely long distances between their breeding grounds and the non-breeding ranges. During these large-scale migrations, the entire species, or subpopulations of it, use specific routes.

■ In some species, the annual migratory cycle is not completed by one individual, but by several generations instead.

■ Crossbreeding experiments using populations of birds differing in their migratory behavior have shown that the urge to migrate, as well as the migratory direction, are under direct genetic control.

■ Some animals exhibit a remarkable homing ability, which enables them to return home after foraging excursions or (natural or experimental) displacement.

■ Migratory behavior and homing can be studied in the field by a variety of methods, including analysis of distribution data and tagging, as well as by use of radar and microtransmitters.

■ In birds, several mechanisms have been shown, or are likely to be involved in navigation. They include the use of a time-compensated sun compass in diurnal migrants and in homing pigeons, and employment of a star compass in nocturnal migrants. Olfactory cues have been hypothesized to play a role in homing of pigeons reared under certain conditions.

■ Some birds, including homing pigeons, are also able to detect cues arising from the earth's geomagnetic field, presumably through specific magnetoreceptors. One of these proposed receptor types appears to be associated with photoreceptors, while the other is photoreceptor independent and probably based on magnetite. In the bobolink, a migratory New World bird, such magnetite deposits have been found in the upper beak region where they appear to be innervated by the ophthalmic branch of the trigeminal nerve. Electrophysiological recordings from this nerve have revealed units responding to changes in the ambient magnetic field with high sensitivity. Magnetite-based magnetoreceptors could provide directional information (compass function), whereas photoreceptor-associated magnetoreceptors could mediate analysis of local information in relation to the bird's destination (map function).

▨ Similarly well examined as the homing of pigeons is the ability of salmon to return, as adults, to their natal stream after extended periods in the sea. One of the best-examined species is the coho salmon. Its life cycle consists of the following stages. The eggs are laid in upstream areas of rivers. The hatching alevins, as well as the later emerging parrs, remain in the natal stream until the fish are approximately 18 months old. They transform into smolts that swim down the river toward the sea. In the sea, the salmon grow, until, after another 18 months, when they reach adulthood, they migrate back to their natal stream to reproduce.

▨ Tagging experiments have shown that roughly 95% of the salmon that survive into adulthood return to their natal stream. The remaining 5% stray and spawn in non-natal rivers.

▨ Experiments pioneered by the laboratory of Arthur Davis Hasler have suggested that the homing ability of salmon is based on imprinting to olfactory cues of the natal stream during the smolt transformation. Strong evidence in favor of this hypothesis has been provided by imprinting of salmon to artificial chemicals, for example morpholine and β-phenylethyl-alcohol.

▨ Smolt transformation, and thus olfactory imprinting, is under control of thyroid hormones. Since these hormones also induce migration, smolting salmon learn within a rather short sensitive period a series of olfactory way points. As adults, they trace this odor sequence to find the correct natal stream. At the neuronal level, the process of imprinting may be mediated by thyroid-hormone controlled postnatal neurogenesis in the olfactory epithelium.

▨ Hypothetical mechanisms involved in navigation in the open sea are sun-compass orientation and the use of magnetic information. The existence of a magnetic sense in salmonids has been suggested by the results of behavioral experiments. Possible structural correlates of this capability are magnetite-containing cells in the olfactory epithelium. These cells are linked to the brain via processes of the superficial ophthalmic ramus of the trigeminal nerve. Electrophysiological recordings from this nerve have revealed responses to changes in the magnetic field.

■ The third group of animals discussed in detail were sea turtles. The adults migrate over enormous distances across the ocean to nest on their natal beaches. Hatchlings, on the other hand, immediately after emerging from the nest, crawl to the ocean where they establish a course toward the open sea.

■ To find their way to the ocean, sea turtle hatchlings appear to employ at least three types of orientation mechanisms: First, immediately after hatching, they move toward the lowest illuminated horizons, which normally establishes a seaward course. Second, the direction of the water waves is used to orient toward the open sea within the wave refraction zone. Third, beyond this zone, magnetic cues are probably involved in maintaining an offshore course. The magnetic directional preference that leads turtles seaward appears to be acquired during the initial phase of movement on land, when the hatchlings crawl toward the sea, and also as they orient offshore by swimming into waves.

Recommended reading

Berthold, P. (1996). *Control of bird migration*. Chapman & Hall, London/Glasgow/Weinheim.

A brief, but good introduction to the genetic analysis of migratory behavior in birds.

Berthold, P. (1991). Genetic control of migratory behaviour in birds. *Trends in Ecology and Evolution*, 6, 254–257.

A comprehensive review of ecophysiological and orientation mechanisms, as well as of microevolutionary processes controlling migration in birds.

Dittman, A. W., and Quinn, T. P. (1996). Homing in Pacific salmon: mechanisms and ecological basis. *Journal of Experimental Biology*, 199, 83–91.

A more recent review of homing in salmon, supplementing well the earlier book by Hasler and Scholz.

Hasler, A. D., and Scholz, A. T. (1983). *Olfactory imprinting and homing in salmon: investigations into the mechanism of the imprinting process*. Springer-Verlag, Berlin/Heidelberg/New York/Tokyo.

Both a review of the research conducted by the laboratory of Arthur Hasler, and a summary of the doctoral thesis of Allan Scholz.

Lohmann, K. J. (1992). How sea turtles navigate. *Scientific American*, 266(1), 82–88.

A popular account of the biology of sea turtles, with special emphasis on their navigational capabilities.

Lohmann, K. J., and Johnson, S. (2000). The neurobiology of magnetoreception in vertebrate animals. *Trends in Neurosciences*, 23, 153–159.

An excellent, brief review of sensory and neural mechanisms possibly involved in magnetoreception.

Lohmann, K. J., and Lohmann, C. M. F. (1996). Orientation and open-sea navigation in sea turtles. *Journal of Experimental Biology*, **199**, 73–81.

A good review of orientation mechanisms mediating the finding of the sea by the sea turtle hatchlings, and open-sea orientation by the adults.

Schmidt-Koenig, K. (1979). *Avian orientation and navigation*. Academic Press, London/New York/San Francisco.

Although somewhat out-of-date, still a useful review discussing critically the results of the research on avian migration and pigeon homing.

Walker, M. M., Diebel, C. E., Haugh, C. V., Pankhurst, P. M., Montgomery, J. C., and Green, C. R. (1997). Structure and function of the vertebrate magnetic sense. *Nature*, **390**, 371–376.

This original research article provides an excellent example of how behuvioral, anatomical, and physiological experiments can be combined to gain an integrative understanding of an important biological phenomenon.

Wiltschko, R., and Wiltschko, W. (1995). *Magnetic orientation in animals*. Springer-Verlag, Berlin/Heidelberg/New York.

The most comprehensive review available on magnetoreception and magnetic orientation in animals, written by two of the pioneers in the field.

Wiltschko, W., and Wiltschko, R. (1996). Magnetic orientation in birds. *Journal of Experimental Biology*, **199**, 29–38.

Excellent for those who seek an authoritative review of the use of magnetic compass in birds, without having to read through hundreds of pages, like the above book by the same authors.

Questions

9.1 What sensory cues are involved in the homing of pigeons and in the annual migrations of migratory birds? What role does an internal clock play in this ability?

9.2 Experiments aimed at verifying the involvement of magnetoreception in the navigation of migratory birds and homing pigeons have, sometimes, been difficult to reproduce. Discuss possible reasons for this difficulty.

9.3 Sketch the life cycle of salmon. How do the adult fish find the way back to their place of birth? What sensory modalities are involved in homing, and what environmental cues are used for this capability?

9.4 Great scientific discoveries rarely occur in isolation. Illustrate this statement by describing how the ethological work of Konrad Lorenz and the behavioral physiological work of Karl von Frisch have influenced the development of the olfactory imprinting hypothesis formulated by the school of Arthur Davis Hasler.

9.5 A magnetic sense likely to exist in many, if not all salmonids has been demonstrated in rainbow trout. Although it has frequently been suggested that such a sensory capability may be involved in long-distance navigation of salmon during the open sea stage, actual proof of this hypothesis is still lacking. Propose experiments through which you could verify the use of magnetoreception by salmon for open sea navigation.

9.6 How do sea turtle hatchlings establish a course to the open sea immediately after emerging from the nest? What sensory mechanisms underlay this navigational capability?

9.7 It has been suggested that sea turtles may use chemosensory cues for navigation in the sea. Discuss how this may be achieved. Design experiments to verify this hypothesis.

Communication: the neuroethology of cricket song

10

- Introduction
- Biophysics of cricket songs
- Mechanism of sound production
- Neural control of sound production
- Behavioral analysis of auditory communication
- Perception of auditory signals
- Song recognition by auditory interneurons
- Temperature coupling
- Genetic coupling in cricket song communication
- Summary
- Recommended reading
- Questions

Introduction

Communication involves the transmission of information from a sender to a receiver (see Chapter 3). The information is specific to a given behavioral situation. This is illustrated in Fig. 10.1, which shows three different songs of male crickets produced in different behavioral situations. **Calling songs** are generated to attract sexually receptive females. **Courtship songs** entice the female to mate. **Aggressive songs** form part of the aggressive behavior during encounters with other males. The vehicle used to convey the information is called the **signal**. In the case of cricket communication, a good portion of the information transmitted between males and females, as well as among males, is encoded by **acoustic signals**.

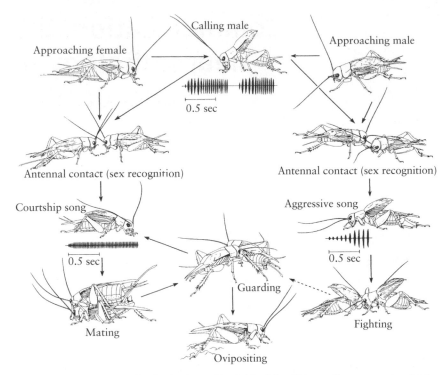

Figure 10.1 Communication in the Australian field cricket *Teleogryllus commodus*. Males attract receptive females by producing a calling song. Upon arrival of the female, the two mating partners touch antennae, a behavior involved in sex recognition. Then, the male generates a courtship song to entice the female to mate. During copulation, the female mounts the male, and the male transfers a small bag of sperm (called the spermatophore) to the end of the female's abdomen. Subsequently, the male exhibits mate guarding, a behavior thought to ensure that the female does not remove and eat the externally attached spermatophore before the sperm are passed into the internal sperm receptacle. The female permits her eggs to be fertilized and, using her long ovipositor, places fertilized eggs into proper substrate. Other males which might also have been attracted by the calling songs of the singing male are inspected by tactile contact via the antennae. Typically the production of aggressive song and fighting between the two males follows. (After **Loher, W., and Dambach, M.** (1989).)

In any communication system, the transmitted signal has to be designed in such a way that it can be detected and 'understood' by the receiver. Using the conveyed information, the receiver makes a decision about how to respond. The response of the receiver affects the fitness of both the sender and the receiver in a beneficial way. This may not necessarily be the case in every instance, but it happens on average over many interactions between the two individuals. The requirement that the receiver understands the signal produced by the sender implies that there must be certain degree of agreement between the sender and the receiver.

In this chapter, we will show how this agreement is morphologically and physiologically implemented by having a closer look at the auditory communication of crickets—one of the first neuroethological model systems in which a detailed and comprehensive analysis of the neurobiological basis of communication was conducted. We will start with a biophysical analysis of the calling songs. Then, we will examine how these signals are produced by the male, and how they are perceived and processed by the female. Finally, we will discuss mechanisms ensuring unambiguous transmission of information in situations when the communication system is challenged by environmentally induced variability.

Biophysics of cricket songs

Like any sound, the songs of crickets can be displayed on an oscilloscope. This is possible, because the microphone converts the sound pressure into voltage. Thus, the oscillogram represents a plot of instantaneous sound pressure over time. The greater the sound pressure (subjectively experienced by humans as an increase in **loudness**), the greater the amplitude of the signal.

Figure 10.2 is an oscillogram of a **calling song**. At a more compressed time scale (Fig. 10.2(a)), it becomes evident that this type of song consists of brief bursts of pulses, followed by intervals of silence. The individual pulses are called **syllables**. One burst of syllables is called a **chirp**. At an expanded time scale (Fig. 10.2(b)), which allows the researcher to resolve the finer details of the sound, the oscillogram reveals that each syllable is comprised of an uninterrupted train of sound waves. These waves vary in amplitude over the duration of the syllable, but are rather constant in frequency. The latter parameter defines the **carrier frequency** of the song and is determined as the reciprocal of the duration of one period of the wave in seconds. The unit of the carrier frequency is, thus, 1/sec, also referred to as Hertz, abbreviated Hz.

The biophysical structure of the calling songs is highly stereotyped and very constant among members of the same species. By contrast, the calling songs of different species may differ considerably. These differences encompass mainly the **temporal structure** of the songs, for example the number of syllables produced within a chirp and the arrangement of chirps, but to a lesser extent the carrier frequency.

The calling songs of different species may differ considerably. These differences encompass mainly the temporal structure of the songs.

Mechanism of sound production

The male cricket uses a **file-and-scraper-mechanism** to produce his songs. During the so-called **stridulation**, the cricket moves one front wing over

Male crickets employ a file-and-scraper stridulation mechanism to produce their songs.

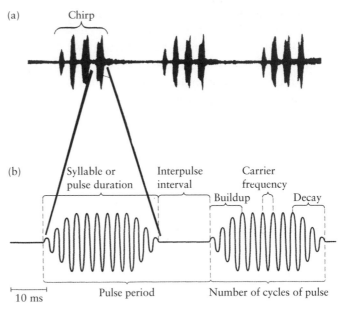

Figure 10.2 Oscillogram of a calling song produced by a male cricket. (a) At a more compressed time scale, the gross structure of the song becomes evident. Each song bout consists of chirps, which in turn are made up of syllables. (b) At an expanded time scale, the oscillogram reveals that the individual syllables are comprised of uninterrupted trains of sound waves. While the amplitude of the syllable varies, particularly at the beginning and the end of each syllable, the duration of the individual wave periods, and thus the carrier frequency of the song, is rather constant. (After **Huber, F., and Thorson, J.** (1985), and **Huber, F., Moore, T. E., and Loher, W.** (ed.) (1989).)

Resonance: oscillation of a system upon supply of energy from an external force at the correct frequency. This frequency is usually close to the so-called resonant frequency of the oscillator with no external force. After supply of the external energy at the appropriate frequency, the resulting amplitude is particularly large, if the damping of the oscillating system is low.

the other in a motion resembling the closing of a pair of scissors. Thereby, the **scraper,** a protuberance of cuticle on the inner edge of the wing, is drawn over a row of regularly spaced teeth, referred to as the **file,** on the underside of the other wing. Movement of the scraper over one tooth of the file produces a single sound pulse. A complete closing stroke of the wings results in a single syllable. This mechanism is schematically shown in Fig. 10.3.

The striking of the scraper over the teeth of the file sets the surface of the wings into oscillation. This vibration is enhanced by the **harp,** a thin triangular area of the wing next to the file, which **resonates** at the carrier frequency of the sound pulse. At the end of the closing strike, the wings are slightly separated, so that the reopening of the wings is not accompanied by any sound production.

In two of the most commonly examined crickets, the European field crickets *Gryllus campestris* and *Gryllus bimaculatus*, each syllable lasts for approximately 15–20 msec. Most commonly, these species produce trains of four syllables, which follow each other at intervals of

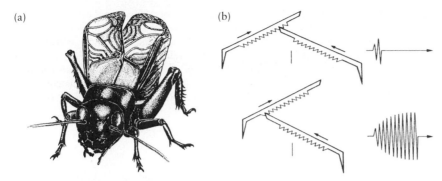

Figure 10.3 Song production by field crickets. (a) Male adopting singing position. (b) Stridulation mechanism. The animal, whose forewings are schematically shown, is oriented as in (a). Both forewings move toward the center, indicated by the vertical line. The scraper of the left (lower) forewing moves across the file of the right (upper) forewing. The movement across one tooth of the file results in the production of a single sound pulse. A complete closing strike of the wings leads to the generation of a syllable. (After **Dambach, M.** (1988).)

about 35 msec. This set of four syllables—the chirp—is followed by a brief period of silence. The chirps are characteristically repeated at a rate of two to four chirps per second.

Neural control of sound production

The closing and opening of the wings during sound production is achieved by the same set of 'twitch' muscles that move the wings during flight, namely **opener** and **closer muscles** of the second thoracic segment. The rhythm of opening and closing, and thus of the production of sound, is controlled by the two thoracic ganglia. Experiments in which these ganglia were deprived from sensory input by cutting of the peripheral nerves have shown that the cricket's central nervous system continues to produce the normal motor pattern. This suggests that the neural control of the calling song pattern is independent of sensory input.

Each muscle of the singing and flight muscle system is driven either by a single motor neuron or by up to five such neurons. This action potential triggers a **muscle impulse** in the bundle of muscle fibers innervated by the respective motor neuron(s). This action potential results in the **contraction** of one of the wing-closing muscles. At the same time, the probability of firing in some neighboring wing-closing motor neurons is enhanced, leading to a powerful contraction of the entire set of muscles.

On the other hand, during discharge of the wing-opening motor neurons, the wing-closing neurons are inhibited. The latter neurons fire

The cricket's central nervous system consists of a chain of 10 knots of neurons called ganglia. Two, including the brain ganglion, are in the head, three in the thorax, and five in the abdomen.

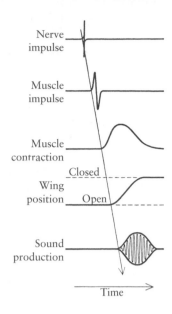

Nerve impulse

Muscle impulse

Muscle contraction

Closed

Wing position   Open

Sound production

Time

Figure 10.4 Sequence of events leading to a single pulse of the male cricket song. At the beginning, a motor neuron in the thoracic ganglion produces an action potential. Upon arrival at the wing muscle fibers, this nerve impulse causes a muscle impulse, which, in turn, results in the contraction of the wing-closing muscles. The closing of the wing draws the scraper of one wing over the file of the other wing. This rubbing of the scraper across the teeth of the file sets the surface of the wing into vibration at a frequency matching the tooth-strike rate. As a final result, a tone-like sound pulse is produced. (After **Bentley, D., and Hoy, R. R. (1974).**)

again after cessation of this inhibition. This alternating activation of wing-closing and wing-opener motor neurons leads to the coordinated alternate closing and opening of the forewing, and, thus, to the production of sound, followed by a short period of silence. The sequence from nerve impulse generation to sound production is summarized in Fig. 10.4.

Behavioral analysis of auditory communication

A female cricket that is in the state of copulatory readiness responds to the male calling songs by flying or walking toward the source of the sound, until she reaches the male. This behavioral response is called **positive phonotaxis**. That the female is guided solely by auditory stimuli was demonstrated in an experiment conducted as early as 1913 by Johann Regen, a high school teacher in Vienna. Making use of the then newly developed telephone, he transmitted the calling songs of a male to a female cricket. The result speaks for itself: the female approached the receiver as soon as the male songs were broadcast.

An approach to analyze the features used by the female to detect and localize the male's calling songs reliably and with high accuracy was developed by Ernst Kramer and Peter Heinecke at the Max Planck Institute for Behavioral Physiology in Seewiesen, Germany (Box 10.1). The central device employed in this approach is a polystyrene sphere 50 cm in diameter. The cricket walks freely on top of this sphere. Any movement of the animal is tracked by an infrared-sensitive device consisting of a small disk of light-reflecting foil glued to the back of the cricket, an infrared light source, and photodetectors sensible in the infrared range of light. When the cricket attempts to move away from the top of the sphere, the photodetectors sense the motion by means of the reflected light. The corresponding signals are fed into a motor system that can compensate the movement by rotating the sphere in opposite direction, thus forcing the animal to walk near the top of the sphere. The signals encoding the movement of the cricket are stored in a computer, thereby providing a precise record of both direction and speed. In recognition of

Positive phonotaxis: movement of an animal toward the source of a sound.

Kramer locomotion compensator: a polystyrene sphere on top of which the cricket can walk freely. A control device monitors the movements of the cricket and counter-rotates the sphere to keep the cricket near the top.

BOX 10.1   Principle of Kramer locomotion compensator

The Kramer locomotion compenstor is a device for the analysis of the features used by the female cricket to detect and localize the male's calling songs. (a) Schematic drawing of the compensator. The cricket walks freely on top of a polystyrene sphere. Sound is broadcasted from a loudspeaker placed on the left (L) and right (R) side, respectively. The cricket's movements are scanned by a camera sensible to infrared light produced by a circular lamp and reflected by a small disk of foil glued on the back of the animal. To exclude visual cues to have an effect upon the outcome of the experiment, the cricket is surrounded by a fabric cylinder. The camera's signals are fed into a computer-controlled motor system (designated M_x and M_y) that compensates the movements of the cricket by rotating the sphere in opposite direction. This compensation mechanism forces the cricket to walk in place. (b) Time profiles obtained in a particular experiments by monitoring a cricket's movement with the locomotion compensator. The top trace shows the velocity, the middle trace the direction of walking of a female during a phonotaxis experiment, and the bottom trace is a polar plot of all the locomotion vectors using the same data obtained during the two stimulus presentations. Each locomotion vector represents the angle of the walking direction relative to the coordinates of the Kramer locomotion compensator, and a length that is proportional to the walking velocity, both for that 1 sec interval. The trial started with 0.5 min of silence, followed by stimulation with a synthetic calling song through a loudspeaker placed at an angle of 270°. At 3.5 min, this loudspeaker fell silent, and the stimulus was switched to a second speaker placed at a 90° angle. Note the tracking of the female, which meanders about the direction of the active speaker.

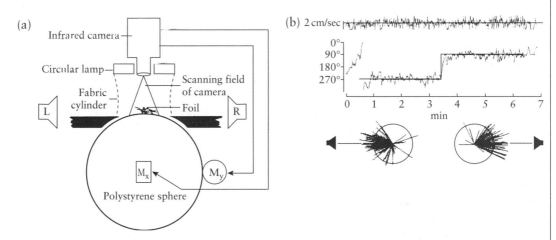

(a) Kramer locomotion compensator (b) Time profiles obtained by monitoring a cricket's movement with the Kramer locomotion compensator. ((a) After **Schmitz, B., Scharstein, H., and Wendler, G.** (1982). (b) After **Pires, A., and Hoy, R. R.** (1992).)

the merits of one of the inventors, this device is often referred to as the **Kramer locomotion compensator** or Kramer treadmill.

Experiments based on the Kramer locomotion compensator, conducted at the Max Planck Institute for Behavioral Physiology in the laboratory of Franz Huber (Box 10.2), have shown that the following three parameters

---

BOX 10.2   Franz Huber

Franz Huber in 2003. (Courtesy: G.K.H. Zupanc.)

Regarded by many as a 'father of insect neuroethology', Franz Huber became particularly known for his research on cricket behavior. Huber, born in Nussdorf (Germany) in 1925, grew up on a Bavarian farm where animals were an integral part of his life. In 1947, he began to study biology, chemistry, and physics at the University of Munich. His mentors were the entomologist Werner Jacobs and the discoverer of the bee dance language, Karl von Frisch (see Chapter 2). Others who had a major influence on the young biologist were Konrad Lorenz, Nico Tinbergen, Erich von Holst (for more information on these three scientists, see Chapter 2) and especially the American Kenneth Roeder, whose work on insects paved the road to relate function of the nervous system to behavior (see Chapter 4).

Huber completed his doctoral thesis in 1953 with an anatomical investigation of the orthopteran nervous system. It was also animals from this order which he chose as a suitable model system to study the neural mechanisms of behavior. These animals,

crickets in particular, remained the focus of his scientific work throughout most of his life. During his subsequent position at the University of Tübingen, Franz Huber was the first to conduct focal brain stimulation experiments in insects—a then novel approach he had learned from Walter Rudolf Hess in Zurich (see Chapter 2).

It is characteristic of Huber to have kept his enthusiasm to learn new techniques and concepts throughout his life. He was among the first who applied intracellular recording techniques to neuroethological model systems—an approach he learned in the laboratory of Theodore Bullock at the University of California at Los Angeles (see Chapter 2), which he joined as a visiting scientist during 1961 to 1962. In 1963, Huber took over a chair at the University of Cologne. Finally, from 1973 until his retirement in 1993, he was one of the directors at the Max Planck Institute for Behavioral Physiology in Seewiesen.

By employing an integrative approach, he and his group pioneered the study of behavior at the level of single nerve cells and neural networks in insects. This research culminated in the establishment of one of the first neuroethological model systems for which a comprehensive, biological understanding was achieved.

Franz Huber's work has been recognized by the award of many honors, including the Karl-Ritter-von-Frisch Medal of the German Zoological Society and honorary doctorates of distinguished universities, both in Germany and abroad. Even more important than the recognition by his peers is the satisfaction he gained through his research. As Franz Huber puts it: "One should search for a suitable model system, study its behavioral tactics in the field, select those that can be treated under controlled conditions with no hesitation to adopt a variety of methods to solve riddles at molecular, cellular, and network levels. But behind all is the curiosity for the living world and how it evolved."

of the male song are important in determining the female's phonotaxic response:

- syllable rate
- carrier frequency
- intensity.

In the European field cricket, *Gryllus campestris*, the best response is evoked by acoustic stimuli comprised of approximately 30 syllables per second—a **syllable rate** found in natural chirps produced by the males. Somewhat surprisingly, the number of syllables per chirp is not important, as long as three or more syllables are generated in a row. The female will even track a continuous **trill** consisting of a continuous repetition of synthesized syllables, provided the stimulus is presented at the correct syllable rate. Also, the duration of the syllable relative to the following interval of silence is not crucial in eliciting a response.

In addition to the syllable rate, the **carrier frequency** also plays an important role. In *Gryllus campestris*, the most effective stimulus to trigger a phonotaxic response by the female is a 5 kHz sound (a frequency found in natural songs) modulated at a syllable rate of approximately 30 Hz.

If a song optimized in such a way is simultaneously broadcast through two differently positioned loudspeakers at different **intensities**, the females consistently choose the louder sound. Such an experimental situation might imitate two singing males at different positions in the field. At equal intensity, females walk between the two sound sources.

The features important in the male's calling song to trigger a phonotaxic response in females are: intensity, carrier frequency of the syllables, and syllable rate.

Perception of auditory signals

When a female cricket detects a calling song, she has to solve the following problems to successfully approach the singing male:

- **localization** of the sound source
- **differentiation** of the calling song produced by conspecific males from sound generated by other animals or from different types of song used by conspecific males in different behavioral situations.

These two tasks are solved both at the level of the ear and by higher auditory processing stations.

In field crickets, the **ears** are encased in special structures associated with the tibiae of the forelegs (**prothoracic legs**) on each side of the body. Externally, the ear is bordered by an eardrum, the **tympanum** (there is one tympanum on the anterior and on the posterior surface of the tibia, but only the latter appears to be involved in auditory perception). The ear on

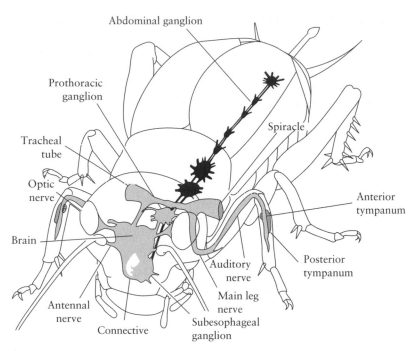

Figure 10.5 The central nervous system and tracheal tube arrangement of the cricket. The ears are situated in the prothoracic legs and bordered by the anterior and posterior tympana. The ears on the two sides of the body are connected via the tracheal tube. Each of the two upper tracheal branches ends in a spiracle. Auditory information is conveyed from the ear to the prothoracic ganglion via the auditory nerve. On top of the hierarchy of central structures processing auditory information is the brain, the frontmost ganglion of the cricket's central nervous system. (After **Huber, F., and Thorson, J.** (1985).)

one side is coupled to the ear on the other side of the body by the lower branches of an air-filled **tracheal tube**, which forms part of the respiratory system. Each of the two upper tracheal branches ends in a **spiracle**, an opening connecting the tracheae with the outside world. Figure 10.5 provides a schematic overview of the tracheal tube system.

The primary auditory organ is the **tympanal organ**. It is attached to the tympanal trachea. To each tympanal trachea, a few dozens of auditory receptor cells (**scolopidia**) are attached. Their axons constitute the **auditory nerve**. Its fibers run through the knee joint of the foreleg up the leg and terminate, by forming numerous branches, in the prothoracic ganglion of the central nervous system on the same side that the leg from which they arise is located (Fig. 10.5). This terminal field is known as the **auditory neuropil**. Tracing experiments have shown that the fiber branches of the auditory nerve are entirely restricted to the ipsilateral side; they do not cross the midline. In the region of the auditory neuropil, the terminal branches of the auditory nerve make synaptic contact with interneurons, which relay the information further within the central nervous system.

As a result of the coupling of the two ears via the tracheae, sound pressure reaches each tympanum not only at its outer surface, but also at its inner surface. To examine the possible contribution of stimulation of the inner tympanal surface, **laser vibrometry measurements** were conducted by Axel Michelsen and Ole Naesbye-Larsen of the University of Southern Denmark, Odense (Denmark). In such experiments, the surface of the tympanum is illuminated by a laser beam, and the velocity of any movement of the surface is computed based on the Doppler shift of the frequency of the light that is reflected from the surface.

These measurements have revealed a major contribution to the deflection of the tympanum by sound traveling via the internal pathway. This is probably due to the tracheae having resonance near the dominant frequency component of the male's calling song. Moreover, it has been shown that, when an amplitude peak of a sinusoidal signal corresponding in its frequency to the carrier frequency of the calling song arrives at the outer tympanal surface, the inner tympanal surface of the same ear experiences more or less a pressure trough. This is due to the length of the internal pathway that sound has to take to arrive at the inner tympanal surface. Measurements have shown that this pathway is approximately half-a-wavelength longer than the way sound has to travel to reach the outer surface. Thus, the tympanum is simultaneously pushed inward from the outside (due to the pressure peak) and pulled inward from the inside (due to the pressure trough). Correspondingly, one-half period later, there is a pressure trough at the outside and a pressure peak at the inside of the tympanal membrane. This causes the tympanum to be simultaneously pulled outward at the outside and pushed outward at the inside. This synergistic effect is particularly pronounced when the male calling song is directly coming from the side of the ear that is closer to the sound source.

On the other hand, if the calling song originates directly from the right side, for example, sound has to travel about the same distance to reach the inner surface of the left tympanum that it takes to arrive at the outer surface of the left tympanum. As a result, an inward push, caused by a pressure peak at the outside, is counterbalanced by an outward push, caused by the pressure peak at the inside. Now, the overall effect is a largely reduced net movement of the tympanal membrane.

Ears, like that of crickets, which are designed in such a way that sound pressure can reach both the inner and the outer surface of the tympanum, are called **pressure gradient ears**. They allow the animal to determine the direction where sound comes from by following a simple behavioral rule: turn toward the ear that receives the highest sound pressure. This is, indeed, what a female cricket does when approaching a singing male. Characteristically, she follows a zigzag path. This meandering enables her,

The cricket's ear functions as a pressure gradient ear: sound arrives at both the outer and the inner surface of the tympanum to stimulate the primary auditory organ, the tympanal organ.

on a continuous basis, to compare the levels of sound pressure arriving at the two ears.

Song recognition by auditory interneurons

The phonotaxic response of female crickets is best elicited by modulation of a 5 kHz sound at a syllable rate matching the one of natural calling songs produced by the male (see above). Based on these behavioral observations, one would expect to find properties either at the level of the ear, or at sensory processing stations within the cricket's central nervous system, that correspond to these features. Indeed, it has been shown that, due to its mechanical resonance properties, the tympanal membrane is tuned to the carrier frequency of the male calling song. A similar frequency tuning is exhibited by the auditory fibers. However, neither the ear nor the primary receptors are sensitive in terms of the temporal parameters of the song. Preference to a particular syllable rate must, therefore, be a property of higher levels of sensory processing.

As mentioned above, the auditory fibers terminate in a restricted region of the prothoracic ganglion called the auditory neuropil. There, the terminal branches make synaptic contact with second-order neurons. A prominent class of these interneurons is formed by the 'omega-1 neurons', commonly referred to as **ON-1 cells**. These neurons have received their name because their shape resembles the capital Greek letter omega (Ω). Axons of the auditory fibers terminate at the numerous, heavily branched dendrites of the ON-1 cells. This input is confined to the side occupied by the cell bodies of the ON-1 cells. In other words, each ON-1 neuron receives input only from the ipsilateral ear. When the ear is stimulated with a train of sound pulses, the ON-1 neurons are rhythmically depolarized such that they closely copy the temporal pattern of the song. This property is shown over a wide range of pulse repetition rates.

The impulses generated by an ON-1 neuron are conducted to the other (contralateral) side of the prothoracic ganglion via the omega-shaped trajectory of its axon, where they cause inhibition. Thus, stimulation of the ear causes excitation in ipsilateral ON-1 neurons and inhibition in contralateral ON-1 neurons. This inhibition is reciprocal—ON-1 neurons in the left prothoracic ganglion cause inhibition on the right side, and ON-1 neurons in the right ganglion do the same on the left side. This mechanism is likely to be used to sharpen **directional sensitivity** of the neural circuit to sound.

Acoustic information encoded by neurons within the prothoracic ganglion leaves this structure via several types of ascending neurons whose axons travel upward to the brain. Among them are the so-called 'ascending neurons of type 1' (**AN-1 cells**). Their dendrites overlap with the axonal arborization of the contralateral ON-1 cells, suggesting that the ON-1 neurons transmit the information directly to AN-1 cells on the contralateral side of the prothoracic ganglion. Although the various types of ascending neurons differ in terms of their sensitivity to different carrier frequencies of sound, all of them, including the AN-1 cells, still truly copy the temporal pattern of sound signals.

Auditory information is relayed to the brain via the projection of the AN-1 cells. Their dense terminal arborizations overlap with dendritic arborizations of 'brain neurons of class 1', known as **BNC-1 cells**. The terminal arborizations of these cells, in turn, overlap with the dendritic arbor of a second class of brain interneurons called **BNC-2 cells**. This distribution has led to the following proposal: auditory information reaches the brain via AN-1 cells, from which it is first relayed to BNC-1 cells and then to BNC-2 cells.

Simplified circuit relaying auditory information from the ear to the brain: scolopidia → ON-1 cells → AN-1 cells → BNC-1 cells → BNC-2 cells.

The physiological properties of the BNC-1 and BNC-2 cells were examined in detail by Klaus Schildberger in the laboratory of Franz Huber. As his studies have shown, the accuracy of copying of the temporal pattern of the calling songs by the BNC cells is significantly reduced compared to neurons of the prothoracic ganglion. On the other hand, they appear to filter out certain features relevant in the phonotaxic response of female crickets. This is evident from the following three functional types among the BNC-1 and BNC-2 (Fig. 10.6):

- One type responds best to syllable rates effective in phonotaxis, plus to higher syllable rates; these cells act like high-pass filters used in electronic circuits, hence called **high-pass cells**.

- A second type responds best to syllable rates effective in phonotaxis, plus to lower syllable rates; these cells act like low-pass filters, hence called **low-pass cells**.

- A third type, found in a subpopulation of BNC-2 cells, responds only to syllable repetition rates in the range that best elicits phonotaxis in females in behavioral tests; these cells appear to function as band-pass filters, and have hence been termed **band-pass cells**. These cells can be regarded as **recognition neurons**.

BNC-2 cells appear to act as pattern recognition neurons: they respond to songs in terms of the number of action potentials elicited by chirps very similar to the way the whole animal behaves, in terms of the phonotaxic response.

According to the model proposed by Schildberger, the high-pass neurons and the low-pass neurons act on the band-pass neurons in a fashion resembling logic AND gates (cf. Chapter 6). In other words, only if input is received simultaneously both from the low-pass and the high-pass neurons, will the band-pass neurons respond. This model is schematically summarized in Fig. 10.7.

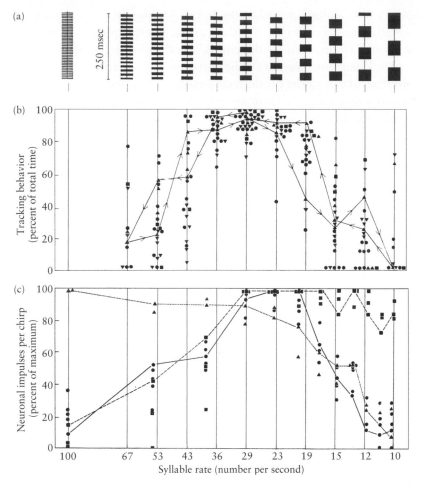

Figure 10.6 Response characteristics of female crickets and various types of BNC cells to synthesized calling songs of varied syllable rates. (a) For stimulation, a series of chirps were used that varied in terms of their syllable rate. (b) Several female crickets (represented by different geometric symbols) were exposed to these different series of chirps. Their behavioral response was monitored using the Kramer locomotion compensator. The data from one of these experiments are connected by arrows to show the sequence of the individual tests. The results of the behavioral experiments indicate a preference for syllable rates near 30 per second, which corresponds to 7–8 syllables over the 250 msec period shown. (c) Response of three types of BNC cells, as determined by the number of action potentials elicited by each chirp. Some cells (circles) exhibit a band-pass response; they respond best to syllable rate around 30 per second. Other (squares) have a low-pass response; they respond best to rates at and below 30 syllables per second. A third class of cells show a high-pass response; they show the strongest responses to syllable rate at and above 30 per second. The data obtained in one experiment each on these three types of BNC neurons are connected by solid line (band-pass response), broken line (low-pass response), and dotted line (high-pass response). Note that, in addition to the syllable rate, other parameters, such as the number of syllables per chirp and the duration of the individual syllables, co-varied in the stimulation experiment shown in (a). However, as has been demonstrated in other investigations, these latter parameters do not affect the response of the neurons. (After **Huber, F., and Thorson, J.** (1985).)

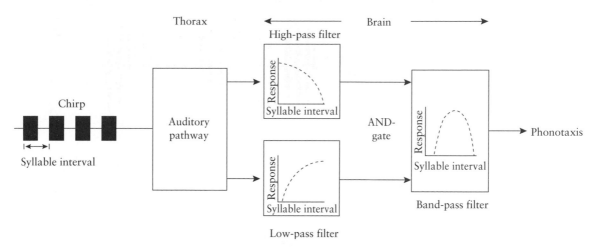

Figure 10.7 Model to explain the properties of band-pass neurons among BNC-2 cells. Within the cricket brain, cells with high-pass filter characteristics respond to syllable rates effective in phonotaxis, plus to higher rates. Similarly, cells with low-pass filter characteristics respond to syllable rates effective in phonotaxis; however, in contrast to the high-pass cells, they also show a preference for lower syllable rates. Both the high-pass and low-pass cells provide input to the band-pass neurons. According to the proposed model, the properties of this latter cell type arise from AND-gating of the input of the high-pass and the low-pass neurons. In other words, the band-pass cells respond only if both the high-pass cells and the low-pass cells detect syllable rates in the range effective to elicit phonotaxis. (After **Huber, F.** (1990).)

Temperature coupling

The problem

Communication is possible only if the receiver 'understands' the signal produced by the sender. The fact that the temporal properties of communication signals in **ectothermic animals** are often under the influence of the **ambient temperature** poses a potential challenge to the integrity of the communication system. The following three solutions to this problem have been found:

- the response criteria of the signal receiver are broadly specified so as to encompass the range of variation of the signal
- the response criteria of the signal receiver rely, at least to some extent, on temperature-invariant properties of the signal
- the response criteria of the signal receiver change parallel to temperature

Among these three solutions, the last one, commonly referred to as **temperature coupling**, has been particularly well studied. The subject of

Temperature coupling: certain properties of a signal produced by the sender and the response criteria of the receiver change parallel to the ambient temperature changes.

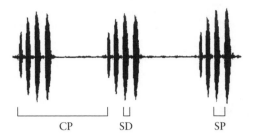

Figure 10.8 Oscillogram of a calling song of the field cricket *Gryllus firmus*. As in any oscillogram, the amplitude of the signal is plotted as a function of time. For further analysis, three temporal parameters have been examined: chirp period (CP), syllable period (SP), and syllable duration (SD). (After **Pires, A., and Hoy, R. R.** (1992).)

a good number of these investigations has been the acoustic communication system of field crickets. Most of this research was carried out by the laboratory of Ron Hoy of Cornell University in Ithaca, New York, using the field cricket *Gryllus firmus*. This North American species lives on the East Coast. In the field, males produce calling songs at ambient temperatures ranging from 12 to 30 °C.

The songs

Figure 10.8 shows an oscillogram of the natural calling song of *Gryllus firmus*. Similar to the European field crickets, *Gryllus campestris* and *Gryllus bimaculatus*, the songs are organized into groups of chirps separated by brief periods of silence. Most commonly, each chirp contains four syllables, although occasionally chirps also consisting of three or five syllables are generated. Figure 10.8 also indicates the temporal parameters used to characterize the song:

- the **chirp period** is the time elapsed between the onset of the first syllable of one chirp and the onset of the first syllable of the next chirp; its reciprocal (in 1/sec) is the chirp rate.

- the **syllable duration** is the time elapsed between the beginning and the end of one syllable

- the **syllable period** is the time elapsed between the onset of one syllable and the onset of the next syllable; its reciprocal (in 1/sec) is the syllable rate.

Temporal patterns used to characterize cricket song: chirp rate (reciprocal of chirp period), syllable rate (reciprocal of syllable period), and syllable duration

Effect of temperature on calling song

In the first set of experiments, Anthony Pires and Ron Hoy made recordings of calling songs in the field. Immediately after each recording, they

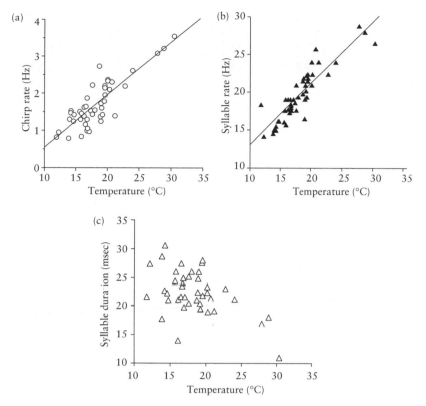

Figure 10.9 Three temporal parameters of the calling song of the field cricket *Gryllus firmus* as a function of temperature. The songs were recorded in the field, and the temperature measured at the calling site. Each point is the mean of one hundred intervals of the respective parameter determined from the song of one individual. (a) Chirp rate. (b) Syllable rate. (c) Syllable duration. (After **Pires, A., and Hoy, R. R. (1992).**)

measured the temperature at the calling site. Back in the laboratory, the calls were played back on an oscilloscope, and the temperature dependence of the three parameters 'chirp period', 'syllable duration', and 'syllable period' was analyzed.

Figure 10.9 shows the results of this analysis. Both chirp rate and syllable rate increase, in a linear fashion, with temperature. From 12–30 °C, chirp rate increases by a factor of 4, from about 0.8 chirps produced per second (thus corresponding to a chirp rate of 0.8 Hz) to 3.3 chirps per second (thus corresponding to a chirp rate of 3.3 Hz). Over the same temperature range, the syllable rate increases by a factor of 2, from approximately 15 to 29 Hz. Conversely, syllable duration is negatively correlated with ambient temperature. From 17 to 30 °C, the mean syllable duration decreases by about 40%.

Chirp rate and syllable rate are positively correlated with temperature, whereas syllable duration is negatively correlated with ambient temperature.

These results might give the impression that each parameter associated with the calling songs changes with temperature. This is, however, not the case. In the population of field crickets studied by Pires and Hoy, the carrier frequency of the calling song ranged from 3.6 to 4.6 kHz. Despite this variability, a plot of the individual carrier frequencies *versus* the temperature at which the corresponding songs were produced indicated that this parameter is not affected by temperature.

Effect of temperature on calling song recognition

As shown earlier in this chapter (see 'Behavioral analysis of auditory communication'), sexually receptive females respond to the calling song produced by the males with positive phonotaxis. On the other hand, the experiments by Pires and Hoy have demonstrated that the temporal properties of the male calling songs are strongly affected by temperature. How do the females cope with this environmentally induced variability in the signal?

Model songs can be synthesized so that their temporal characteristics correspond to the differences found in natural songs under different temperatures.

To answer this question, Anthony Pires and Ron Hoy synthesized model songs with temporal patterns corresponding to those recorded in the field at different temperatures. To model a song produced at 15 °C, a chirp rate of 1.3 Hz and a syllable rate of 14.3 Hz was used. Similarly, songs modeled after those generated at 21 °C exhibited a chirp rate of 1.8 Hz and a syllable rate of 20.0 Hz. The values of these parameters were further increased in synthetic songs imitating natural songs produced at 30 °C, namely to a chirp rate of 3.3 Hz and a syllable rate of 25.0 Hz.

To avoid a possible effect of previous experience, crickets were reared in the laboratory, and only female nymphs isolated from males were allowed in the experiments. To quantify phonotaxis, the Kramer locomotion compensator was used. The experiments were conducted at three different ambient temperatures, 15, 21, and 30 °C. The two loudspeakers were placed at 180° angles from each other. For stimulation, the female crickets were placed on the spherical treadmill and the synthetic males songs modeled after temperatures of 15, 21, and 30 °C were presented.

> *Question: Using the above framework, how would you design experiments to find out whether or not females respond to stimulation by male calling songs produced under different temperatures with a correct phonotaxic response?*

Pires and Hoy conducted, in the laboratory, the following two types of experiments:

- single-stimulus sequential experiments
- two-stimuli-simultaneous choice experiments.

In the single-stimulus sequential experiments, each trial lasted for 7 min. In the first 30 sec, there was silence. Then, the female was stimulated for 3 min with song from one speaker, followed immediately by another 3 min of stimulation from the opposite speaker, and ending with 30 sec of silence. To analyze the female's response, a **locomotion vector** was calculated. This vector, sampled once every second, was defined by an angle describing the cricket's walking direction relative to the coordinates of the Kramer locomotion compensator, and a length proportional to her walking velocity. To screen out trivial movements, only the 30 'best' responses in each 3-min presentation, of magnitude greater than the mean velocity for the stimulus presentation, were analyzed. In the next step, these 30 locomotion vectors were transformed relative to the direction of the speaker active during the experiment and pooled with the 30 corresponding vectors from the other stimulus presentation. As a result, a mean vector characterizing an individual animal's response to each trial was obtained.

In the two-stimuli simultaneous choice experiments, a song was matched to a given temperature at which the experiment was performed. This song was presented through one loudspeaker, while simultaneously through the opposite speaker one of the other two synthetic songs was played back. This paired presentation lasted 8 min, with a switch of the songs between the two speakers after the first 4 min. Similar to the single-stimulation sequential experiments, the locomotor activity of the female on the Kramer locomotion compensator was sampled and mean vectors calculated.

The results of the single-stimulus sequential experiments conducted under the three ambient temperatures—15, 21, and 30 °C—are shown in Fig. 10.10. This figure demonstrates that at all three temperatures the strongest responses, characterized by the largest mean speaker components and the highest mean walking velocities, were obtained to the model song matching the ambient temperature. Thus, females tested at 15 °C showed the strongest responses to songs modeled after natural songs produced at 15 °C (Fig. 10.10(a)). Females tested at 21 and 30 °C responded best to 21 °C songs and 30 °C songs, respectively (Fig. 10.10(b,c)). Similar results were obtained in the two-stimuli choice experiments.

Taken together with the data obtained in the experiments where the effect of temperature on the calling songs was examined, these findings demonstrate that the signal and response properties change parallel to the temperature changes. In other words, the cricket's song communication system is temperature coupled.

Vector: a physical quantity, such as force, which possesses both 'magnitude' and 'direction'. This quantity can be represented by an arrow having appropriate length and direction and emanating from a given reference point. Correspondingly, the 'locomotion vector' in cricket research is defined by the speed and the direction of movement.

Two approaches to test a female cricket's response to different songs involve the conduction of single-stimulus sequential experiments and two-stimuli simultaneous choice experiments.

Female crickets exhibit the strongest phonotactic response to model songs matching the ambient temperature.

Genetic coupling in cricket song communication

The coupling between sender and receiver in the cricket's song communication system is likely to have a **genetic basis**. This is suggested by experiments

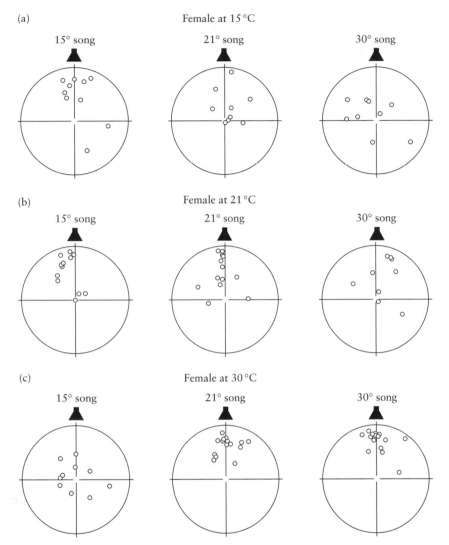

Figure 10.10 Orientation responses of female field crickets, *Gryllus firmus*, to synthetic calling songs modeled after songs produced at three different ambient temperatures, 15, 21, and 30 °C. In each of the single-stimulus sequential experiments, the females were tested at three different ambient temperatures. Each point represents the coordinates of the mean vector calculated from a single trial. (a) Females kept at 15 °C ambient temperature during the experiment. (b) Females kept at 21 °C ambient temperature during the experiment. (c) Females kept at 30 °C ambient temperature during the experiment. The direction of the loudspeaker relative to the coordinates of the Kramer locomotion compensator is indicated by the loudspeaker icons. (After **Pires, A., and Hoy, R. R.** (1992).)

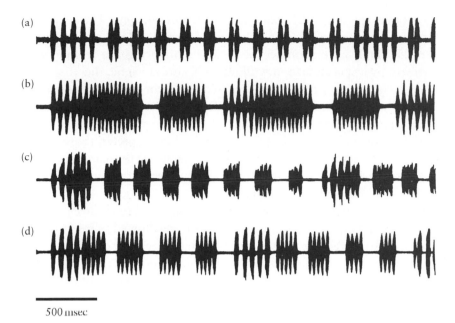

(a)

(b)

(c)

(d)

500 msec

Figure 10.11 Male calling songs of two species of Australian field crickets and their hybrids. (a) *Teleogryllus oceanicus*. (b) *Teleogryllus commodus*. (c) F_1 hybrid of *Teleogryllus oceanicus* female × *Teleogryllus commodus* male. (d) F_1 hybrid of *Teleogryllus commodus* female × *Teleogryllus oceanicus* male. Note that, although the two hybrid songs are somewhat intermediate to the two parental songs, they differ in their fine structure. (After **Bentley, D. R., and Hoy, R. R. (1972).**)

performed by Ron Hoy, Janet Hahn, and Robert Paul of Cornell University. For their study, these scientists used two species of Australian field crickets, *Teleogryllus commodus* and *Teleogryllus oceanicus*. The male songs of these two species are clearly different. The songs of hybrid males are somewhat intermediate between *Teleogryllus commodus* and *Teleogryllus oceanicus*, but the exact temporal structure of the song depends on the parental animals used for the crossing. Sons of the cross between a *Teleogryllus oceanicus* female and a *Teleogryllus commodus* male produce songs different from those generated by males whose mother is *Teleogryllus commodus* and whose father is *Teleogryllus oceanicus*. Oscillograms of the male songs of the two parental species and the hybrids are shown in Fig. 10.11.

The reason for the existence of the two types of cross is that certain elements of the song pattern are controlled by genes on the X chromosome. Female crickets have two X chromosomes (XX), whereas males have only one X chromosome and lack the Y chromosome found in many other animals (X0; the '0' indicates the absence of the chromosome). The maternal parent always contributes one X chromosome, whereas only

Two types of cross can be made between two cricket species, termed A and B, one using males from species A and females from species B, and the other (reciprocal) using males from species B and females from species A.

50% of the sperms of the males have an X chromosome. The other 50% lack an X chromosome. When an egg is fertilized by a spermatozoon carrying the X chromosome, the resulting offspring is a female; when fertilized by a spermatozoon without an X chromosome, the offspring is a male. Male offspring from the two different hybrid crosses will, therefore, be genetically alike, except that they have X chromosomes from different maternal parents.

Which songs do females of the two hybrid types prefer, those of their 'siblings' (the term is here used to indicate males of the same cross type, but not necessarily of the same crossing event), or of 'reciprocal' individuals (males of the other cross type)?

To test the females' preference, they were subjected to auditory discrimination tests. In these experiments, females were allowed to freely walk to choose between the two hybrid songs. To minimize prior auditory experience, the females were isolated from the males in postembryonic nymphal stages before auditory behavior develops. At the beginning of the experiment, the cricket was placed into a small screened container and set on the gravel-covered floor of an arena in a soundproof anechoic room. The room was illuminated only by dim red light to observe the cricket's behavior. Then the female was stimulated by tape-recorded calling songs of the two hybrids. These songs were played simultaneously through separate loudspeakers placed at adjacent corners of the arena so that the speakers and the cricket formed the corners of an equilateral triangle, 1 m on a side. This set-up is schematically shown in Fig. 10.12.

Each female was allowed 5 min to leave the container and an additional 5 min to make a choice. A choice was defined as walking to within 2.5 cm of the speaker or, in a second series of experiments, as actually touching the speaker.

The results of the two series of experiments are summarized in Table 10.1. As these findings demonstrate, the hybrid females show a phonotaxis preference for the calls of their 'sibling' males over those of males from the reciprocal cross. This genetic linkage of the neural structures involved in

Hybrid female crickets show a phonotaxic preference for calls of males from the same type of cross from which they originate.

Table 10.1 Phonotaxic choice of hybrid female crickets in response to calling songs of males from sibling crosses and reciprocal crosses

Hybrid type performing discrimination	Response to *T. oceanicus* female × *T. commodus* male	Response to *T. commodus* female × *T. oceanicus* male	Relative response to sibling cross
T. oceanicus female × *T. commodus* male	97	50	66%
T. commodus female × *T. oceanicus* male	40	125	76%

Note: The figures in the first two columns indicate the total number of females that made a choice in two series of phonotaxis experiments. As each female was tested only once, the data represent different individuals. The figures in the third column represent the percentage of responses of females to male songs from the same type of cross.

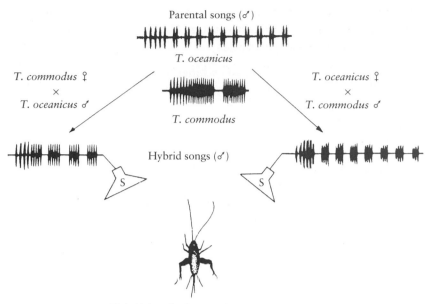

Parental songs (♂)

T. oceanicus

T. commodus ♀
×
T. oceanicus ♂

T. commodus

T. oceanicus ♀
×
T. commodus ♂

Hybrid songs (♂)

S

S

Hybrid female phonotaxic response

Figure 10.12 Schematic diagram of set-up of auditory discrimination experiment. A female cricket and two loudspeakers are placed such that they form the corners of an equilateral triangle, 1 m on each side. During the experiments, the cricket has to choose between two hybrid calling songs played simultaneously through the speakers. The female cricket has made a choice as soon as she walks to within 2.5 cm of a speaker—as used as a criterion in one series of experiments—or actually touches the speaker, as defined in another series of experiments. (After **Hoy, R. R., Hahn, J., and Paul, R. C.** (1977).)

song production by the males and of the neural structures involved in song receptivity by the females ensures that the corresponding genes are transmitted together, and that females will preferentially respond to males of their own kind with phonotaxic behavior.

Summary

▪ Male crickets produce different types of sound. One type, commonly referred to as calling song, triggers positive phonotaxis in females. This approaching behavior is followed by courtship and mating.

▪ The calling songs consist of individual pulses called syllables. Each syllable is composed of an uninterrupted train of sound waves. The duration of the individual waves determines the carrier frequency of the song. The intervals between the onset of one syllable and the onset of the next syllable defines the syllable rate. Syllables occurring in repeated short sequences are called chirps.

■ The calling songs of different species may differ considerably. This is mainly due to differences in the temporal structure of the songs, such as the syllable rate and the arrangement of syllables.

■ The songs are produced by a file-and-scraper stridulation mechanism localized to the forewings. The harp acts as a resonator. The movements of the forewings are achieved through the action of the wing-opener and wing-closer muscles of the second thoracic segment.

■ A commonly used approach to collect quantitative behavioral data on the phonotaxic response of receptive females to sound broadcast from loudspeakers employs the Kramer locomotion compensator consisting of a spherical treadmill and an infrared monitoring device. This set-up forces the cricket to walk on top of the sphere throughout the experiment, while recording the direction and velocity of the cricket's movement.

■ Results of such behavioral experiments have shown that the features important for eliciting phonotaxis in the female are the carrier frequency and the syllable repetition rate of the songs.

■ The ears of crickets act as pressure gradient ears. Sound pressure reaches both the outer and the inner surface of the tympanum. This, as well as the physiological properties of the primary auditory organ, the tympanal organ, enable the cricket to localize sound. The directional sensitivity is sharpened by omega neurons, auditory interneurons in the first thoracic ganglion.

■ Neither at the level of the ear and the primary receptors, nor at the level of the neurons in the prothoracic ganglion, is a preference for species-specific syllable rates found. Such a preference occurs only in higher-order brain neurons, especially in a subpopulation of the so-called BNC-2 cells.

■ Comparison of the biophysical structure of the male calling songs with the behavioral and physiological response of the females has demonstrated a close match between sender and receiver in this communication system.

■ The sender–receiver matching is challenged by the fact that, in the field, male crickets produce calling songs over a wide range of temperatures. The temporal structure of the songs may vary considerably, as chirp rate and syllable rate increase linearly with temperature. However both sequential and two-choice paradigms have shown that the strongest phonotaxic response can be evoked in females to model songs appropriate for the temperature at which they are kept ('temperature coupling' of sender and receiver).

■ The coupling of sender–receiver systems is likely to have a genetic basis. This could be shown in crossing experiments using two species of Australian field crickets, *Teleogryllus commodus* and *Teleogryllus oceanicus*, which produce remarkably different calling songs. The structure of the calling songs of the hybrid males is somewhat intermediate between those of the two parental species, but differs depending on which species was used as the paternal and maternal animal, respectively. F_1 hybrid females show a phonotaxis preference for the calls of males of the same type of cross from which they originate, over songs of males of the reciprocal cross (genetic coupling of sender and receiver).

Recommended reading

Bentley, D., and Hoy, R. R. (1974). The neurobiology of cricket song. *Scientific American*, **231**(8), 34–44.

The focus of this easy-to-read article is on the production of cricket songs: how the muscles moving the forewings with the file and scraper are controlled by command interneurons, when this neuronal network is built and becomes physiologically active during development, and how the genetic basis of particular characteristics of the cricket song can be explored using hybrids of different species.

Gerhardt, H.C., and Huber, F. (2002). *Acoustic communication in insects and anurans: common problems and diverse solutions*. University of Chicago Press, Chicago.

A marvelous synthesis of how frogs and insects produce their calls, what messages are encoded within the sounds, and how the intended recipients receive and decode these signals. The two authors have placed special emphasis on a discussion of the common solutions that the different animal groups have evolved to shared challenges, such as ectothermy, as well as on a presentation of the diversity of solutions that reflect the differences in evolutionary history. A must for any student of animal communication.

Huber, F. (1990). Cricket neuroethology: neuronal basis of intraspecific acoustic communication. *Advances in the Study of Behavior*, **19**, 299–356.

A comprehensive review of the neuroethology of acoustic communication in crickets, written by the founding father of cricket neuroethology and targeted at the informed reader.

Huber, F., Moore, T. E., and Loher, W. (ed.) (1989). *Cricket behavior and neurobiology*. Cornell University Press, Ithaca/London.

The 'bible' of the cricket neuroethologist. 565 pages and 15 chapters, mostly on various aspects of auditory communication, written by many of the leading figures in the field.

Huber, F., and Thorson, J. (1985). Cricket auditory communication. *Scientific American*, **253**(6), 46–54.

In this article, Franz Huber and his collaborator John Thorson give a stimulating account of how sounds used for auditory communication are produced by male crickets, how these songs are perceived, and how, at a neuronal level, the female extracts features relevant to produce the phonotaxic response.

Questions

10.1. Starting with a biophysical analysis of the calling song, describe how male field crickets produce signals to attract females, and how the female perceives and processes features relevant to elicit phonotaxis.

10.2. How are crickets able to localize sound? Discuss the role of both the ears and the central structures devoted to the analysis of auditory information in this process.

10.3. Changes in ambient temperature lead to alterations in certain parameters of the calling song of field crickets. How do the female crickets cope with this potential problem to ensure a correct phonotaxic response?

Cellular mechanisms of learning and memory

11

- Introduction
- Explicit and implicit memory systems
- The cell biology of an implicit memory system: sensitization in *Aplysia*
- The cell biology of an explicit memory system: the hippocampus of mammals
- New neurons for new memories
- Summary
- Recommended reading
- Questions

Introduction

Whereas a large part of this book has focused on neural correlates of simple behaviors, the reductionist approach of behavioral neurobiology can equally well be applied to the study of more complex behavioral and cognitive processes. As we will demonstrate in this chapter, one of these processes involves the way memory is stored in the brain. Research in this area has especially been stimulated by the work of the eminent Canadian psychobiologist Donald O. Hebb, who, in his book *The Organization of Behavior: A Neuropsychological Theory*, published in 1949, made concrete proposals on how the physiological properties of neural networks might contribute to perception and memory formation (see Box 11.1). Since then, an ever-growing number of researchers have tested and expanded the ideas of Hebb using a variety of empirical approaches. Out of the numerous aspects studied over the past half a century, we will cover mainly one—the cellular mechanisms that underlie memory storage. Before we can take a closer look at the major findings in this area, we first need to discuss the evidence pointing to different memory systems.

BOX 11.1   Donald O. Hebb

Donald O. Hebb (Courtesy: McGill University Archives.)

A rather modest and self-critical man during his lifetime, Donald Hebb has become a legacy especially after his death. His theories on the self-assembly of neurons and on the relation between brain and behavior, particularly, revolutionized psychology, then dominated by Freudian views or radical behaviorism, and made Hebb the father of cognitive psychobiology.

Donald Olding Hebb was born in Chester, Nova Scotia (Canada) in 1904 and graduated from Dalhousie University in 1925. He first went into education and became a school principal in the Province of Quebec. In 1936, he obtained his Ph.D. from Harvard with a thesis on the effects of early visual deprivation upon size and brightness perception in rats. Following an invitation by Wilder Penfield, he accepted a fellowship at the Montreal Neurological Institute to examine the impact of brain injury and surgery on human behavior and intelligence. In 1942, Hebb joined Karl Lashley (see Chapter 2) at the Yerkes Laboratory of Primate Behavior, where he explored fear, anger, and other emotional processes in chimpanzees. Particularly stimulated by the intellectual climate of this laboratory, Donald Hebb made a successful attempt to bridge, in the first half of the twentieth century, the wide gap between neurophysiology and psychology through the publication of *The Organization of Behavior: A Neuropsychological Theory* in 1949. In this influential book, Hebb postulated pivotal ideas, which have laid the path for numerous empirical studies. Among them is what has become famous as the 'Hebb synapse'—the prediction that the efficacy of connections between neurons increases with repeated stimulation of the postsynaptic neuron by the presynaptic neuron. Experimental confirmation of this postulate was obtained only decades later when long-term potentiation and the generation of new synapses, as a result of increased neuronal activity, was discovered (see this chapter).

Hebb returned to McGill as Professor of Psychology and was appointed chair in this institution in 1948. In this position, he attracted many outstanding scientists and made McGill University one of the premier centers of psychobiology in North America. He died in 1985.

Explicit and implicit memory systems

In 1957, a seminal discovery was reported by the American neurosurgeon William Scoville, and the Canadian psychologist Brenda Milner who had done her Ph.D. thesis under the supervision of Donald Hebb at McGill University in Montreal. They reported the case of a patient who later became famous as 'H.M.'. As a child, H.M. had a head injury that eventually led to epilepsy. Over the years, his seizures worsened to the point that he became severely incapacitated. As a last resort to control the epileptic seizures, in 1953, when H.M. was 27, Scoville removed his medial temporal lobe, including the **hippocampus**, on both sides of the brain. While the operation indeed relieved H.M. of seizures, it left him since then with a devastating loss of memory.

Similar profound and irreversible deficits in memory have been found in other patients with lesions of temporal lobe structures, especially of the hippocampus. Clinically, their memory deficit is typically manifested as a severe **anterograde amnesia** and a partial **retrograde amnesia** covering a certain time period preceding the lesion. Following the event that led to the lesion, these patients can still perform learning tasks, and they do these with efficiency similar to that of normal persons. However, when these patients carry out learning tasks, they do well only as long as they focus on a given task. As soon as their attention shifts to a new topic, the whole event is forgotten. On the other hand, old memories from their childhood appear to be intact, which indicates that the hippocampus is not the sole site of memory storage within the brain.

Extensive testing of the learning and memory capabilities has revealed another remarkable phenomenon in patients with temporal lobe lesions. Despite their severe memory impairments, they are able to learn new sensorimotor skills quite well, with stable retention over time. One of these tasks is illustrated in Fig. 11.1(a). A human subject is shown a picture of a double-margin star and asked to draw a line between the two margins. Both a normal subject and a memory-impaired patient can do this very easily. Then, the person is asked to draw the lines while seeing

Amnesia: disturbance in long-term memory, manifested by total or partial inability to recall past experiences. Anterograde amnesia: deficit in memory in reference to events occurring after the traumatic event or disease that caused the condition. Retrograde amnesia: deficit in memory in reference to events that occurred before the trauma or disease that caused the condition.

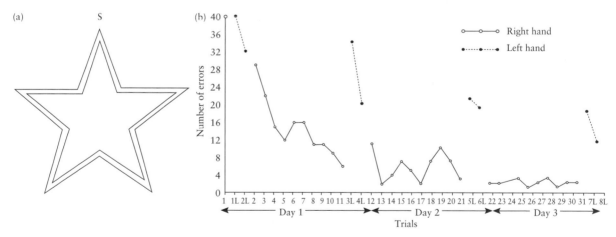

Figure 11.1 Mirror drawing task. (a) The test person is asked to draw a line between the two outlines of the star, starting from point 'S', while viewing the star and the hand in a mirror. (b) Learning curve of a patient who underwent temporal-lobe surgery. Over the three days of testing, he steadily improved in performing the mirror drawing task. The number of errors indicates the number of times in each trial he strayed outside the boundaries of the star while drawing. (After **Milner, B., Squire, L. R., and Kandel, E. R. (1998)**.)

the hand and the star in a mirror. This task is quite difficult, as one tends to draw the line from the points in wrong direction. However, one can learn this skill with practice. Interestingly, the memory-impaired patients exhibited similar learning curves as normal subjects (Fig. 11.1(b)). Yet, while they show steady improvement over the training sessions, and retain this motor skill beyond the training phase, in the end they are unable to remember that they had ever done this task before.

This observation made by Brenda Milner was one of the first pieces of experimental evidence indicating that more than one memory system exists in the brain. Together with similar findings in other amnesic patients, this led to the distinction of the following two types of memory:

- **Explicit memory** (also called **declarative** or **episodic memory**) indicates what is ordinarily associated with the term 'memory'. In humans, it requires conscious recollection of people, places, objects, and events. This type of memory is dependent on the integrity of regions within the medial temporal lobe of the cerebral cortex, including the hippocampus.

- **Implicit memory** (also called **nondeclarative** or **procedural memory**) forms the basis for perceptual and motor skills. In humans, it does not involve a conscious recall of the past. Further, it does not require an intact temporal lobe.

Implicit memory includes not only classical conditioning, but also forms of behavioral changes traditionally not considered types of learning by ethologists, such as habituation and sensitization. In both explicit and implicit memory, a short-term form of information storage can be distinguished from a long-term form of memory. Most importantly, and as will be shown in the following, both explicit and implicit memory make use of similar molecular strategies to implement short-term and long-term behavioral changes.

The cell biology of an implicit memory system: sensitization in *Aplysia*

At about the same time when Brenda Milner's behavioral observations suggested an association of the hippocampus with memory function, scientists made the first attempts to apply a cellular approach to the study of learning and memory. However, soon it became clear that, with the techniques then available, a focus on explicit memory processes was unlikely to succeed. Therefore, scientists turned to implicit memory and searched extensively for simple model systems, particularly among invertebrates. One of these scientists, who pioneered this research like no one else, was the American Eric Kandel. A portrait of his life and work is presented in Box 11.2. Kandel focused on the marine snail *Aplysia*, which, since then, has become one of the major model systems in neurobiology.

BOX 11.2   Eric R. Kandel

Eric R. Kandel. (Courtesy: The Nobel Foundation.)

Like no other scientist before him, Eric Kandel pioneered a reductionist approach to the study of learning and memory. His merit is based on successfully linking various forms of behavioral changes, including simple types of learning, to specific subcellular processes and synaptic plasticity.

Eric Kandel was born in Vienna, Austria, in 1929. In 1939, he emigrated with his family to the USA after Nazi-Germany's annexation of his country. Following graduation from Harvard College with a degree in history and literature, he studied medicine at New York University School of Medicine. During his postdoctoral training with Wade Marshall in the Laboratory of Neurophysiology of the National Institute of Mental Health in Bethesda from 1957 to 1960, he studied the cellular properties of the hippocampus, the part of the mammalian brain closely associated with complex memory. Contrary to the initial expectation, he found that the intrinsic signaling properties of the hippocampus are not much different from neurons in other areas of the brain. Rather, it appeared to be the pattern of functional interconnections that determined the unique functions of the hippocampus. However, due to the immense number of neurons and interconnections in the hippocampus and the limitations of the techniques available at that time, this aspect was not approachable. Therefore, an intensive search for a simple model system began. Finally, during a postdoctoral fellowship with Ladislav Tauc at the Institut Morey in Paris, he found such a suitable experimental animal—the marine snail *Aplysia*.

In the following decades, *Aplysia* formed the basis for Kandel's research. After completing his residency in clinical psychiatry at the Massachusetts Mental Health Center of the Harvard Medical School, Kandel held faculty positions at the Harvard Medical School and the New York University School of Medicine. In 1974, he joined Columbia University in New York as founding director of the Center for Neurophysiology and Behavior, which was established with the aim to understand the biological basis of behavior by employing an integrated approach. Since then, he has remained on the faculty of Columbia. Among the numerous honors received for his research are such prestigious awards as the Wolf prize, the Lasker Award, the National Medal of Science, and the Nobel Prize for Medicine and Physiology, which he shared in the year 2000 with Arvid Carlsson of the University of Gothenburg (Sweden) and Paul Greengard of Rockefeller University in New York.

After three decades of work on *Aplysia*, during which he and his laboratory discovered key molecular mechanisms underlying habituation, sensitization, and classical conditioning, Erich Kandel returned, at the age of 60, to the hippocampus. Again, he pioneered research. By using genetically modified mice, the work of his group provided now the framework to understand the cellular and subcellular mechanisms of long-term potentiation and its relation to spatial memory.

The sea hare (Aplysia californica) is a gastropod mollusk that lives along the coastal waters of southern California.

Why *Aplysia*?

As an experimental animal, *Aplysia* offers several important advantages:

1. It exhibits several simple reflex behaviors that can be modified by different forms of learning.

2. The total number of neurons in the nervous system of *Aplysia* is low (approximately 20 000, compared to a trillion or so in the central nervous system of mammals), and less than 100 are involved in the reflex pattern studied by Kandel and his group.

3. The neurons are among the largest known, some reaching diameters of up to 1 000 μm, and thus are visible to the naked eye. This makes it relatively easy to record from these cells, even over long periods of time. Moreover, thanks to their large size, individual cells can be dissected out of the nervous system and used for biochemical analysis. With the modern techniques available, one can obtain sufficient messanger RNA (mRNA) from a single cell to make a cDNA library!

4. Many of the nerve cells are uniquely identifiable. This makes it possible to record from the same type of cell in different individuals, or to return to the same cell in a later recording session. Moreover, dyes or molecular constructs can be injected into these identified cells, which provides the researcher with a very powerful tool to perform a molecular analysis.

The gill-withdrawal reflex

The behavior Kandel and his group chose for their investigations consists of a defensive reflex. When the siphon of the snail is touched, both the siphon and the gill are withdrawn into the mantle cavity under the mantle shelf (Fig. 11.2). This response is referred to as the **gill-withdrawal reflex**. In spite of its simplicity, this reflex can be modified by several forms of learning, including habituation, dishabituation, sensitization, classical conditioning, and operant conditioning. Most of the work conducted by Kandel and his group has focused on sensitization. The effect of this simple type of learning can best be illustrated using the response of *Aplysia* to external stimuli. Without sensitization, a weak tactile stimulus applied to the siphon evokes a weak response, and the siphon and the gill withdraw only briefly. However, after an aversive shock has been applied to another part of the body, such as the tail, the same weak touch increases both the size and the duration of the reflex response. This effect is called **sensitization**.

Sensitization: enhancement of a behavioral response to a stimulus after application of a different, noxious, stimulus.

The duration of this memory depends on the number of repetitions of this noxious stimulation. If a single shock is applied to the tail, then the

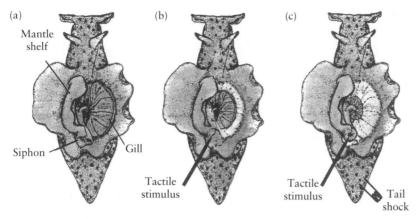

Figure 11.2 The gill-withdrawal reflex. (a) Dorsal view of *Aplysia*. The mantle shelf has been retracted to expose the siphon and the gill. (b) Upon applying a light touch to the siphon with a fine probe, the siphon contracts, and the gill is withdrawn into the mantle cavity for protection. (c) Application of a noxious stimulus, such as an electric shock, to the tail leads to sensitization of the gill-withdrawal reflex. Following such sensitizing training, the same light touch as applied in (b) causes now a much larger siphon- and gill-withdrawal response. Also, siphon and gill are withdrawn for much longer periods of time than before sensitization. (After **Kandel, E. R.** (2001).)

sensitization is effective for only a few minutes. If, on the other hand, four to five spaced shocks are applied to the tail, then a modulation of the gill-withdrawal reflex can be observed for several days. A typical result of such experiments is shown in Fig. 11.3. The duration of the sensitization effect does allow the investigator to distinguish between **short-term memory** and **long-term memory**. These two types of memory are paralleled by two different cellular mechanisms mediating their behavioral outcome. As will become evident below, the essential difference between these two mechanisms is that short-term memory does not require the synthesis of new protein, whereas long-term memory does.

Neural circuit of the gill-withdrawal reflex

Before we describe in detail the molecular mechanism of sensitization, we need to take a look at the neural substrate underlying the gill withdrawal reflex. Figure 11.4 sketches the major components of this neural network, which is located in the abdominal ganglion. It consists of six motor neurons that make direct synaptic connections onto the gill. These motor neurons are connected with 24 sensory neurons that innervate the siphon. The sensory neurons also connect to excitatory and inhibitory inter-neurons that, in turn, are connected to the motor neurons. Stimulation of the tail activates three types of modulatory interneurons: the so-called

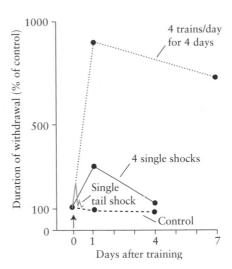

Figure 11.3 Conversion of short-term to long-term memory storage in *Aplysia*. As an indicator of the behavioral response, the duration of the withdrawal of the siphon and gill was measured and expressed relative to unsensitized (control) reflex responses arbitrarily set to 100%. Unsensitized animals respond to a light touch of the siphon with a weak siphon- and gill-withdrawal reflex ('control' curve). Application of a single electric shock to the tail causes the same light tactile stimulus to produce a much larger siphon- and gill-withdrawal reflex; however, this enhancement lasts for only about 1 h ('single tail shock' curve). Four spaced shocks applied to the tail result in an enhanced gill- and siphon-withdrawal reflex lasting for several days ('4 single shocks' curve). More intense training, consisting of four brief trains of shocks a day applied for four days in a row, increases the size and duration of the reflex response even more; the underlying memory now lasts for weeks ('4 trains/day for 4 days' curve). (After **Kandel, E. R.** (2001).)

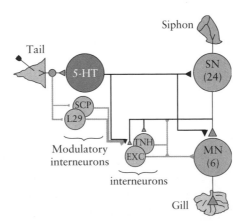

Figure 11.4 Neural circuit of the gill-withdrawal reflex. Synaptic terminals are symbolized by triangles. EXC, excitatory interneurons; INH, inhibitory interneurons; L29, interneurons that release an unidentified modulatory neurotransmitter; MN, motor neurons; SCP, neurons that release small cardioactive peptide; SN, sensory neurons; 5-HT, neurons that release serotonin. For further explanations, see text. (After **Kandel, E. R.** (2001).)

L29 cell, cells that release small cardioactive peptide at their terminal site, and cells that release serotonin. These interneurons connect to the sensory neurons and the excitatory interneurons. The serotonergic pathway is particularly important for modulation of the gill-withdrawal reflex, as blocking of the serotonergic cells abolishes the effect of the sensitizing tail stimulus.

Molecular biology of short-term sensitization

In the initial stage of the molecular analysis of the mechanisms of short-term memory, Kandel and his group focused on short-term sensitization. They found that the associated behavioral changes result from alterations in the strength of the synaptic connections between neurons, a phenomenon referred to as **facilitation**. Their investigations, furthermore, provided clear evidence that both the synaptic changes and the short-term behavioral changes are expressed even when protein synthesis is inhibited. This suggested that this type of behavioral and neuronal plasticity is mediated by a second-messenger system.

Facilitation: Enhancement of synaptic transmission.

Subsequent studies confirmed this hypothesis and revealed the following mechanism schematically shown in Fig. 11.5: A single sensitizing stimulus to the tail of *Aplysia* leads to activation of a modulatory interneuron that releases serotonin as its transmitter. Serotonin acts on a transmembrane serotonin receptor, which belongs to the G protein-coupled receptor superfamily, on a sensory neuron. This activates the enzyme adenylyl cyclase, which converts adenosine triphosphate (ATP) to cyclic adenosine-3′,5′-monophosphate (cyclic AMP or cAMP), thereby increasing the level of this second messenger. In turn, the cAMP recruits the cAMP-dependent protein kinase A (PKA). This leads to an enhancement in transmitter release from the sensory neuron by the following two mechanisms.

1. The activation of PKA by cAMP causes phosphorylation of specific K^+ channels. This results in a closure of these channels, and, thus, in a reduction of repolarizing K^+ currents. As a consequence, action potentials are broadened and the influx of Ca^{2+}, which occurs as a normal event during the action potential, is prolonged. The increased concentration of Ca^{2+} ions, which are necessary for vesicle exocytosis (see Chapter 3), contributes to a greater transmitter release and, thus, to an enhanced activation of the motor neuron, as observed during sensitization (pathway 1).

2. The activation of the PKA directly affects, through a mechanism not yet understood, one or more steps in vesicle mobilization and exocytotic release, thereby also enhancing transmitter release (pathway 2).

These effects triggered by stimulation of the tail can be simulated by application of serotonin to the sensory neuron or by injection of cAMP directly into this cell, thus providing further evidence for the involvement of these molecules in controlling synaptic strength.

The importance of cAMP for the establishment of short-term memories is underlined by results obtained in another model system, the fruit fly *Drosophila*. This work, started by Seymour Benzer at the California Institute of Technology in the early 1970s, is based on a genetic screen in *Drosophila* for mutants that affect learning and memory. Benzer and his group used a simple conditioning paradigm in which the flies had to learn to distinguish between two odors—one associated with an electric shock, the other not paired with an electric stimulus. Flies that had been successfully conditioned avoided the odor associated with the shock. Using this behavioral assay, Benzer and his group found, among others, one mutant called *dunce* that has a defect specifically in short-term memory storage, but not in other behaviors or sensory capabilities, such as locomotion or odor detection, necessary to perform this task. Subsequently, it was shown that the mutant gene encodes a cAMP-dependent phosphodiesterase. This enzyme degrades cAMP. Mutant flies, therefore, accumulate too much cAMP, and this interferes with the ability to establish olfactory memories.

> Genetic approaches to learning and memory storage frequently make use of mutagenized animals that are screened by employing learning assays. The genetic modifications associated with the mutants are then analyzed at the molecular level.

Molecular biology of long-term sensitization

In contrast to short-term memory storage, the establishment of long-term memories is blocked when protein synthesis is inhibited. This suggests that the expression of certain genes is induced in the process of transferring information from the short-term memory to the long-term memory.

What are these genes, and how is their expression regulated? Studies on long-term sensitization of the gill-withdrawal reflex in *Aplysia* have revealed the following mechanism turned on by repeated tail shocks or several puffs of serotonin to the terminal region of the sensory neuron (Fig. 11.5). First, as during short-term sensitization, PKA is activated. However, now the PKA recruits another kinase, the **mitogen-activated protein kinase** (MAPK). Both of these kinases translocate into the cell's nucleus where they activate a cascade of transcriptional processes. At the beginning of this cascade, the transcriptional activator CREB-1a (cAMP response element binding protein-1a) is activated. This protein binds to a cAMP response element (CRE) in the promoters of target genes.

The critical involvement of CREB-1a in the conversion of short-term memory to long-term memory is demonstrated by injection of oligonucleotides carrying the CRE DNA sequence into the nucleus of a sensory neuron. These oligonucleotides bind CREB-1a, as does the endogenous CRE. Thus, if their titer is high enough, they bind most of the CREB-1a

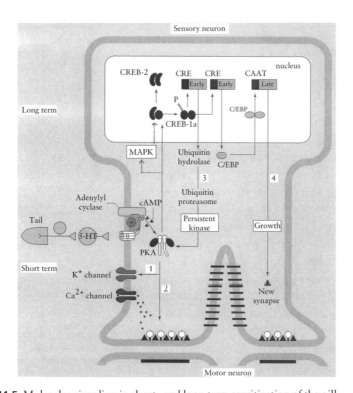

Figure 11.5 Molecular signaling in short- and long-term sensitization of the gill-withdrawal reflex in *Aplysia*. A noxious stimulus, such as an electric shock, applied to the tail activates sensory neurons that excite facilitatory interneurons. These serotonergic (5-HT) interneurons synapse onto the terminals of sensory neurons, which detect tactile stimuli given to the siphon skin and synapse onto motor neurons controlling the withdrawal of the siphon and the gill. At the axo-axonic synapses, the facilitatory interneurons are able to enhance transmitter release from the sensory neurons. This so-called presynaptic facilitation is mediated by serotonin released by the facilitatory interneurons upon stimulation of the tail. Depending on the mode of serotonergic activation, the synapse of the sensory neuron can undergo short-term facilitation, lasting for minutes, or long-term facilitation, lasting for days. Short-term facilitation: Presynaptic facilitation induced by a single tail shock (or a single pulse of serotonin) activates pathways (1) and (2). Serotonin acts on a transmembrane serotonin receptor that activates the enzyme adenylyl cyclase to convert ATP to the second messenger cAMP. The cAMP, in turn, activates PKA, which phosphorylates a number of target proteins involved in channel functions (pathway 1) and exocytosis of synaptic vesicles (pathway 2). As a result, transmitter availability and release are enhanced. These molecular modifications last for minutes and, thus, underlay short-term memory storage. Long-term facilitation: Presynaptic facilitation induced by repeated-shock stimulation of the tail (or multiple pulses of serotonin) activates pathways (3) and (4). The switch to these mechanisms is initiated by PKA. This kinase recruits another kinase, MAPK. Both kinases translocate to the nucleus where they phosphorylate the CREB protein and remove the repressive action of CREB-2, which inhibits CREB-1a. CREB-1a activates several immediate-response genes. One of them encodes for ubiquitin hydrolase, which is a crucial factor in the production of a persistently active PKA. The other immediate-response gene activated by CREB-1a encodes for the transcription factor C/EBP, which is involved in the activation of downstream genes. Their expression ultimately leads to the generation of new synapses. (After **Mayford, M., and Kandel, E. R.** (1999).)

molecules, which are at this point no longer available for binding to the CRE sequence in the promoter regions. In other words, the function of CREB-1a is inhibited through competitive action. As a result of such experiments, long-term facilitation is blocked, but short-term facilitation remains unaltered. Conversely, injection of a phosphorylated form of the cloned *Aplysia* CREB-1a activator can cause long-term facilitation of synaptic transmission, without any further behavioral or serotonergic stimulation.

The CREB-1a exerts its function by activating a cascade of immediate-early genes. One of these genes encodes for ubiquitin hydrolase, which activates ubiquitin proteosome. The proteosome, in turn, cleaves the regulatory subunit that inhibits the catalytic subunit of PKA. This action frees the catalytic subunit of PKA, which then phosphorylates the same substrates as done during short-term facilitation. However, now neither cAMP nor serotonin are necessary—the PKA exhibits persistent activity. This carries on for approximately 12 h.

The memory of this long-term facilitation process is stabilized by the establishment of new synaptic connections. The growth of these synapses is achieved through the activation by CREB-1a of a second immediate-early gene encoding the transcriptional factor C/EBP. This factor acts on downstream genes that give rise to the growth of new synapses. If the expression of the C/EBP gene is blocked, then the growth of new synaptic connections is suppressed and the establishment of the long-term memory inhibited.

These results of molecular biology experiments expanded earlier electron microscopic findings made by Craig Bailey and Mary Chen, both, like Kandel, from Columbia University. They combined intracellular labeling techniques (see Chapter 3) with the ultrastrucutral analysis of completely reconstructed identified sensory neurons of animals from different behavioral groups. Their results demonstrate that the establishment of long-term memory is accompanied by the following two major types of changes in synaptic organization:

1. The number, size, and vesicle complement of the active zones of the presynaptic varicosities of sensory neurons are larger in sensitized animals than in controls, and smaller in habituated individuals.

2. The total number of presynaptic varicosities per sensory neuron depends on the behavioral status of the animal. Control animals have about 1200 synaptic varicosities. Sensory neurons of long-term habituated animals have, on average, 35% fewer varicosities than controls, whereas in animals that underwent long-term sensitization the total number of synaptic varicosities doubled compared to controls (Fig. 11.6).

The alterations in both the number of sensory neuron varicosities and the active zones persist in parallel with the behavioral retention of the memory. This correspondence in the time course of the two events

Immediate-early genes: a class of genes whose expression is low or undetectable in quiescent cells, but whose transcription is activated within minutes after proper stimulation.

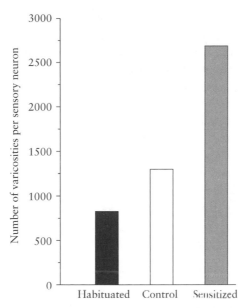

Figure 11.6 Structural changes in synaptic organization involved in long-term habituation and sensitization. The histogram shows the total number of presynaptic varicosities per completely reconstructed sensory neuron in *Aplysia*. This number is highest in sensitized animals and lowest in habituated animals. (After **Bailey, C. H., and Chen, M.** (1988).)

provides additional evidence for the notion that the structural modifications are an important causal factor in the maintenance of long-term sensitization or long-term habituation, respectively.

An interesting discovery was made in the mid 1990s by the groups of Eric Kandel and Craig Bailey. They found that there are not only positive regulators of the conversion of short-term memory into long-term memory, but also inhibitory constraints. One of these **memory suppressor genes** encodes for the inhibitory transcription factor CREB-2. Overexpression of this repressor selectively blocks long-term facilitation in *Aplysia*. Conversely, blocking of CREB-2 results in a single pulse of serotonin (that normally produces only short-term facilitation) to induce long-term facilitation and the growth of new synaptic connections.

The conversion of short-term memory storage into long-term memory storage is characterized by the synthesis of new proteins and the establishment of structural changes at the synaptic level.

The cell biology of an explicit memory system: the hippocampus of mammals

As suggested by the observations of temporal lobe patients mentioned at the beginning of this chapter, explicit memory in mammals requires specialized structures within this part of the cortex, including the hippocampus. This is a feature fundamentally different from implicit memory, which does not involve specialized anatomical systems devoted to memory storage. Rather, as the above detailed molecular analysis of sensitization has shown, implicit memory storage is achieved by modification of the synaptic properties of the neural circuitry concerned with processing of the information relevant to the execution of the respective behavior (e.g. of the gill-withdrawal reflex in *Aplysia*). Despite this difference, research has demonstrated that the molecular mechanisms for converting the labile short-term memory into a stable long-term memory are, at least to a certain degree, remarkably similar among implicit and explicit forms of learning. In the following, we will discuss, in some detail, the biochemical pathways involved in this process in the mammalian hippocampus.

Nonhuman models to study explicit memory

The knowledge gained through the study of human patients with severe memory loss after brain surgery pointed to the importance of structures within the medial temporal lobe for memory function. The identity of these structures and their precise role in memory formation, however, remained enigmatic due to the low number of such patients available, and the frequently poor definition of their brain lesions. A better definition was possible only when the patient's brain was available for autopsy after death. (Today, these difficulties can be overcome by the use of magnetic resonance imaging of the patient's brain during lifetime.)

Due to these difficulties in humans, efforts began, in the late 1950s, to establish animal models in which defined lesions could be correlated with specific memory deficits. Such model systems include mice, rats, rabbits, and monkeys, all of which have a medial temporal lobe system, including a hippocampus.

The hippocampal formation is comprised of the following structures of the neocortex in mammals: hippocampus proper, dentate gyrus, subicular complex, and entorhinal cortex.

Lesioning studies, combined with behavioral testing, in such mammalian model systems have suggested that the key structures of the medial temporal-lobe system are the hippocampus proper, the dentate gyrus, the subicular complex, and the entorhinal cortex (collectively referred to as the **hippocampal formation**), as well as two adjacent parts of the cortex, the perirhinal and parahippocampal cortices. Lesions of any of these structures impair explicit memory.

The function of the medial temporal-lobe memory system appears to be to direct changes in the organization of cortical areas that represent stored memory. These areas within the cortex are typically geographically separate and afford to be bound together in order to fulfill their function in permanent memory storage. In the process of this cortical reorganization, information transiently stored in the medial temporal-lobe system is passed on to the neocortex. As a consequence, the medial temporal-lobe system is needed for memory storage only for a limited amount of time after learning. This explains why patients with temporal lobe ablations had intact memory of events that had occurred years before the surgery, but—in addition to the anterograde amnesia—they also exhibited retrograde amnesia covering a period between several months and a few years prior to the removal of parts of the temporal lobe.

The structure of the hippocampus

A remarkable feature of the hippocampus is its highly regular cellular organization and connectivity. This regularity makes it possible to obtain a number of cross-sections, each of which is very similar to the others, in terms of the morphological and physiological properties. Figure 11.7 shows such a cross-section through the hippocampus on one side of the brain. As this simplified diagram reveals, two principal neuronal fields— the granule cells of the dentate gyrus and the pyramidal cells of areas CA3 and CA1—and the following three principal neural pathways exchange information within the hippocampus.

1. The **perforant pathway**; it originates in the entorhinal cortex and forms excitatory connections with the granule cells of the dentate gyrus.

2. The **mossy fiber pathway**; it is formed by the axons of the granule cells of the dentate gyrus, which make synaptic contact with the pyramidal cells in the CA3 area.

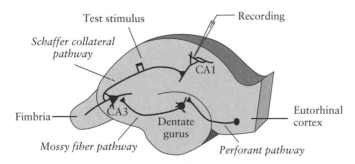

Figure 11.7 The hippocampus and its major neural pathways. The perforant pathway originates in the entorhinal cortex and forms excitatory connections with the granule cells of the dentate gyrus. The mossy fiber pathway is formed by the axons of the granule cells of the dentate gyrus and terminates onto pyramidal cells in the CA3 area of the hippocampus. The Schaffer collateral pathway is formed by the pyramidal cells of the CA3 area projecting to the pyramidal cells of the CA1 area. In addition to these three pathways, the drawing shows a typical experimental arrangement to demonstrate LTP. A stimulation electrode is placed next to the Schaffer collateral fibers. The postsynaptic potentials evoked by stimulation of this fiber path are recorded from the cell bodies of the pyramidal cells of the CA1 area. (After **Kandel, E. R.** (2001).)

3. The **Schaffer collateral pathway**, which connects the pyramidal cells of the CA3 area with the pyramidal cells in the CA1 region.

Taken together, the pattern of connectivity exhibited by these three pathways is often referred to as the **trisynaptic circuitry**.

The regular organization of the hippocampus repeats itself along the longitudinal axis. Therefore, each cross-section taken through the hippocampus and kept alive in artificial cerebrospinal fluid (yielding a so-called **slice preparation**) allows the researcher to study many physiological properties of the principal cell groups and connections *in vitro*.

The three major pathways within the hippocampal formation are the perforant pathway, the mossy fiber pathway, and the Schaffer collateral pathway. Together, they form the trisynaptic circuitry.

Place cells in the hippocampus

As soon as lesioning studies provided evidence of a role of the hippocampal formation in learning and memory, an extensive search began to attribute more specific functions to this part of the brain. A seminal discovery was published in 1971 when John O'Keefe and John Dostrovsky of the University College London reported the results of their study in a brief communication in the journal *Brain Research*. They made extracellular recordings from pyramidal cells of the hippocampus of rats, while the animals could freely move around in the experimental chamber. What O'Keefe and Dostrovsky found was a peculiar firing pattern of some of these cells: A specific cell increased the rate of spike generation when the rat moved into a particular part of the chamber. When the rat left this area, the discharge rate decreased, but as soon as the rat returned to the area, the cell increased firing again. Other cells behaved in a very similar way, with the only difference that they increased firing in a different section of the environment. This remarkable discharge pattern of such cells is shown in Fig. 11.8.

The particular area that evokes firing by the pyramidal cells is referred to as the cell's **place field** or **firing field**. The cells exhibiting this kind of behavior are called **place cells**. Each place cell has a somewhat different spatial preference. This preference is quite stable in a given environment, but may be different in a different environment. Place cells can, therefore, encode place fields in more than one environment. When the animal is introduced to a new environment, new place fields are formed within minutes, but, as soon as the cell's preference is established, the characteristic firing pattern becomes and remains stable over weeks or even months. Collectively, the place fields of the different place cells form an internal representation of the space—a so-called **spatial map**. The establishment of these maps involves a learning process and a conversion of the acquired space information from a short-term into a long-term memory.

These results demonstrate that the hippocampus exerts a crucial function in establishing spatial memory. On the other hand, this does not

Place cells are a subpopulation of hippocampal pyramidal cells that encode place fields in an environment. Collectively, the place fields of the different place cells form a spatial map of the space.

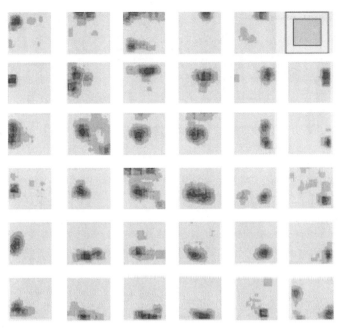

Figure 11.8 Place fields of 35 simultaneously recorded pyramidal place cells in the CA1 region of the rat hippocampus. During the recordings, the freely moving rat searched for grains of rice on a small 40×40 cm^2 open platform. The surface area of the holding box relative to the overall camera view is shown in the upper right-hand corner of the picture. The recordings were obtained by four implanted microelectrodes. During foraging, the rat's location and the pyramidal cells' activities from each electrode were recorded. The firing rates within the place fields are represented by four gray-scale levels, each representing 20% of the peak firing rate. In the picture, the place fields of the individual place cells are arranged according to field location. Fields toward the northwest of the box are shown in the upper left portion of the picture, those toward the southeast in the lower right part, and so on. Note that the fields of the 35 cells collectively cover a considerable area of the total space defined by the box. (After **O'Keefe, J., Burgess, N., Donnett, J. G., Jeffery, K. J., and Maguire, E. A.** (1998).)

imply that the hippocampus is involved just in spatial memory, or that no other brain structures contribute to the long-term storage of environmental information. Clearly, an immense amount of work still lies ahead to unravel the structure–function relationship of brain areas involved in learning and memory.

Long-term potentiation

In each of these three hippocampal pathways, a type of synaptic plasticity occurs that has become known as **long-term potentiation**, commonly abbreviated **LTP**. This phenomenon was discovered in 1973 by Timothy Bliss and Terje Lømo in the Laboratory of Per Andersen at the

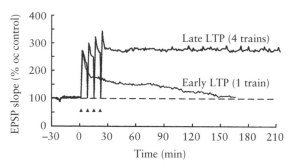

Figure 11.9 Early and late phases of LTP in the Schaffer collateral pathway of the hippocampus. A response was evoked by applying constant test stimuli before and after induction of LTP. For the induction, a single train, or four trains at 10-min intervals, of 100 Hz stimuli were delivered for 1 sec each (the time points of stimulation are indicated by triangles below the plot). The response is indicated by a plot of the slope of the rising phase of the evoked postsynaptic potential (EPSP). As revealed by the diagram, one train of this high-frequency stimulation elicits early LTP, whereas four trains evoke the late phase of LTP. (After **Kandel, E. R.** (2001).)

Long-term potentiation: abrupt and sustained increase in the efficiency of synaptic transmission after high-frequency stimulation of a monosynaptic excitatory pathway.

University of Oslo, Norway. With brief volleys of constant-voltage pulses, lasting for a few seconds and presented at frequencies ranging between 10 and 100 Hz, they stimulated the perforant path fibers in rabbits. The stimulations resulted in an increase in the efficiency of synaptic transmission for periods ranging from 30 min to 10 h after delivering the electric pulses. Since the original discovery made by Bliss and Lømo, LTP has also been found in the other two hippocampal pathways. Moreover, it has also been shown that, under proper conditions, LTP can last for days, and even for weeks. Thus, LTP exhibits a property that makes it suitable for memory storage.

As with facilitation of the gill-withdrawal reflex in *Aplysia*, LTP in the hippocampus exhibits short-term and long-term phases. A typical experimental result revealing these two phases is shown in Fig. 11.9. A single train of stimuli produces a short-term phase of LTP, called **early phase of LTP**. This phase lasts for 1–3 h and does not require synthesis of new protein. By contrast, four or more trains of electrical stimuli induce a long-term phase of LTP, called **late phase of LTP**. This phase lasts for at least one day and requires the synthesis of both new mRNA and new protein.

Molecular biology of LTP in the mammalian hippocampus

The late phase LTP has been particularly well studied in the Schaffer collateral pathway. Glutamate released from the presynaptic terminals of the Schaffer collateral fibers binds to N-methyl-D-aspartate (NMDA) receptors on the pyramidal cells of the CA1 region. As shown in Fig. 11.10, this triggers an influx of Ca^{2+} ions into the postsynaptic cell. Upon repeated stimulation, the Ca^{2+} influx recruits an adenylyl cyclase,

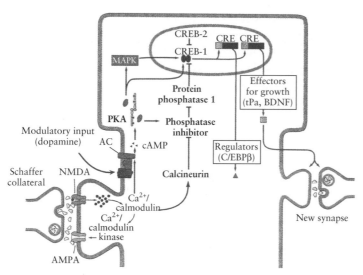

Figure 11.10 Cellular mechanisms of the late phase of LTP in the Schaffer collateral pathway. Repeated stimulation of the Schaffer collateral fibers leads to activation of NMDA receptors and influx of Ca^{2+} ions into the postsynaptic cell. The Ca^{2+} recruits an adenylyl cyclase (AC), which activates a cAMP-dependent PKA. The PKA recruits the MAPK, and both kinases translocate into the nucleus, where they phosphorylate the CREB-1. The latter protein, which is inhibited by CREB-2, binds to CREs, in the promoter regions of target genes. These genes are thought to express growth factors involved in mediating structural changes in synaptic organization, including the growth of new synapses. This pathway resulting in the establishment of the late phase of LTP is regulated by constraints (printed in **bold**), which inhibit long-term memory storage. Abbreviations not explained in the text: AMPA, α-amino-3-hydroxy-5-methyl-4-isoxazole-propionic acid; BDNF, brain-derived neurotropic factor; tPa, tissue plasminogen activator. (After **Kandel, E. R.** (2001).)

which, in turn, activates the cAMP-dependent PKA. Like in long-term facilitation in *Aplysia*, the activated PKA recruits the MAPK, and both translocate to the nucleus, where they activate a transcriptional cascade regulated by CREB proteins. The target genes activated by CREB-1 include those that express growth-promoting effectors, such as brain-derived neurotrophic factor. There is, indeed, some evidence that the late phase LTP finally leads to the formation of new synapses, which could make the synaptic changes persistent.

To study the role of PKA in the establishment of LTP and long-term memory, Ted Abel, Peter Nguyen, Rusiko Bourtchouladze, and Eric Kandel employed a transgenic approach. They generated transgenic mice with a reduced activity of PKA in the hippocampus. These mice expressed a dominant negative form of the regulatory subunit of PKA. This so-called R(AB) protein carries mutations in both cAMP-binding sites and acts as an inhibitor of enzymatic activity. To make the inhibition region-specific, the researchers used a promoter from the gene expressing the α subunit of the Ca^{2+}/calmodulin-dependent protein kinase II.

As expression of the latter gene is specific to somata and dendrites of neurons of the forebrain in adult mice, the expression of the transgene was also limited to this brain region.

As expected, in the R(AB) transgenic mice the hippocampal PKA activity was significantly reduced. This did, however, not affect the early phase of LTP, which was normal. Also, the learning process itself, as well as short-term memory, were not impaired. However, a marked decrease was found in the late phase of LTP, and this defect was paralleled by deficits in long-term spatial memory.

A more detailed analysis of the R(AB) transgenic mice showed that the diminished PKA activity caused a reduced stability of the place cells in the CA1 region of the hippocampus. When R(AB) mice were placed in a new environment, the place cells were able to establish new place fields, and the resulting map was stable when tested 1 h after the exposure to the new environment. However, these place fields were not stable when the mice were tested at 24 h.

Taken together, these results indicate that the CA1 area of the hippocampus plays an important role in the conversion of short-term memory into long-term memory. In this process of memory consolidation, PKA is thought to induce the transcription of genes encoding proteins that are required to make the synaptic potentiation long-lasting.

New neurons for new memories

Another new and exciting research direction in the neurobiological study of learning and memory has been established in recent years. Based on the discovery that new neurons are continuously born in the adult hippocampus, a number of investigations have provided evidence of a link between adult neurogenesis and memory formation. The two model systems used in these studies are the hippocampus of rodents and the hippocampus of birds.

Adult neurogenesis and memory formation in the hippocampus of rodents

In a series of studies conducted by the laboratory of Fred ('Rusty') Gage at the Salk Institute for Biological Sciences in La Jolla, California, the effect of different environmental conditions on the rate of cell proliferation and survival in the **dentate gyrus** of the hippocampus of mice was examined. As shown in Fig. 11.11, one group of mice were kept in standard housing cages, which were rather small and did not provide many opportunities for exploratory behavior and social interactions. A second group of mice lived in an **enriched environment**, which consists of

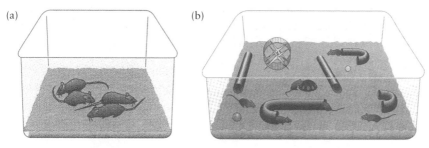

Figure 11.11 Living conditions of two experimental groups of mice. (a) Mice living in a standard housing cage. (b) Mice living in a large cage that provides an enriched environment. The enrichment offers plenty of opportunities for social interaction (14 mice were kept in this type of cage instead of 4 mice kept in the standard cage); for exploration of the environment by providing appropriate objects, such as toys and a set of tunnels; and for physical exercise by being equipped with a running wheel. (After **Van Praag, H., Kempermann, G., and Gage, F. H.** (2000).)

large cages with a variety of objects, such as tunnels, nesting material, and toys. Furthermore, the groups of mice kept under the latter condition were larger (14 mice vs. 3–4 mice under standard conditions). Thus, the enriched environment offers plenty of opportunities for exploration and for social interaction among the other individuals, compared to the relatively impoverished standard laboratory environment.

Behavioral experiments demonstrated that environmental enrichment enhances memory functions in various learning tasks. To examine the possible involvement of neurogenesis in this behavioral effect, newborn cells were identified by the BrdU technique. This method involves (usually intraperitoneal) administration of 5-bromo-2′-deoxyuridine, commonly abbreviated 'BrdU'. This molecule is an analogue of thymidine that naturally does not exist in the body. Similar to thymidine, BrdU is incorporated into replicating DNA during the S-phase of mitosis. Since BrdU is foreign to the organism, it can be recognized by antibodies raised against it. Further immunohistochemical processing with secondary antibodies (see Chapter 3), therefore, enables the researcher to visualize the BrdU label in those cell nuclei that undergo mitosis at the time of, or shortly after, the administration of BrdU.

Histological investigations employing the BrdU method revealed that mice kept under such conditions exhibit significantly more neurons in the dentate gyrus than mice kept under standard laboratory conditions. In principle, the larger number of neurons in the dentate gyrus of mice living in an enriched environment could be due either to a higher proliferative activity in mice exposed to this environmental condition, or to better survival of the new cells. As studies of Gage and his group have suggested, it is mainly the higher survival rate of the newborn cells that leads to more

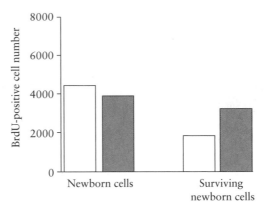

Figure 11.12 Total number of newborn cells and surviving newborn cells per dentate gyrus of mice kept under standard housing conditions (open bars) and in an enriched environment (closed bars). The number of newborn cells was estimated using the BrdU technique one day after the last administration of a series of BrdU injections. The number of surviving new cells was determined four weeks following the last administration of BrdU. (After **Van Praag, H., Kempermann, G., and Gage, F. H.** (1999).)

neurons in the dentate gyrus of mice kept under enriched conditions (Fig. 11.12). Further experiments, in which the scientists investigated aspects of the enriched environment that exert the behavioral improvements and the increase in cell numbers, suggest that the increased running activity plays a role in these effects.

A crucial question is whether or not the new cells become **functional**. Various lines of evidence suggest that this is, indeed, the case. Retrograde tracing experiments have shown that the newly generated neurons send axons to the normal projection area of dentate gyrus neurons, namely the so-called CA_3 field of the hippocampus. Furthermore, the young neurons develop immunoreactivity against the calcium-binding protein calbindin-28 kD, and they are surrounded by synaptic profiles containing the synaptic vesicle protein synaptophysin—both features that are characteristic of functional dentate gyrus neurons.

Since the hippocampus is involved in memory formation, it is tempting to speculate that the changes in the number of hippocampal neurons are causally related to the alterations in learning behavior and memory. This notion is, indeed, supported by experiments conducted by the research groups of Tracey Shors of Rutgers University in Piscataway, New Jersey, and of Elizabeth Gould of Princeton University, New Jersey. They treated rats with methylazoxymethanol acetate (MAM), a DNA-methylating agent that diminishes the number of proliferating cells in the hippocampus (Fig. 11.13(a)). To test the effect of reduction of cell proliferation on

An enriched environment promotes survival of neurons produced in the dentate gyrus of the hippocampus and induces behavioral improvement in spatial learning tasks of adult mice.

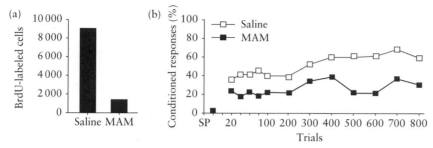

Figure 11.13 Effect of treatment of rats with MAM on the number of adult-generated neurons in the dentate gyrus (a) and on trace conditioning (b) in rats. The animals were treated with MAM or saline (as a control) daily for 14 days. On days 10, 12, and 14, they received one injection each of BrdU to label newborn cells. After being left untreated for 2 days, they were subjected to an eye-blink conditioning paradigm involving 200 trials per day for 4 days, for a total of 800 trials. As the graph shows, the treatment with MAM reduced both the number of newly generated cells in the dentate gyrus and the overall number of conditioned responses during trace conditioning. (After **Shors, T. J., Miesegaes, G., Beylin, A., Zhao, M., Rydel, T., and Gould, E. (2001).**)

memory formation, two conditioning paradigms were used. One of these paradigms involved delay conditioning, while the other employed trace conditioning.

During **delay conditioning**, the conditioned stimulus and the unconditioned stimulus show a temporal overlap. Acquisition of a conditioned response of this type does not require an intact hippocampus. In contrast, during **trace conditioning**, there is a temporal gap between the conditioned stimulus and the unconditioned stimulus. Thus, the animal has to resurrect a memory 'trace' of the conditioned stimulus to associate this stimulus with the unconditioned stimulus. Acquisition of this type of conditioning requires the hippocampus.

Rats treated with the cell proliferation-suppressing agent showed a reduction of the number of conditioned responses during trace conditioning, but not during delay conditioning (Fig. 11.13(b)). This result is consistent with the idea that generation of new cells in the hippocampus during adulthood is causally linked to the formation of hippocampus-dependent memories. As further experiments have suggested, the cells become critical about one to two weeks after their generation. This is the time when the young cells become incorporated into the granule cell layer of the hippocampus, form dendrites, and extend axons into their projection area, the CA_3 region. This highly plastic phase appears to make the young neurons ideally suited to be involved in the formation of associations between stimuli.

The generation of new cells in the adult hippocampus may be causally linked to the formation of hippocampus-dependent memories.

Adult neurogenesis and spatial learning in the avian hippocampus

A close relationship between adult neurogenesis in the hippocampus, on the one side, and learning and memory performance, on the other, has also been suggested by experiments conducted on birds. Most of these studies were carried out in the laboratories of John Krebs at Oxford University and of Fernando Nottebohm at Rockefeller University in New York. Hand-reared juvenile marsh tits (*Parus palustris*) allowed to store sunflower seeds in storage sites and to retrieve this food showed a significantly higher rate of cell proliferation in the ventricular zone bordering the hippocampus than did controls. The control group consisted of age-matched birds which were treated in identical way as the test animals, except that they could not store or retrieve food. Also, after several weeks of food storing and retrieval activity, the experienced birds exhibited a larger number of cells in the hippocampus than did the control birds.

Similar results were obtained in free-ranging black-capped chickadees (*Parus atricapillus*), songbirds which are very common in forests of temperate North America. In the late summer and early fall, their diet shifts from insects to seeds. At that time, they start to hide part of the seeds they find in storing sites, with typically only a few items per site. In the wild, they retrieve these seeds after a period of hours or days. Experiments on captive birds have, however, demonstrated that the memory of cache sites can last much longer, namely for several weeks.

Although determination of neuronal birth dates with radioactively labeled thymidine showed that chickadees form new neurons during all times of the year, these experiments also revealed a marked peak in the fall, as shown in Fig. 11.14. These neurons live for a few months and then appear to die. Moreover, the rate of cell proliferation in the region of the hippocampus is notably higher in juveniles than in adults.

Taken together, both the experiments on tits and chickadees suggest that the recruitment of new neurons into the hippocampus plays a role for the acquisition of new spatial memories. In chickadees, the need for such an acquisition is particularly acute in the fall and in juveniles. It is especially the young birds which encounter a wealth of novel information in the fall. Then, the birds cease to defend territories and form flocks, change their diet, and experience a rapidly changing environment, such as the trees changing color and losing leaves, and the later arrival of the snow.

On the other hand, sustained learning appears to require to periodically replace neurons within the hippocampus. It is thought that, in order to avoid running out of memory space, memory information encoded by older neurons is transferred from the hippocampus to sites elsewhere in the brain. Since the older neurons are likely to have lost their plastic properties essential for the formation of memories, these neurons are finally replaced by newly generated ones.

Experiments on birds suggest a close relationship between recruitment of new neurons in the hippocampus and acquisition of spatial memories.

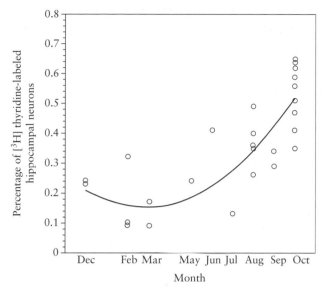

Figure 11.14 Percentage of new neurons in the hippocampus of adult chickadees at different times of the year. The free-ranging birds were captured and received a single injection of [³H]thymidine (i.e. thymidine radioactively tagged by containing, in one position, tritium instead of non-radioactive hydrogen) to label cells undergoing mitosis at or shortly after the time of administration. Then, the birds were released. Six weeks later, they were recaptured, and the relative proportion of radioactively labeled neurons in the hippocampus was determined. The graph reveals that, although new neurons are produced throughout the year, their proportion is significantly higher in the fall than at other times of the year. (After **Barnea, A., and Nottebohm, F.** (1994).)

Summary

▓ Studies on human patients with memory deficits have suggested two types of memory. First, an explicit or declarative memory, which in humans requires conscious recollection of places, objects, and events, and is dependent upon the integrity of the medial temporal lobe, including the hippocampus. Second, an implicit or nondeclarative memory, which forms the basis for sensorimotor skills, and does not require an intact medial temporal lobe.

▓ One of the prime model systems to study the cellular basis of implicit memory has been the gill-withdrawal reflex in the marine snail *Aplysia*. This defensive reflex involves a withdrawal of the siphon and the gill into the mantle cavity when the siphon of the snail is touched.

▓ The gill-withdrawal reflex can be modified by several forms of learning, including sensitization, which occurs when an aversive shock is applied to another part of the body. Then, a weak tactile stimulus applied to the siphon evokes a much greater reflex response than it normally would.

■ If a single shock is applied to the tail, then the sensitization of the gill-withdrawal reflex lasts for only a few minutes, and thus exhibits features of short-term memory. If shocks are applied repeatedly, then the gill-withdrawal reflex remains sensitized for several days, thus displaying properties of long-term memory.

■ At the neuronal level, the sensitization effect is mediated by modulatory interneurons in the abdominal ganglion which make synaptic contact with sensory neurons. The latter neurons innervate the siphon and make direct synaptic contact with motoneurons innervating the gill.

■ Cellular analysis of short-term sensitization revealed an increase in the strength of the synaptic connection between the sensory neuron and the motoneuron. This so-called facilitation is initiated by activation of the modulatory interneuron, which releases serotonin at the synapse making contact with the sensory neuron. The serotonin acts on a transmembrane serotonin receptor on the sensory neuron which activates adenylyl cyclase, thereby leading to an increase in the intracellular level of cAMP. The cAMP, in turn, recruits PKA, which results in an enhancement of transmitter release from the sensory neuron.

■ Repeated tail shocks, or several puffs of serotonin to the terminal region of the sensory neuron, leading to long-term sensitization, activate initially a similar biochemical pathway as during short-term sensitization. However, now the PKA recruits MAPK. Both kinases translocate into the cell's nucleus, where they activate a cascade of transcriptional processes.

■ The activation of several target genes results in two major effects: First, the PKA exhibits persistent activity. Second, the synaptic organization of the connection between the sensory and motor neurons is altered; this includes the growth of new synaptic connections.

■ Nonhuman models to study explicit memory have been established in a number of mammals which have a medial temporal-lobe, including a hippocampus.

■ These studies have revealed place cells among hippocampal pyramidal cells which collectively form spatial maps of different environments of the animal. The establishment of these maps resembles learning processes and the conversion of short-term spatial memory into long-term memory.

■ A second phenomenon discovered in the hippocampus is LTP, which is reflected by an abrupt and sustained increase in efficiency of synaptic transmission after high-frequency stimulation of a monosynaptic excitatory pathway. Due to its long-lasting effect, LTP is thought to be involved in memory storage.

▣ A detailed molecular analysis of the late phase of LTP, which lasts for at least one day, in the Schaffer collateral pathway of the hippocampus has implicated PKA as a key molecule. The final step in this process appears to be the formation of new synapses.

▣ Reduction of the activity of PKA by employing a transgenic approach causes a marked decrease of the late phase of LTP, and a reduced stability of place cells in the CA1 region of the hippocampus. This result underlines the importance of PKA in the process of memory consolidation.

▣ Generation of new neurons appears to be involved in hippocampus-dependent learning behavior and memory formation of both rodents and birds.

Recommended reading

Bailey, C. H., and Kandel, E. R. (1993). Structural changes accompanying memory storage. *Annual Review of Neuroscience*, 55, 397–426.

This review examines the formation of new synapses as a highly conserved feature of long-lasting behavioral memory in a variety of vertebrate and invertebrate species.

Bliss, T. V. P., and Collingridge, G. L. (1993). A synaptic model of memory: long-term potentiation in the hippocampus. *Nature*, 361, 31–39.

A comprehensive review of the physiological and cellular mechanisms of LTP in the hippocampus. Primarily targeted for readers at a more advanced level.

Kandel, E. R. (2001). The molecular biology of memory storage: a dialogue between genes and synapses. *Science*, 294, 1030–1038.

Based on the author's address to the Nobel Foundation, this essay provides an excellent review of the work of Eric Kandel.

Mayford, M., and Kandel, E. R. (1999). Genetic approaches to memory storage. *Trends in Genetics*, 15, 463–470.

This article reviews some key advances made through genetic approaches in the study of learning and memory, and discusses possible future directions.

O'Keefe, J., Burgess, N., Donnett, J. G., Jeffery, K. J., and Maguire, E. A. (1998). Place cells, navigational accuracy, and the human hippocampus. *Philosophical Transactions of the Royal Society of London B*, 353, 1333–1340.

A useful summary of how the hippocampal formation is involved in spatial navigation, mainly based on the work of the group headed by John O'Keefe and Neil Burgess.

Squire, L. R., and Kandel, E. R. (1999). *Memory: from mind to molecules*. Scientific American Library, New York.

A highly stimulating book, written by two of the world-leaders in memory research. Easy to read, without oversimplifying any facts. The ideal start!

Yuste, R., and Bonhoeffer, T. (2001). Morphological changes in dendritic spines associated with long-term synaptic plasticity. *Annual Review of Neuroscience*, 24, 1071–1089.

An authoritative review of the role of morphological changes of dendritic spines in hippocampal LTP in rodents, with special emphasis on a discussion of 'online' studies using time-lapse imaging to visualize the generation of new spines.

Questions

11.1 What is the difference between implicit and explicit memory? What experimental approaches have led to the distinction between these two types of memory?

11.2 How do the behavioral effects of short- and long-term sensitization on the gill-withdrawal reflex of *Aplysia* differ? What are the molecular mechanisms mediating these two effects?

11.3 What role has been proposed for the mammalian hippocampus in the establishment of long-term memory?

11.4 What is long-term potentiation? Describe the molecular mechanisms underlying the late phase of long-term potentiation in the Schaffer collateral pathway of the mammalian hippocampus. How can a transgenic approach assist the researcher in defining the role of molecular factors involved in this process?

11.5 Summarize the experimental evidence suggesting a link between the generation of new neurons in the adult hippocampus and the formation of memories.

References

1: Introduction

Baerends, G. P., and Baerends-van Roon, J. M. (1950). An introduction to the study of the ethology of cichlid fishes. *Behaviour*, **1**(Suppl.), 1–242.

Zupanc, G. K. H., and Banks, J. R. (1998). Electric fish: animals with a sixth sense. *Biological Sciences Review*, **11**(2), 23–27.

2: The study of animal behavior: a brief history

Autrum, H. (1996). *Mein Leben: Wie sich Glück und Verdienst verketten*. Springer-Verlag, Berlin.

Becker, J. B., Breedlove, S. M., and Crews, D. (ed.) (1993). *Behavioral endocrinology*. MIT Press, Cambridge, Massachusetts/London.

Bullock, T. H. (1996). Theodore H. Bullock. In *The history of neuroscience in autobiography. Volume 1* (ed. L. R. Squire), pp. 110–156. Society for Neuroscience, Washington, DC.

Bullock, T. H. (1999). Neuroethology has pregnant agendas. *Journal of Comparative Physiology A*, **185**, 291–295.

Bullock, T. H., and Horridge, G. A. (1965). *Structure and function in the nervous system of invertebrates. Volumes I and II*. W.H. Freeman and Company, San Francisco/London.

Darwin, C. (1892). *The expression of the emotions in man and animals*. John Murray, London.

De Bono, M., and Bargmann, C. I. (1998). Natural variation in a neuropeptide Y receptor homolog modifies social behavior and food response in *C. elegans. Cell*, **94**, 679–689.

Dewsbury, D. A. (1989). A brief history of the study of animal behavior in North America. In *Perspectives in ethology. Volume 8: whither ethology?* (ed. P. P. G. Bateson and P. H. Klopfer), pp. 85–122. Plenum Press, New York.

Ewert, J.-P. (1980). *Neuroethology: an introduction to the neurophysiological fundamentals of behavior*. Springer-Verlag, Berlin.

Frisch, K. v. (1950). *Bees: their vision, chemical senses, and language*. Cornell University Press, Ithaca, New York.

Frisch, K. v. (1966). *The dancing bees: an account of the life and senses of the honey bee*. Methuen & Co., Ltd, London.

Frisch, K. v. (1967). *A biologist remembers*. Pergamon Press, Oxford.

Frisch, K. v. (1967). *The dance language and orientation of bees*. The Belknap Press of Harvard University Press/Oxford University Press, Cambridge, Massachusetts/London.

Hassenstein, B. (1964). Erich von Holst †. *Zoologischer Anzeiger*, **27**(Suppl.), 676–682.

Heisenberg, M. (1997). Genetic approach to neuroethology. *BioEssays*, **19**, 1065–1073.

Holst, E. v. (1973). *The behavioural physiology of animals and man. Volume 1*. Methuen & Co. Ltd, London.

Jaynes, J. (1969). The historical origins of 'ethology' and 'comparative psychology'. *Animal Behaviour*, **17**, 601–606.

Lorenz, K. (1966). *On aggression*. Methuen & Co., Ltd, London.

Lorenz, K. (1977). *Behind the mirror: a search for a natural history of human knowledge*. Methuen & Co., Ltd, London.

Lorenz, K. Z. (1954). *Man meets dog*. Methuen & Co., Ltd, London.

Lorenz, K. Z. (1964). *King Solomon's ring: new light on animal ways*. Methuen & Co., Ltd, London.

Lorenz, K. Z. (1981). *The foundations of ethology*. Springer-Verlag, New York/Wien.

Morgan, C. L. (1900). *Animal behaviour*. Edward Arnold, London.

Pavlov, I. P. (1960). *Conditioned reflexes: an investigation of the physiological activity of the cerebral cortex*. Dover Publications, Inc., New York.

Pflüger, H.-J., and Menzel, R. (1999). Neuroethology, its roots and future. *Journal of Comparative Physiology A*, 185, 389–392.

Tinbergen, N. (1951). *The study of instinct*. Oxford University Press, London.

Tinbergen, N. (1965). *Social behaviour in animals: with special reference to vertebrates*. Methuen & Co., Ltd, London.

Warden, C. J. (1927). The historical development of comparative psychology. *Psychological Review*, 34, 57–168.

Watson, J. B. (1912). Instinctive activity in animals: some recent experiments and observations. *Harper's Magazine*, 124, 376–382.

Wilson, E. O. (1975). *Sociobiology: the new synthesis*. Belknap Press of Harvard University Press, Cambridge, Massachusetts/London.

3:   The tools and concepts of behavioral neurobiology

Bower, T. G. R. (1966). Heterogeneous summation in human infants. *Animal Behaviour*, 14, 395–398.

Drees, O. (1952). Untersuchungen über die angeborenen Verhaltensweisen bei Springspinnen (Salticidae). *Zeitschrift für Tierpsychologie*, 9, 169–207.

Heiligenberg, W., and Kramer, U. (1972). Aggressiveness as a function of external stimulation. *Journal of Comparative Physiology*, 77, 332–340.

Lack, D. (1939). The behaviour of the robin. Part I. The life-history, with special reference to aggressive behaviour, sexual behaviour, and territory. Part II. A partial analysis of aggressive and recognitional behaviour. *Proceedings of the Zoological Society A*, 109, 169–219.

Lack, D. (1953). *The life of the robin*. Penguin Books, Melbourne/London/Baltimore.

Leong, C.-Y. (1969). The quantitative effect of releasers on the attack readiness of the fish *Haplochromis burtoni* (Cichlidae, Pisces). *Zeitschrift für Vergleichende Physiologie*, 65, 29–50.

Michelsen, A., Andersen, B. B., Storm, J., Kirchner, W. H., and Lindauer, M. (1992). How honeybees perceive communication dances, studied by means of a mechanical model. *Behavioral Ecology and Sociobiology*, 30, 143–150.

Miller, J. P., and Selverston, A. I. (1979). Rapid killing of single neurons by irradiation of intracellularly injected dye. *Science*, 206, 702–704.

Seitz, A. (1940). Die Paarbildung bei einigen Cichliden: I. Die Paarbildung bei *Astatotilapia strigigena* Pfeffer. *Zeitschrift für Tierpsychologie*, 4, 40–84.

Tinbergen, N., and Perdeck, A. C. (1950). On the stimulus situation releasing the begging response in the newly hatched herring gull chick (*Larus argentatus argentatus* Pont.). *Behaviour*, 3, 1–39.

4: Spatial orientation and sensory guidance

Barnes, W. J. P. (1993). Sensory guidance in arthropod behaviour: common principles and experimental approaches. *Comparative Biochemistry and Physiology*, 104A, 625–632.

Corey, D. P., and Hudspeth, A. J. (1979). Ionic basis of the receptor potential in a vertebrate hair cell. *Nature*, 281, 675–677.

Flock, Å. (1965). Transducing mechanisms in the lateral line canal organ receptors. *Cold Spring Harbor Symposia on Quantitative Biology*, 30, 133–145.

Fraenkel, G. S., and Gunn, D. L. (1940). *The orientation of animals: kineses, taxes and compass reactions*. Clarendon Press, Oxford.

Fullard, J. H., Simmons, J. A., and Saillant, P. A. (1994). Jamming bat echolocation: the dogbane tiger moth *Cycnia tenera* times its clicks to the terminal attack calls of the big brown bat *Eptesicus fuscus*. *Journal of Experimental Biology*, 194, 285–298.

Holst, E. v. (1950). Die Tätigkeit des Statolithenapparates im Wirbeltierlabyrinth. *Naturwissenschaften*, 37, 265–272.

Howard, J., Roberts, W. M., and Hudspeth, A. J. (1988). Mechanoelectrical transduction by hair cells. *Annual Review in Biophysics and Biophysical Chemistry*, 17, 99–124.

Hudspeth, A. J., and Corey, D. P. (1977). Sensitivity, polarity, and conductance change in the response of vertebrate hair cells to controlled mechanical stimuli. *Proceedings of the National Academy of Sciences, USA*, 74, 2407–2411.

Jennings, H. S. (1962). *Behavior of the lower organisms*. Indiana University Press, Bloomington.

Kühn, A. (1919). *Die Orientierung der Tiere im Raum*. Fischer Verlag, Jena.

Machemer, H. (1988). Galvanotaxis: Grundlagen der elektromechanischen Kopplung und Orientierung bei

Paramecium. In *Praktische Verhaltensbiologie* (ed.
G. K. H. Zupanc), pp. 60–82. Verlag Paul Parey,
Berlin/Hamburg.

Mogami, Y., Pernberg, J., and Machemer, H. (1990).
Messenger role of calcium in ciliary electromotor
coupling: a reassessment. *Cell Calcium*, 11, 665–673.

Pollak, G. D., Marsh, D. S., Bodenhamer, R., and
Souther, A. (1977). Characteristics of phasic on neurons
in inferior colliculus of unanaesthetized bats with
observations relating to mechanisms of echo ranging.
Journal of Neurophysiology, 40, 926–942.

Schnitzler, H.-U., and Ostwald, J. (1983). Adaptations for
the detection of fluttering insects by echolocation in
horseshoe bats. In *Advances in vertebrate neuroethology*
(ed. J.-P. Ewert, R. R. Capranica, and
D. J. Ingle), pp. 801–827. Plenum Press, New York.

Schöne, H. (1984). *Spatial orientation: the spatial control
of behavior in animals and man.* Princeton University
Press, Princeton.

Simmons, J. A., Fenton, M. B., and O'Farrell, M. J. (1979).
Echolocation and pursuit of prey by bats. *Science*, 203,
16–21.

Suga, N. (1990). Bisonar and neural computation in bats.
Scientific American, 262(6), 60–68.

5: Neuronal control of motor output: swimming in toad tadpoles

Brown, T. G. (1911). The intrinsic factors in the act of
progression in the mammal. *Proceedings of the Royal
Society of London Series B*, 84, 308–319.

Brown, T. G. (1914). On the nature of the fundamental
activity of the nervous centres; together with an analysis
of the conditioning of rhythmic activity in progression
and a theory of the evolution of function in the nervous
system. *Journal of Physiology (London)*, 48, 18–46.

Marder, E. (2001). Moving rhythms. *Nature*, 410, 755.

Roberts, A. (1990). How does a nervous system produce
behaviour? A case study in neurobiology. *Science
Progress*, 74, 31–51.

Roberts, A., Soffe, S. R., and Perrins, R. (1997). Spinal
networks controlling swimming in hatchling *Xenopus*
tadpoles. In *Neurons, networks, and motor behavior*
(ed. P. S. G. Stein, S. Grillner, A. I. Selverston, and
D. G. Stuart), pp. 83–89. MIT Press, Cambridge, MA.

Roberts, A., Soffe, S. R., Wolf, E. S., Yoshida, M., and
Zhao, F.-Y. (1998). Central circuits controlling
locomotion in young frog tadpoles. *Annals of the
New York Academy of Sciences*, 860, 19–34.

Sherrington, C. S. (1906). *The integrative action of the
nervous system.* Yale University Press, New Haven, CT.

6: Neuronal processing of sensory information

Carr, C. E., and Konishi, M. (1990). A circuit for detection
of interaural time differences in the brain stem of the
barn owl. *Journal of Neuroscience*, 10, 3227–3246.

Ewert, J.-P. (1967). Aktivierung der Verhaltensfolge beim
Beutefang der Erdkröte (*Bufo bufo* L.) durch elektrische
Mittelhirn-Reizung. *Zeitschrift für Vergleichende
Physiologie*, 54, 455–481.

Ewert, J.-P. (1968). Der Einfluß von Zwischenhirndefekten
auf die Visuomotorik im Beute- und Fluchtverhalten der
Erdkröte (*Bufo bufo* L.). *Zeitschrift für Vergleichende
Physiologie*, 61, 41–70.

Ewert, J.-P. (1969). Quantitative Analyse von Reiz
Reaktionsbeziehungen bei visuellem Auslösen der
Beutefang-Wendereaktion der Erdkröte *Bufo bufo* (L.).
Pflügers Archiv, 308, 225–243.

Ewert, J.-P., Arend, B., Becker, V., and Borchers, H.-W.
(1979). Invariants in configurational prey selection by
Bufo bufo (L.). *Brain, Behavior and Evolution*,
16, 38–51.

Ewert, J.-P., and Borchers, H.-W. (1971).
Reaktionscharakteristik von Neuronen aus dem
Tectum opticum und Subtectum der Erdkröte (*Bufo bufo*
L.). *Zeitschrift für Vergleichende Physiologie*,
71, 165–189.

Ewert, J.-P., and Hock, F. (1972). Movement-sensitive
neurones in the toad's retina. *Experimental Brain
Research*, 16, 41–59.

Ewert, J.-P., and Traud, R. (1979). Releasing stimuli for
antipredator behaviour in the common toad *Bufo bufo*
(L.). *Behaviour*, 68, 170–180.

Ewert, J.-P., and Wietersheim, A. v. (1974).
Musterauswertung durch Tectum- und Thalamus/
Praetectum-Neurone im visuellen System der Kröte *Bufo
bufo* (L.). *Journal of Comparative Physiology*, 92,
131–148.

Ewert, J.-P., and Wietersheim, A. v. (1974). Einfluß von
Thalamus/Praetectum-Defekten auf die Antwort von
Tectum-Neuronen gegenüber bewegten visuellen
Mustern bei der Kröte *Bufo bufo* (L.). *Journal of
Comparative Physiology*, 92, 149–160.

Grüsser-Cornehls, U., Grüsser, O.-J., and Bullock, T. H.
(1963). Unit responses in the frog's tectum to moving
and nonmoving visual stimuli. *Science*, 141, 820–822.

Jeffress, L. A. (1948). A place theory of sound localization. *Journal of Comparative and Physiological Psychology*, **41**, 35–39.

Knudsen, E. I. (1982). Auditory and visual maps of space in the optic tectum of the owl. *Journal of Neurosicence*, **2**, 1177–1194.

Knudsen, E. I., and Brainard, M. S. (1991). Visual instruction of the neural map of auditory space in the developing optic tectum. *Science*, **253**, 85–87.

Knudsen, E. I., and Konishi, M. (1979). Mechanisms of sound localization in the barn owl (*Tyto alba*). *Journal of Comparative Physiology A*, **133**, 13–21.

Knudsen, E. I., Blasdel, G. G., and Konishi, M. (1979). Sound localization by the barn owl (*Tyto alba*) measured with the search coil technique. *Journal of Comparative Physiology A*, **133**, 1–11.

Konishi, M. (1990). Similar algorithms in different sensory systems and animals. *Cold Spring Harbor Symposia on Quantitative Biology*, **55**, 575–584.

Lettvin, J. Y., Maturana, H. R., McCulloch, W. S., and Pitts, W. H. (1959). What the frog's eye tells the frog's brain. *Proceedings of the Institute of Radio Engineers*, **47**, 1940–1951.

Manley, G. A., Köppl, C., and Konishi, M. (1988). A neural map of interaural intensity differences in the brain stem of the barn owl. *Journal of Neuroscience*, **8**, 2665–2676.

Payne, R. S. (1971). Acoustic location of prey by barn owls (*Tyto alba*). *Journal of Experimental Biology*, **54**, 535–573.

7:   Sensori-motor integration: the jamming avoidance response of the weakly electric fish, *Eigenmannia*

Bastian, J., and Yuthas, J. (1984). The jamming avoidance response of *Eigenmannia*: properties of a diencephalic link between sensory processing and motor output. *Journal of Comparative Physiology A*, **154**, 895–908.

Bullock, T. H., and Heiligenberg, W. (ed.) (1986). *Electroreception*. John Wiley & Sons, New York/Chichester/Brisbane/Toronto/Singapore.

Heiligenberg, W. (1977). *Principles of electrolocation and jamming avoidance response in electric fish*. Springer-Verlag, Berlin/Heidelberg/New York.

Heiligenberg, W. (1991). The jamming avoidance response of the elecric fish, *Eigenmannia*: computational rules and their neuronal implementation. *Seminars in the Neurosciences*, **3**, 3–18.

Heiligenberg, W., and Rose, G. (1985). Phase and amplitude computations in the midbrain of an electric fish: intracellular studies of neurons participating in the jamming avaoidance response of *Eigenmannia*. *Journal of Neuroscience*, **5**, 515–531.

Heiligenberg, W., Baker, C., and Matsubara, J. (1978). The jamming avoidance response in *Eigenmannia* revisited: the structure of a neuronal democracy. *Journal of Comparative Physiology A*, **127**, 267–286.

Heiligenberg, W., Keller, C. H., Metzner, W., and Kawasaki, M. (1991). Structure and function of neurons in the complex of the nucleus electrosensorius of the gymnotiform fish *Eigenmannia*: detection and processing of electric signals in social communication. *Journal of Comparative Physiology A*, **169**, 151–164.

Heiligenberg, W., Metzner, W., Wong, C. J. H., and Keller, C. H. (1996). Motor control of the jamming avoidance response of *Apteronotus leptorhynchus*: evolutionary changes of a behavior and its neuronal substrates. *Journal of Comparative Physiology A*, **179**, 653–674.

Kawasaki, M., Maler, L., Rose, G. J., and Heiligenberg, W. (1988). Anatomical and functional organization of the prepacemaker nucleus in gymnotiform electric fish: the accommodation of two behaviors in one nucleus. *Journal of Comparative Neurology*, **276**, 113–131.

Keller, C. H. (1988). Stimulus discrimination in the diencephalon of *Eigenmannia*: the emergence and sharpening of a sensory filter. *Journal of Comparative Physiology A*, **162**, 747–757.

Keller, C. H., Maler, L., and Heiligenberg, W. (1990). Structural and functional organization of a diencephalic sensory–motor interface in the gymnotiform fish, *Eigenmannia*. *Journal of Comparative Neurology*, **293**, 347–376.

Lissmann, H. W. (1958) On the function and evolution of electric organs in fish. *Journal of Experimental Biology*, **35**, 156–191.

Lissmann, H. W., and Machin, K. E. (1958) The mechanism of object location in *Gymnarchus niloticus* and similar fish. *Journal of Experimental Biology*, **35**, 451–486.

Metzner, W. (1993). The jamming avoidance response in *Eigenmannia* is controlled by two separate motor pathways. *Journal of Neuroscience*, **13**, 1862–1878.

Rose, G. J., and Heiligenberg, W. (1985). Temporal hyperacuity in the electric sense of fish. *Nature*, **318**, 178–180.

Rose, G. J., and Heiligenberg, W. (1986). Neural coding of difference frequencies in the midbrain of the electric fish *Eigenmannia*: reading the sense of rotation in an amplitude–phase plane. *Journal of Comparative Physiology A*, **158**, 613–624.

Watanabe, A., and Takeda, K. (1963). The change of discharge frequency by ac stimulus in a weak electric fish. *Journal of Experimental Biology*, **40**, 57–66.

Zupanc, G. K. H., and Lamprecht, J. (1994). Walter Heiligenberg (1938–1994). *Trends in Neurosciences*, **17**, 507–508.

8: Neuromodulation: the accommodation of motivational changes in behavior

Dickinson, P. S., Mecsas, C., and Marder, E. (1990). Neuropeptide fusion of two motor-pattern generator circuits. *Nature*, **344**, 155–157.

Forger, N. G., and Breedlove, S. M. (1987). Seasonal variation in mammalian striated muscle mass and motoneuron morphology. *Journal of Neurobiology*, **18**, 155–165.

Getting, P. A., and Dekin, M. S. (1985). *Tritonia* swimming: a model system for integration within rhythmic motor systems. In *Model neural networks and behavior* (ed. A. I. Selverston), pp. 3–20. Plenum Press, New York/London.

Hatton, G. I. (1997). Function-related plasticity in hypothalamus. *Annual Review of Neuroscience*, **20**, 375–397.

Herkenham, M. (1987). Mismatches between neurotransmitter and receptor localizations in brain: observations and implications. *Neuroscience*, **23**, 1–38.

Huber, R., and Delago, A. (1998). Serotonin alters decisions to withdraw in fighting crayfish, *Astacus astacus*: the motivational concept revisited. *Journal of Comparative Physiology A*, **182**, 573–583.

Huber, R., and Kravitz, E. A. (1995). A quantitative analysis of agonistic behavior in juvenile American lobsters (*Homarus americanus* L.). *Brain, Behavior and Evolution*, **46**, 72–83.

Huber, R., Smith, K., Delago, A., Isaksson, K., and Kravitz, E. A. (1997). Serotonin and aggressive motivation in crustaceans: altering the decision to

retreat. *Proceedings of the National Academy of Sciences, USA*, **94**, 5939–5942.

Kawasaki, M., Maler, L., Rose, G. J., and Heiligenberg, W. (1988). Anatomical and functional organization of the prepacemaker nucleus in gymnotiform electric fish: the accommodation of two behaviors in one nucleus. *Journal of Comparative Neurology*, **276**, 113–131.

Keller, C. H., Maler, L., and Heiligenberg, W. (1990). Structural and functional organization of a diencephalic sensory–motor interface in the gymnotiform fish, *Eigenmannia*. *Journal of Comparative Neurology*, **293**, 347–376.308

Kurz, E. M., Sengelaub, D. R., and Arnold, A. P. (1986). Androgens regulate the dendritic length of mammalian motoneurons in adulthood. *Science*, **232**, 395–398.

Marder, E., and Richards, K. S. (1999). Development of the peptidergic modulation of a rhythmic pattern generating network. *Brain Research*, **848**, 35–44.

Swanson, L. W. (1988/89). The neural basis of motivated behavior. *Acta Morphologica Neerlando-Scandinavica*, **26**, 165–176.

Yeh, S.-R., Musolf, B. E., and Edwards, D. H. (1997). Neuronal adaptations to changes in the social dominance status of crayfish. *Journal of Neuroscience*, **17**, 697–708.

Zupanc, G. K. H. (1991). The synaptic organization of the prepacemaker nucleus in weakly electric knifefish, *Eigenmannia*: a quantitative ultrastructural study. *Journal of Neurocytology*, **20**, 818–833.

Zupanc, G. K. H. (2001). Adult neurogenesis and neuronal regeneration in the central nervous system of teleost fish. *Brain, Behavior and Evolution*, **58**, 250–275.

Zupanc, G. K. H., and Heiligenberg, W. (1989). Sexual maturity-dependent changes in neuronal morphology in the prepacemaker nucleus of adult weakly electric knifefish, *Eigenmannia*. *Journal of Neuroscience*, **9**, 3816–3827.

Zupanc, G. K. H., and Heiligenberg, W. (1992). The structure of the diencephalic prepacemaker nucleus revisited: light microscopic and ultrastructural studies. *Journal of Comparative Neurology*, **323**, 558–569.

Zupanc, G. K. H., and Lamprecht, J. (2000). Towards a cellular understanding of motivation: structural reorganization and biochemical switching as key mechanisms of behavioral plasticity. *Ethology*, 106, 467–477.

9:   Large-scale navigation: migration and homing

Beason, R. C., and Nichols, J. E. (1984). Magnetic orientation and magnetically sensitive material in a transequatorial migratory birds. *Nature*, 309, 151–153.

Blakemore, R. P. (1982). Magnetotactic bacteria. *Annual Review of Microbiology*, 36, 217–238.

Bowen, B. W., Abreu-Grobois, F. A., Balazs, G. H., Kamezaki, N., Limpus, C. J., and Ferl, R. J. (1995). Trans-Pacific migrations of the loggerhead turtle (*Caretta caretta*) demonstrated with mitochondrial DNA markers. *Proceedings of the National Academy of Sciences, USA*, 92, 3731–3734.

Brower, L. P. (1996). Monarch butterfly orientation: missing pieces of a magnificent puzzle. *Journal of Experimental Biology*, 199, 93–103.

Cooper, J. C., Scholz, A. T., Horrall, R. M., Hasler, A. D., and Madison, D. M. (1976). Experimental confirmation of the olfactory hypothesis with artificially imprinted homing coho salmon (*Oncorhynchus kisutch*). *Journal of the Fisheries Research Board of Canada*, 33, 703–710.

Dickhoff, W. W., Folmar, L. C., and Gorbman, A. (1978). Changes in plasma thyroxine during the smoltification of coho salmon, *Oncorhynchus kisutch*. *General and Comparative Endocrinology*, 36, 229–232.

Emlen, S. T. (1967). Migratory orientation in the Indigo Bunting, *Passerina cyanea*. Part I. Evidence for use of celestial cues. *Auk*, 84, 309–342.

Emlen, S. T. (1967). Migratory orientation in the Indigo bunting, *Passerina cyanea*. Part II. Mechanisms of celestial orientation. *Auk*, 84, 463–489.

Emlen, S. T., and Emlen, J. T. (1966). A technique for recording migratory orientation of captive birds. *Auk*, 83, 361–367.

Frisch, K. v. (1941). Die Bedeutung des Geruchssinnes im Leben der Fische. *Naturwissenschaften*, 29, 321–333.

Hasler, A. D. (1966). *Underwater guideposts: homing of salmon*. University of Wisconsin Press, Madison/Milwaukee/London.

Hasler, A. D., and Wisby, W. J. (1951). Discrimination of stream odors by fishes and relation to parent stream behavior. *American Naturalist*, 85, 223–238.

Helbig, A. J. (1991). Inheritance of migratory direction in a bird species: a cross-breeding experiment with SE- and SW-migrating blackcaps (*Sylvia atricapilla*). *Behavioral Ecology and Sociobiology*, 28, 9–12.

Hoar, W. S. (1976). Smolt transformation: evolution, behavior and physiology. *Journal of the Fish Research Board of Canada*, 33, 1234–1252.

Hoffmann, K. (1960). Experimental manipulation of the orientational clock in birds. *Cold Spring Harbor Symposia on Quantitative Biology*, 25, 379–387.

Kalmijn, A. J. (1982). Electric and magnetic field detection in elasmobranch fishes. *Science*, 218, 916–918.

Keeton, W. T. (1971). Magnets interfer with pigeon homing. *Proceedings of the National Academy of Sciences USA*, 68, 102–106.

Kramer, G. (1950). Orientierte Zugaktivität gekäfigter Singvögel. *Naturwissenschaften*, 37, 188.

Kramer, G. (1950). Weitere Analyse der Faktoren, welche die Zugaktivität des gekäfigten Vogels orientieren. *Naturwissenschaften*, 37, 377–378.

Kramer, G., and St. Paul, U. v. (1950). Stare (*Sturnus vulgaris* L.) lassen sich auf Himmelsrichtungen dressieren. *Naturwissenschaften*, 37, 526–527.

Kreithen, M. L. (1979). The sensory world of the homing pigeon. In *Neural mechanisms of behavior in the pigeon* (ed. A. M. Granda and J. H. Maxwell), pp. 21–33. Plenum Press, New York/London.

Kreithen, M. L., and Keeton, W. T. (1974). Detection of changes in atmospheric pressure by the homing pigeon, *Columba livia*. *Journal of Comparative Physiology*, 89, 73–82.

Lohmann, K. J., and Fittinghoff Lohmann, C. M. (1992). Orientation to oceanic waves by green turtle hatchlings. *Journal of Experimental Biology*, 171, 1–13.

Lohmann, K. J., and Fittinghoff Lohmann, C. M. (1994). Acquisition of magnetic directional preference in hatchling loggerhead sea turtles. *Journal of Experimental Biology*, 190, 1–8.

Lohmann, K. J., and Lohmann, C. M. F. (1998). Migratory guidance in marine turtles. *Journal of Avian Biology*, 29, 585–596.

Lohmann, K. J., Swartz, A. W., and Lohmann, C. M. F. (1995). Perception of ocean wave direction by sea turtles. *Journal of Experimental Biology*, 198, 1079–1085.

Matthews, G. V. T. (1953). Navigation in the Manx shearwater. *Journal of Experimental Biology*, 30, 370–396.

Papi, F. (1982). Olfaction and homing in pigeons: ten years of experiments. In *Avian navigation* (ed. F. Papi and H. G. Wallraff), pp. 149–159. Springer-Verlag, Berlin/Heidelberg/New York.

Quinn, T. P., and Dittman, A. H. (1990). Pacific salmon migrations and homing: mechanisms and adaptive significance. *Trends in Ecology and Evolution*, 5, 174–177.

Ritz, T., Adem, S., and Schulten, K. (2000). A model for photoreceptor-based magnetoreception in birds. *Biophysical Journal*, 78, 707–718.

Sauer, F. (1957). Die Sternorientierung nächtlich ziehender Grasmücken (*Sylvia atricapilla, borin* und *curruca*). *Zeitschrift für Tierpsychologie*, 14, 29–70.

Schmidt-Koenig, K. (1960). Internal clocks and homing. *Cold Spring Harbor Symposia on Quantitative Biology*, 25, 389–393.

Scholz, A. T., Horrall, R. M., Cooper, J. C., and Hasler, A. D. (1976). Imprinting to chemical cues: the basis for homestream selection in salmon. *Science*, 196, 1247–1249.

Scholz, A. T., Cooper, J. C., Madison, D. M., Horrall, R. M., Hasler, A. D., Dizon, A. E., and Poff, R. J. (1973). Olfactory imprinting in coho salmon: behavioral and electrophysiological evidence. *Proceedings of the Conference of the Great Lakes Research*, 16, 143–153.

Semm, P., and Beason, R. C. (1990). Responses to small magnetic variations by the trigeminal system of the bobolink. *Brain Research Bulletin*, 25, 735–740.

Teichmann, H. (1959). Über die Leistung des Geruchssinnes beim Aal (*Anguilla anguilla*). *Zeitschrift für Vergleichende Physiologie*, 42, 206–254.

Walcott, C. (1996). Pigeon homing: observations, experiments and confusions. *Journal of Experimental Biology*, 199, 21–27.

Walcott, C., and Green, R. P. (1974). Orientation of homing pigeons altered by a change in the direction of an applied magnetic field. *Science*, 184, 180–182.

Wiltschko, R. (1996). The function of olfactory input in pigeon orientation: does it provide navigational information or play another role? *Journal of Experimental Biology*, 199, 113–119.

Wiltschko, R., and Wiltschko, W. (1978). Evidence for the use of magnetic outward-journey information in homing pigeons. *Naturwissenschaften*, 65, 112–113.

Wiltschko, W. (1968). Über den Einfluß statischer Magnetfelder auf die Zugorientierung der Rotkehlchen (*Erithacus rubecula*). *Zeitschrift für Tierpsychologie*, 25, 537–558.

10: Communication: the neuroethology of cricket song

Bentley, D. R., and Hoy, R. R. (1972). Genetic control of the neuronal network generating cricket (*Teleogryllus gryllus*) song patterns. *Animal Behaviour*, 20, 478–492.

Horseman, G., and Huber, F. (1994). Sound localisation in crickets: I. Contralateral inhibition of an ascending auditory interneuron (AN1) in the cricket *Gryllus bimaculatus*. *Journal of Comparative Physiology A*, 175, 389–398.

Horseman, G., and Huber, F. (1994). Sound localisation in crickets: II. Modelling the role of a simple neural network in the prothoracic ganglion. *Journal of Comparative Physiology A*, 175, 399–413.

Hoy, R. R., and Paul, R. C. (1973). Genetic control of song specificity in crickets. *Science*, 180, 82–83.

Hoy, R. R., Hahn, J., and Paul, R. C. (1977). Hybrid cricket auditory behavior: evidence for genetic coupling in animal communication. *Science*, 195, 82–84.

Huber, F. (1990). Cricket neuroethology: neuronal basis of intraspecific acoustic communication. *Advances in the Study of Behavior*, 19, 299–356.

Huber, F. (1990). Nerve cells and insect behavior: studies on crickets. *American Zoologist*, 30, 609–627.

Kleindienst, H.-U., Wohlers, D. W., and Larsen, O. N. (1983). Tympanal membrane motion is necessary for hearing in crickets. *Journal of Comparative Physiology A*, 151, 397–400.

Larsen, O. N., and Michelsen, A. (1978). Biophysics of the ensiferan ears: III. The cricket ear as a four-input system. *Journal of Comparative Physiology A*, 123, 217–227.

Loher, W., and Dambach, M. (1989). Reproductive behavior. In *Cricket behavior and neurobiology* (ed. F. Huber, T. E. Moore, and W. Loher), pp. 43–82. Cornell University Press, Ithaca/London.

Pires, A., and Hoy, R. R. (1992). Temperature coupling in cricket acoustic communication: I. Field and laboratory studies of temperature effects on calling song production and recognition in *Gryllus firmus*. *Journal of Comparative Physiology A*, 171, 69–78.

Regen, J. (1914). Über die Anlockung des Weibchens von *Gryllus campestris* L. durch telephonisch übertragene Stridulationslaute des Männchens. *Pflüger's Archiv für die gesamte Physiologie des Menschen und der Tiere*, 155, 193–200.

Schildberger, K. (1984). Temporal selectivity of identified auditory neurons in the cricket brain. *Journal of Comparative Physiology A*, 155, 171–185.

Schmitz, B., Scharstein, H., and Wendler, G. (1982). Phonotaxis in *Gryllus campestris* L. (Orthoptera, Gryllidae): I. Mechanisms of acoustic orientation in intact female crickets. *Journal of Comparative Physiology A*, **148**, 431–444.

Schmitz, B., Scharstein, H., and Wendler, G. (1983). Phonotaxis in *Gryllus campestris* L. (Orthoptera, Gryllidae): II. Acoustic orientation of female crickets after occlusion of single sound entrances. *Journal of Comparative Physiology A*, **152**, 257–264.

Stout, J. F., and Huber, F. (1981). Responses to features of the calling song by ascending auditory interneurons in the cricket *Gryllus campestris. Physiological Entomology*, **6**, 199–212.

Thorson, J., Weber, T., and Huber, F. (1982). Auditory behavior of the cricket: II. Simplicity of calling-song recognition in *Gryllus*, and anomalous phonotaxis at abnormal carrier frequencies. *Journal of Comparative Physiology A*, **146**, 361–378.

Weber, T., Thorson, J., and Huber, F. (1981). Auditory behavior of the cricket: I. Dynamics of compensated walking and discrimination paradigms on the Kramer treadmill. *Journal of Comparative Physiology A*, **141**, 215–232.

Wohlers, D. W., and Huber, F. (1982). Processing of sound signals by six types of neurons in the prothoracic ganglion of the cricket, *Gryllus campestris* L. *Journal of Comparative Physiology A*, **146**, 161–173.

Young, D., and Ball, E. (1974). Structure and development of the auditory system in the prothoracic leg of the cricket *Teleogryllus commodus* (Walker): I. Adult structure. *Zeitschrift für Zellforschung und mikroskopische Anatomie*, **147**, 293–312.

11: Cellular mechanisms of learning and memory

Abel, T., Martin, K. C., Bartsch, D., and Kandel, E. R. (1998). Memory suppressor genes: inhibitory constraints on the storage of long-term memory. *Science*, **279**, 338–341.

Abel, T., Nguyen, P. V., Barad, M., Deuel, T. A. S., Kandel, E. R., and Bourtchouladze, R. (1997). Genetic demonstration of a role for PKA in the late phase of LTP and in hippocampus-based long-term memory. *Cell*, **88**, 615–626.

Bailey, C. H., and Chen, M. (1983). Morphological basis of long-term habituation and sensitization in *Aplysia. Science*, **220**, 91–93.

Bailey, C. H., and Chen, M. (1988). Long-term memory in *Aplysia* modulates the total number of varicosities of single identified sensory neurons. *Proceedings of the National Academy of Sciences, USA*, **85**, 2373–2377.

Barnea, A., and Nottebohm, F. (1994). Seasonal recruitment of hippocampal neurons in adult free-ranging black-capped chickadees. *Proceedings of the National Academy of Sciences, USA*, **91**, 11217–11221.

Barnea, A., and Nottebohm, F. (1996). Recruitment and replacement of hippocampal neurons in young and adult chickadees: an addition to the theory of hippocampal learning. *Proceedings of National Academy of Sciences, USA*, **93**, 714–718.

Bartsch, D., Ghirardi, M., Skehel, P. A., Karl, K. A., Herder, S. P., Chen, M., Bailey, C. H., and Kandel, E. R. (1995). *Aplysia* CREB2 represses long-term facilitation: relief of repression converts transient facilitation into long-term functional and structural change. *Cell*, **83**, 979–992.

Bliss, T. V. P., and Lømo, T. (1973). Long-lasting potentiation of synaptic transmission in the dentate area of the anaesthetized rabbit following stimulation of the perforant path. *Journal of Physiology*, **232**, 331–356.

Byers, D., Davis, R. L., and Kiger Jr., J. A. (1981). Defect in cyclic AMP phosphodiesterase due to the *dunce* mutation of learning in *Drosophila melanogaster. Nature*, **289**, 79–81.

Carew, T. J., and Sahley, C. L. (1986). Invertebrate learning and memory: from behavior to molecules. *Annual Review of Neuroscience*, **9**, 435–487.

Dudai, Y., Jan, Y.-N., Byers, D., Quinn, W. G., and Benzer, S. (1976). *dunce*, a mutant of *Drosophila* deficient in learning. *Proceedings of the National Academy of Sciences, USA*, **73**, 1684–1688.

Engert, F., and Bonhoeffer, T. (1999). Dendritic spine changes associated with hippocampal long-term synaptic plasticity. *Nature*, **399**, 66–70.

Frost, W. N., Castellucci, V. F., Hawkins, R. D., and Kandel, E. R. (1985). Monosynaptic connections made by the sensory neurons of the gill- and siphon-withdrawal reflex in *Aplysia* participate in the storage of long-term memory for sensitization. *Proceedings of the National Academy of Sciences, USA*, **82**, 8266–8269.

Glanzman, D. L., Mackey, S. L., Hawkins, R. D., Dyke, A. M., Lloyd, P. E., and Kandel, E. R. (1989). Depletion of serotonin in the nervous system of

Aplysia reduces the behavioral enhancement of gill withdrawal as well as the heterosynaptic facilitation produced by tail shock. *Journal of Neuroscience*, 9, 4200–4213.

Hebb, D. O. (1949). *The organization of behavior: a neuropsychological theory*. John Wiley & Sons, New York.

Kempermann, G., Kuhn, H. G., and Gage, F. H. (1997). More hippocampal neurons in adult mice living in an enriched environment. *Nature*, 386, 493–495.

Kempermann, G., Kuhn, H. G., and Gage, F. H. (1998). Experience-induced neurogenesis in the senescent dentate gyrus. *Journal of Neuroscience*, 18, 3206–3212.

Lechner, H. A., and Byrne, J. H. (1998). New perspectives on classical conditioning: a synthesis of Hebbian and Non-Hebbian mechanisms. *Neuron*, 20, 355–358.

Markakis, E. A., and Gage, F. H. (1999). Adult-generated neurons in the dentate gyrus send axonal projections to field CA$_3$ and are surrounded by synaptic vesicles. *Journal of Comparative Neurology*, 406, 449–460.

Patel, S. N., Clayton, N. S., and Krebs, J. R. (1997). Spatial learning induces neurogenesis in the avian brain. *Behavioural Brain Research*, 89, 115–128.

Shors, T. J., Miesegaes, G., Beylin, A., Zhao, M., Rydel, T., and Gould, E. (2001). Neurogenesis in the adult is involved in the formation of trace memories. *Nature*, 410, 372–376.

Milner, B., Squire, L. R., and Kandel, E. R. (1998). Cognitive neuroscience and the study of memory. *Neuron*, 20, 445–468.

O'Keefe, J., and Dostrovsky, J. (1971). The hippocampus as a spatial map. Preliminary evidence from unit activity in the freely-moving rat. *Brain Research*, 34, 171–175.

O'Keefe, J., and Nadel, L. (1978). *The hippocampus as a cognitive map*. Clarendon Press, Oxford.

Quinn, W. G., Harris, W. A., and Benzer, S. (1974). Conditioned behavior in *Drosophila melanogaster*. *Proceedings of the National Academy of Sciences, USA*, 71, 708–712.

Rotenberg, A., Abel, T., Hawkins, R. D., Kandel, E. R., and Muller, R. U. (2000). Parallel instabilities of long-term potentiation, place cells, and learning caused by decreased protein kinase A activity. *Journal of Neuroscience*, 20, 8096–8102.

Squire, L. R., and Zola, S. M. (1996). Structure and function of declarative and nondeclarative memory systems. *Proceedings of the National Academy of Sciences, USA*, 93, 13515–13522.

Squire, L. R., and Zola-Morgan, S. (1991). The medial temporal lobe memory system. *Science*, 253, 1380–1386.

Van Praag, H., Kempermann, G., and Gage, F. H. (1999). Running increases cell proliferation and neurogenesis in the adult mouse dentate gyrus. *Nature Neuroscience*, 2, 266–270.

Van Praag, H., Kempermann, G., and Gage, F. H. (2000). Neural consequences of environmental enrichment. *Nature Reviews*, 1, 191–198.

Figure and table acknowledgments

Chapter 1: Introduction

Fig. 1.1. Courtesy: Günther K. H. Zupanc.

Fig. 1.2. After: Wickler, W. (1968). *Das Züchten von Aquarienfischen: Eine Einführung in ihre Fortpflanzungsbiologie*. Franckh'sche Verlagshandlung, Stuttgart. Reprinted by permission of Kosmos Verlag.

Fig. 1.3. Courtesy: Günther K. H. Zupanc.

Fig. 1.4. Courtesy: Günther K. H. Zupanc and Jonathan R. Banks.

Fig. 1.5. Courtesy: Günther K. H. Zupanc.

Chapter 2: The study of animal behavior: a brief history

Fig. 2.1. Courtesy: Kunsthistorisches Museum, Vienna, Austria/Bridgeman Art Library.

Fig. 2.2. Courtesy: Bettmann/CORBIS.

Fig. 2.3. Courtesy: University of Bristol Special Collections. Photograph by Frank Holmes.

Fig. 2.4. Courtesy: National Academy of Sciences of the USA.

Fig. 2.5. Courtesy: AKG, London.

Fig. 2.6. After: McFarland, D. (1993). *Animal behaviour. Second edition*. Addison Wesley Longman, Harlow. Based on photographs by Thomas Eisner.

Fig. 2.7. Courtesy: The Nobel Foundation. © The Nobel Foundation.

Fig. 2.8. After: McFarland, D. (1993). *Animal behaviour. Second edition.* Addison Wesley Longman, Harlow.

Fig. 2.9. Courtesy: Nickolas Muray/Getty Images.

Fig. 2.10. Courtesy: Nina Leen/Timepix.

Fig. 2.11. After: Manning, A. (1972). *An introduction to animal behaviour. Second edition.* Edward Arnold Publishers, London.

Fig. Box 2.1. Courtesy: Nina Leen/Timepix.

Fig. Box 2.2. Courtesy: The Nobel Foundation. © The Nobel Foundation.

Fig. Box 2.3. Courtesy: Gerhard Gronefeld.

Fig. Box 2.4. Courtesy: Bettmann/CORBIS.

Fig. Box 2.5. Courtesy: Scripps Institution of Oceanography.

Chapter 3: The tools and concepts of behavioral neurobiology

Fig. 3.1. After: Marin-Padilla, M. (1987). The Golgi method. In *Encyclopedia of neuroscience, Vol. 1,* (ed. G. Adelman), pp. 470–471. Birkhäuser, Boston/Basel/Stuttgart.

Fig. 3.2. After: Kandel, E. R., and Schwartz, J. H. (1985). *Principles of neural science. Second edition.* Elsevier, New York/Amsterdam/Oxford.

Fig. 3.3. After: Camhi, J. M. (1984). *Neuroethology: nerve cells and the natural behavior of animals.* Sinauer Associates Inc. Publishers, Sunderland, MA; and: Kandel, E. R., and Schwartz, J. H. (1985). *Principles of neural science. Second edition.* Elsevier, New York/Amsterdam/Oxford. Reprinted by permission of Sinauer Associates, Inc.

Fig. 3.4. After: Zigmond, M. J., Bloom, F. E., Landis, S. C., Roberts, J. L., and Squire, L. R. (eds.) (1999). *Fundamental neuroscience.* Academic Press, San Diego/London. Reprinted by permission of Elsevier. © 1999 Elsevier.

Fig. 3.5. Courtesy: Günther K.H. Zupanc.

Fig. 3.6. After: Shepherd, G. M. (1988). *Neurobiology. Second edition.* Oxford University Press, New York/Oxford.

Fig. 3.7. After: Penzlin, H. (1980). *Lehrbuch der Tierphsyiologie. Third edition.* Gustav Fischer Verlag, Stuttgart.

Fig. 3.8. Courtesy: Günther K. H. Zupanc.

Fig. 3.9. Courtesy: Günther K. H. Zupanc.

Fig. 3.10. Courtesy: Günther K. H. Zupanc.

Fig. 3.11. Courtesy: Jonathan Crowe.

Fig. 3.12. After: Lack, D. (1953). *The life of the robin.* Penguin Books, Melbourne/London/Baltimore.

Fig. 3.13. After: Tinbergen, N. (1969). *The study of instinct.* Oxford University Press, London. Reprinted by permission of Oxford University Press.

Fig. 3.14. After: Tinbergen, N. (1969). *The study of instinct.* Oxford University Press, London. Reprinted by permission of Oxford University Press.

Fig. 3.15. After: Tinbergen, N., and Perdeck, A. C. (1950). On the stimulus situation releasing the begging response in the newly hatched herring gull chick (*Larus argentatus argentatus* Pont.). *Behaviour,* 3, 1–39. Reprinted by permission of Brill Academic Publishers.

Fig. 3.16. Courtesy: Jose Luis Pelaez, Inc./CORBIS.

Fig. 3.17. After: Eibl-Eibesfeldt, I. (1974). *Grundrifl der vergleichenden Verhaltensforschung. Fourth edition.* R. Piper & Co Verlag, München/Zürich. Reprinted by permission of Irenäus Eibl-Eibesfeldt.

Fig. 3.18. After: Leong, C.-Y. (1969). The quantitative effect of releasers on the attack readiness of the fish *Haplochromis burtoni* (Cichlidae, Pisces). *Zeitschrift für Vergleichende Physiologie,* 65, 29–50. Reprinted by permission of Springer-Verlag. © 1969 Springer-Verlag.

Fig. 3.19. Courtesy: Günther K. H. Zupanc. Based on data by: Bower, T. G. R. (1966). Heterogeneous summation in human infants. *Animal Behaviour,* 14, 395–398.

Fig. 3.20. Courtesy: Günther K. H. Zupanc. Based on data by: Drees, O. (1952). Untersuchungen über die angeborenen Verhaltensweisen bei Springspinnen (Salticidae). *Zeitschrift für Tierpsychologie,* 9, 169–207.

Fig. 3.21. Courtesy: Günther K. H. Zupanc.

Fig. 3.22. Courtesy: Günther K. H. Zupanc. Based on data by: Heiligenberg, W., and Kramer, U. (1972). Aggressiveness as a function of external stimulation. *Journal of Comparative Physiology,* 77, 332–340. Reprinted by permission of Springer-Verlag. © 1972 Springer-Verlag.

Fig. 3.23. After: Wickler, W. (1978). Das Problem der stammesgeschichtlichen Sackgassen. In *Evolution II: Ein Querschnitt der Forschung*, (ed. H. von Ditfurth), pp. 29–47. Hoffmann und Campe, Hamburg. Reprinted by permission of Hoffmann und Campe Verlag GmbH. © Hoffmann und Campe Verlag.

Fig. 3.24. After: Manning, A., and Stamp Dawkins, M. (1992). *An introduction to animal behaviour. Fourth edition*. Cambridge University Press, Cambridge/New York/Melbourne. Reprinted by permission of Cambridge University Press.

Fig. 3.25. After: Keeton, W. T. (1980). *Biological Sciences*. W.W. Norton & Company, New York/London. Reprinted by permisison of W.W. Norton & Company, Inc. © 1980, 1979, 1978, 1972, 1967 W.W. Norton & Company, Inc.

Fig. 3.26. After: Frisch, K. v. (1977) *Aus dem Leben der Bienen. Ninth edition*. Springer-Verlag, Berlin/Heidelberg/New York. Reprinted by permission of Springer-Verlag. © 1977 Springer-Verlag.

Fig. 3.27. After: Frisch, K. v. (1977) *Aus dem Leben der Bienen. Ninth edition*. Springer-Verlag, Berlin/Heidelberg/New York. Reprinted by permission of Springer-Verlag. © 1977 Springer-Verlag.

Fig. 3.28. After: Frisch, K. v. (1977) *Aus dem Leben der Bienen. Ninth edition*. Springer-Verlag, Berlin/Heidelberg/New York. Reprinted by permission of Springer-Verlag. © 1977 Springer-Verlag.

Fig. 3.29. After: Frisch, K. v. (1977) *Aus dem Leben der Bienen. Ninth edition*. Springer-Verlag, Berlin/Heidelberg/New York. Reprinted by permission of Springer-Verlag. © 1977 Springer-Verlag.

Fig. 3.30. After: Manning, A., and Stamp Dawkins, M. (1998). *An introduction to animal behaviour. Fifth edition*. Cambridge University Press, Cambridge/New York/Oakleigh. Based on a photograph in: Michelsen, A. (1989). Ein mechanisches Modell der tanzenden Honigbiene. *Biologie in unserer Zeit*, 19, 121–126. Reprinted by permission of Cambridge University Press.

Chapter 4: Spatial orientation

Fig. 4.1. After: Keeton, W. T. (1980). *Biological science. Third edition*. W.W. Norton & Company, New York.

Fig. 4.2. After: Machemer, H. (1988). Galvanotaxis: Grundlagen der elektromechanischen Kopplung und Orientierung bei Paramecium. In *Praktische Verhaltensbiologie*, (ed. G. K. H. Zupanc), pp. 60–82. Verlag Paul Parey, Berlin/Hamburg. Reprinted by permission of Blackwell Verlag GmbH.

Fig. 4.3. After: Machemer, H. (1988). Galvanotaxis: Grundlagen der elektromechanischen Kopplung und Orientierung bei Paramecium. In *Praktische Verhaltensbiologie*, (ed. G. K. H. Zupanc), pp. 60–82. Verlag Paul Parey, Berlin/Hamburg. Reprinted by permission of Blackwell Verlag GmbH.

Fig. 4.4. After: Machemer, H. (1988). Galvanotaxis: Grundlagen der elektromechanischen Kopplung und Orientierung bei Paramecium. In *Praktische Verhaltensbiologie*, (ed. G. K. H. Zupanc), pp. 60–82. Verlag Paul Parey, Berlin/Hamburg. Reprinted by permission of Blackwell Verlag GmbH.

Fig. 4.5. After: Machemer, H. (1988). Galvanotaxis: Grundlagen der elektromechanischen Kopplung und Orientierung bei Paramecium. In *Praktische Verhaltensbiologie*, (ed. G. K. H. Zupanc), pp. 60–82. Verlag Paul Parey, Berlin/Hamburg. Reprinted by permission of Blackwell Verlag GmbH.

Fig. 4.6. After: Kandel, E. R., and Schwartz, J. H. (ed.) (1985). *Principles of neural science. Second Edition*. Elsevier, New York.

Fig. 4.7. After: Holst, E. v. (1950). Die Tätigkeit des Statolithenapparates im Wirbeltierlabyrinth. *Naturwissenschaften*, 37, 265–272. Reprinted by permission of Springer-Verlag. © 1950 Springer-Verlag.

Fig. 4.8. After: Holst, E. v. (1950). Die Tätigkeit des Statolithenapparates im Wirbeltierlabyrinth. *Naturwissenschaften*, 37, 265–272. Reprinted by permission of Springer-Verlag. © 1950 Springer-Verlag.

Fig. 4.9. After: Flock, Å. (1965). Transducing mechanisms in the lateral line canal organ receptors. *Cold Spring Harbor Symposia on Quantitative Biology*, 30, 133–145. Reprinted by permission of Cold Springer Harbor Laboratory Press.

Fig. 4.10. After: Pickles, J. O., and Corey, D. P. (1992). Mechanical transduction by hair cells. *Trends in Neurosciences*, 15, 254–259. Reprinted by permission of Elsevier. © 1992 Elsevier.

Fig. 4.11. After: Simmons, J. A., Fenton, M. B., and O'Farrell, M. J. (1979). Echolocation and pursuit of prey by bats. *Science*, 203, 16–21.

Fig. 4.12. After: Pollak, G. D., Marsh, D. S., Bodenhamer, R., and Souther, A. (1977). Characteristics of phasic on neurons in inferior colliculus of unanaesthetized bats with observations relating to mechanisms of echo ranging. *Journal of Neurophysiology*, 40, 926–942. Reprinted by permission of the American Physiological Society.

Fig. 4.13. After: Schnitzler, H.-U., and Ostwald, J. (1983). Adaptations for the detection of fluttering insects by echolocation in horseshoe bats. In *Advances in vertebrate neuroethology*, (eds. J.-P. Ewert, R.R. Capranica, and D.J. Ingle), pp. 801–827. Plenum Press, New York. Reprinted by permission of Kluwer Academic/Plenum Publishers.

Fig. 4.14. After: Schnitzler, H.-U., and Ostwald, J. (1983). Adaptations for the detection of fluttering insects by echolocation in horseshoe bats. In *Advances in vertebrate neuroethology*, (eds. J.-P. Ewert, R.R. Capranica, and D.J. Ingle), pp. 801–827. Plenum Press, New York. Reprinted by permission of Kluwer Academic/Plenum Publishers.

Fig. Box 4.1. Courtesy: Bettmann/CORBIS.

Chapter 5: Neuronal control of motor output: swimming in toad tadpoles

Fig. 5.1. After: Arshavsky, Y. I., Orlovsky, G. N., Panchin, Y. V., Roberts, A., and Soffe, S. R. (1993). Neuronal control of swimming locomotion: analysis of the pteropod mollusc *Clione* and embryos of the amphibian *Xenopus*. *Trends in Neurosciences*, 16, 227–233. Reprinted by permission of Elsevier. © 1993 Elsevier.

Fig. 5.2. After: Roberts, A. (1990). How does a nervous system produce behaviour? Λ case study in neurobiology. *Science Progress*, 74, 31–51.

Fig. 5.3. After: Roberts, A., Soffe, S. R., and Perrins, R. (1997). Spinal networks controlling swimming in hatchling *Xenopus* tadpoles. In *Neurons, networks, and motor behavior*, (eds. P. S. G. Stein, S. Grillner, A. I. Selverston, and D. G. Stuart), pp. 83–89. MIT Press, Cambridge, MA.

Fig. 5.4. After: Arshavsky, Y. I., Orlovsky, G. N., Panchin, Y. V., Roberts, A., and Soffe, S. R. (1993). Neuronal control of swimming locomotion: analysis of the pteropod mollusc *Clione* and embryos of the amphibian *Xenopus*. *Trends in Neurosciences*, 16, 227–233.

Fig. Box 5.1. Courtesy: Alan Roberts.

Chapter 6: Neuronal Processing of Sensory Information

Fig. 6.1. After: Ewert, J.-P. (1980). *Neuroethology: an introduction to the neurophysiological fundamentals of behavior*. Springer-Verlag, Berlin/Heidelberg/New York. Reprinted by permission of Springer-Verlag. © 1980 Springer-Verlag.

Fig. 6.2. After: Ewert, J.-P. (1980). *Neuroethology: an introduction to the neurophysiological fundamentals of behavior*. Springer-Verlag, Berlin/Heidelberg/New York. Reprinted by permission of Springer-Verlag. © 1980 Springer-Verlag.

Fig. 6.3. After: Ewert, J.-P., and Ewert, S. B. (1981). *Warnehmung*. Quelle & Meyer, Heidelberg. Reprinted by permission of Quelle & Meyer Verlag GmbH & Co.

Fig. 6.4. After: Ewert, J.-P. (1980). *Neuroethology: an introduction to the neurophysiological fundamentals of behavior*. Springer-Verlag, Berlin/Heidelberg/New York. Reprinted by permission of Springer-Verlag. © 1980 Springer-Verlag.

Fig. 6.5. After: Ewert, J.-P. (1980). *Neuroethology: an introduction to the neurophysiological fundamentals of behavior*. Springer-Verlag, Berlin/Heidelberg/New York. Reprinted by permission of Springer-Verlag. © 1980 Springer-Verlag.

Fig. 6.6. After: Ewert, J.-P. (1980). *Neuroethology: an introduction to the neurophysiological funda-mentals of behavior*. Springer-Verlag, Berlin/Heidelberg/New York. Reprinted by permission of Springer-Verlag. © 1980 Springer-Verlag.

Fig. 6.7. After: Ewert, J.-P. (1974). The neural basis of visually guided behavior. *Scientific American*, 230(3), 34–42.

Fig. 6.8. After: Ewert, J.-P. (1980). *Neuroethology: an introduction to the neurophysiological funda-mentals of behavior*. Springer-Verlag, Berlin/Heidelberg/New York. Reprinted by permission of Springer-Verlag. © 1980 Springer-Verlag.

Fig. 6.9. After: Ewert, J.-P. (1980). *Neuroethology: an introduction to the neurophysiological funda-mentals of behavior*. Springer-Verlag, Berlin/Heidelberg/New York. Reprinted by permission of Springer-Verlag. © 1980 Springer-Verlag.

Fig. 6.10. After: Ewert, J.-P. (1980). *Neuroethology: an introduction to the neurophysiological funda-mentals of behavior*. Springer-Verlag, Berlin/Heidelberg/New York. Reprinted by permission of Springer-Verlag. © 1980 Springer-Verlag.

Fig. 6.11. After: Knudsen, E. I. (1981). The hearing of the barn owl. *Scientific American*, 245 (6), 82–91.

Fig. 6.12. Courtesy: Günther K. H. Zupanc.

Fig. 6.13. After: Knudsen, E. I., and Konishi, M. (1979). Mechanisms of sound localization in the barn owl (*Tyto alba*). *Journal of Comparative Physiology A*, 133, 13–21. Reprinted by permission of Springer-Verlag. © 1979 Springer-Verlag.

Fig. 6.14. After: Knudsen, E. I. (1981). The hearing of the barn owl. *Scientific American*, 245 (6), 82–91; and Knudsen, E. I., and Konishi, M. (1979). Mechanisms of sound localization in the barn owl (*Tyto alba*). *Journal of Comparative Physiology A*, 133, 13–21. Reprinted by permission of Springer-Verlag. © 1979 Springer-Verlag.

Fig. 6.15. After: Konishi, M. (1990). Similar algorithms in different sensory systems and animals. *Cold Spring Harbor Symposia on Quantitative Biology*, 55, 575–584. Reprinted by permission of Cold Springer Harbor Laboratory Press.

Fig. 6.16. After: Konishi, M. (1993). Listening with two ears. *Scientific American*, 268 (4), 34–41. Reprinted by permission of Scientific American. © 1993 Scientific American, Inc. All rights re-served.

Fig. 6.17. After: Konishi, M. (1992). The neural algorithm for sound localization in the owl. *The Harvey Lectures*, 86, 47–64. Reprinted by permission of Wiley-Liss, Inc., a subsidiary of John Wiley & Sons, Inc. © 1992 Wiley-Liss, Inc.

Fig. 6.18. After: Konishi, M. (1992). The neural algorithm for sound localization in the owl. *The Harvey Lectures*, 86, 47–64; and: Konishi, M. (1993). Listening with two ears. *Scientific American*, 268 (4), 34–41. Reprinted by permission of Scientific American. © 1993 Scientific American, Inc. All rights reserved.

Fig. 6.19. After: Konishi, M. (1993). Listening with two ears. *Scientific American*, 268 (4), 34–41. Reprinted by permission of Scientific American. © 1993 Scientific American, Inc. All rights reserved.

Fig. 6.20. After: Konishi, M. (1993). Listening with two ears. *Scientific American*, 268 (4), 34–41. Reprinted by permission of Scientific American. © 1993 Scientific American, Inc. All rights reserved.

Fig. 6.21. After: Konishi, M. (1993). Listening with two ears. *Scientific American*, 268 (4), 34–41. Reprinted by permission of Scientific American. © 1993 Scientific American, Inc. All rights reserved.

Fig. Box 6.1. Courtesy: Masakazu Konishi.

Chapter 7: Sensori-motor integration: the jamming avoidance response of the weakly electric fish, *Eigenmannia*

Fig. 7.1. Courtesy: Günther K. H. Zupanc.

Fig. 7.2. Courtesy: Günther K. H. Zupanc. Based on a histological preparation by Frank Kirschbaum.

Fig. 7.3. Courtesy: Günther K. H. Zupanc.

Fig. 7.4. Courtesy: Günther K. H. Zupanc. Based on a histological preparation by Frank Kirschbaum.

Fig. 7.5. After: Heiligenberg, W. (1977). *Principles of electrolocation and jamming avoidance response in electric fish*. Springer-Verlag, Berlin/Heidelberg/New York. Reprinted by permission of Springer-Verlag. © 1977 Springer-Verlag.

Fig. 7.6. After: Heiligenberg, W., Baker, C., and Matsubara, J. (1978). The jamming avoidance response in *Eigenmannia* revisited: the structure of a neuronal democracy. *Journal of Comparative Physiology A*, 127, 267–286. Reprinted by permission of Springer-Verlag. © 1978 Springer-Verlag.

Fig. 7.7. After: Heiligenberg, W. (1991). *Neural nets in electric fish*. MIT Press, Cambridge, MA.

Fig. 7.8. Courtesy: Günther K. H. Zupanc.

Fig. 7.9. After: Heiligenberg, W. (1991). *Neural nets in electric fish*. MIT Press, Cambridge, MA.

Fig. 7.10. After: Heiligenberg, W. (1991). *Neural nets in electric fish*. MIT Press, Cambridge, MA.

Fig. 7.11. After: Heiligenberg, W. (1991). *Neural nets in electric fish*. MIT Press, Cambridge, MA.

Fig. 7.12. After: Metzner, W. (1999). Neural circuitry for communication and jamming avoidance in gymnotiform fish. *Journal of Experimental Biology*, 202, 1365–1375. Reprinted by permission of the Company of Biologists, Ltd.

Fig. Box 7.1. Courtesy: Günther K. H. Zupanc.

Chapter 8: Neuromodulation: the accommodation of motivational changes in behavior

Fig. 8.1. Courtesy: Günther K. H. Zupanc.

Fig. 8.2. Courtesy: Günther K. H. Zupanc.

Fig. 8.3. Courtesy: Günther K. H. Zupanc.

Fig. 8.4. After: Zupanc, G. K. H., and Heiligenberg, W. (1989). Sexual maturity-dependent changes in neuronal morphology in the prepacemaker nucleus of adult weakly electric knifefish, *Eigenmannia*. *Journal of Neuroscience*, 9, 3816–3827. Reprinted by permission of The Journal of Neuroscience. © 1989 Society for Neuroscience.

Fig. 8.5. Courtesy: Günther K. H. Zupanc.

Fig. 8.6. After: Kurz, E. M., Sengelaub, D. R., and Arnold, A. P. (1986). Androgens regulate the dendritic length of mammalian motoneurons in adulthood. *Science*, 232, 395–398.

Fig. 8.7. After: Marder, E., and Richards, K. S. (1999). Development of the peptidergic modulation of a rhythmic pattern generating network. *Brain Research*, 848, 35–44. Reprinted by permission of Elsevier. © 1999 Elsevier.

Chapter 9: Large-scale navigation: migration and homing

Fig. 9.1. After: Schüz, E. (1971). *Grundriß der Vogelzugskunde*. Verlag Paul Parey, Berlin/Hamburg. Reprinted by permission of Blackwell Verlag GmbH.

Fig. 9.2. After: Keeton, W. T. (1980). *Biological sciences*. W.W. Norton & Company, New York/London.

Fig. 9.3. After: Brower, L. P. (1996). Monarch butterfly orientation: missing pieces of a magnificent puzzle. *Journal of Experimental Biology*, 199, 93–103. Reprinted by permission of the Company of Biologists, Ltd.

Fig. 9.4. After: Helbig, A. J. (1991). Inheritance of migratory direction in a bird species: a cross-breeding experiment with SE- and SW-migrating blackcaps (*Sylvia atricapilla*). *Behavioral Ecology and Sociobiology*, 28, 9–12. Reprinted by permission of Springer-Verlag. © 1991 Springer-Verlag.

Fig. 9.5. After: Helbig, A. J. (1991). Inheritance of migratory direction in a bird species: a cross-breeding experiment with SE- and SW-migrating blackcaps (*Sylvia atricapilla*). *Behavioral Ecology and Sociobiology*, 28, 9–12. Reprinted by permission of Springer-Verlag. © 1991 Springer-Verlag.

Fig. 9.6. After: Kramer, G. (1950). Weitere Analyse der Faktoren, welche die Zugaktivität des gekäfigten Vogels orientieren. *Naturwissenschaften*, 37, 377–378. Reprinted by permission of Springer-Verlag. © 1950 Springer-Verlag.

Fig. 9.7. After: Hoffmann, K. (1954). Versuche zu der im Richtungsfinden der Vögel enthaltenen Zeiteinschätzung. *Zeitschrift für Tierpsychologie*, 11, 453–475. Reprinted by permission of Blackwell Verlag GmbH.

Fig. 9.8. After: Kreithen, M. L. (1979). The sensory world of the homing pigeon. In *Neural mechanisms of behavior in the pigeon*, (eds. A. M. Granda, and J. H. Maxwell), pp. 21–33. Plenum Press, New York/London. Reprinted by permission of Kluwer Academic/Plenum Publishers.

Fig. 9.9. After: Emlen, S. T., and Emlen, J. T. (1966). A technique for recording migratory orientation of captive birds. *Auk*, 83, 361–367.

Fig. 9.10. After: Emlen, S. T., and Emlen, J. T. (1966). A technique for recording migratory orientation of captive birds. *Auk*, 83, 361–367.

Fig. 9.11. After: Emlen, S. T. (1967). Migratory orientation in the Indigo Bunting, *Passerina cyanea*. Part I: Evidence for use of celestial cues. Auk, 84, 309–342.

Fig. 9.12. After: Merkel, F. W., and Fromme, H. G. (1958). Untersuchungen über das Orientierungsvermögen nächtlich ziehender Rotkehlchen, *Erithacus rubecula*. *Naturwissenschaften*, 45, 499–500. Reprinted by permission of Springer-Verlag. © 1958 Springer-Verlag.

Fig. 9.13. After: Wiltschko, W., and Wiltschko, R. (1996). Magnetic orientation in birds. *Journal of Experimental Biology*, 199, 29–38. Reprinted by permission of the Company of Biologists, Ltd.

Fig. 9.14. After: Wiltschko, W., and Wiltschko, R. (1996). Magnetic orientation in birds. *Journal of Experimental Biology*, 199, 29–38. Reprinted by permission of the Company of Biologists, Ltd.

Fig. 9.15. After: Walcott, C., and Green, R. P. (1974). Orientation of homing pigeons altered by a change in the direction of an applied magnetic field. *Science*, 184, 180–182.

Fig. 9.16. After: Wiltschko, R., and Wiltschko, W. (1978). Evidence for the use of magnetic outward-journey information in homing pigeons. *Naturwissenschaften*, 65, 112–113. Reprinted by permission of Springer-Verlag. © 1978 Springer-Verlag.

Fig. 9.17. After: Semm, P., and Beason, R. C. (1990). Responses to small magnetic variations by the trigeminal system of the bobolink. *Brain Research Bulletin*, 25, 735–740.

Fig. 9.18. Courtesy: Günther K. H. Zupanc. Based on Childerhose, R. J., and Trim, M. (1979). *Pacific salmon and steelhead trout*. Douglas & McIntyre, Vancouver; and: Quinn, T. P., and Dittman, A. H. (1990). Pacific salmon migrations and homing: mechanisms and adaptive significance. *Trends in Ecology and Evolution*, 5, 174–177. Reprinted by permission of Elsevier. © 1990 Elsevier.

Fig. 9.19. After: Quinn, T. P., and Dittman, A. H. (1990). Pacific salmon migrations and homing: mechanisms and adaptive significance. *Trends in Ecology and Evolution*, 5, 174–177. Reprinted by permission of Elsevier. © 1990 Elsevier.

Fig. 9.20. Courtesy: Günther K. H. Zupanc. Based on a histological preparation by Helmut Altner.

Fig. 9.21. Courtesy: Günther K. H. Zupanc. Based on a histological preparation by Helmut Altner.

Fig. 9.23. After: Hasler, A. D., and Scholz, A. T. (1983). *Olfactory imprinting and homing in salmon: investigations into the mechanism of the imprinting process.* Springer-Verlag, Berlin/Heidelberg/New York/Tokyo. Reprinted by permission of Springer-Verlag. © 1983 Springer-Verlag.

Fig. 9.24. After: Dickhoff, W. W., Folmar, L. C., and Gorbman, A. (1978). Changes in plasma thyroxine during the smoltification of coho salmon, *Oncorhynchus kisutch. General and Comparative Endocrinology*, 36, 229–232. Reprinted by permission of Elsevier. © 1978 Elsevier.

Fig. 9.25. After: Hasler, A. D., and Scholz, A. T. (1983). *Olfactory imprinting and homing in salmon: investigations into the mechanism of the imprinting process.* Springer-Verlag, Berlin/Heidelberg/New York/Tokyo. Reprinted by permission of Springer-Verlag. © 1983 Springer-Verlag.

Fig. 9.26. After: Dittman, A. W., and Quinn, T. P. (1996). Homing in Pacific salmon: mechanisms and ecological basis. *Journal of Experimental Biology*, 199, 83–91. Reprinted by permission of the Company of Biologists, Ltd.

Fig. 9.27. After: Walker, M. M., Diebel, C. E., Haugh, C. V., Pankhurst, P. M., Montgomery, J. C., and Green, C. R. (1997). Structure and function of the vertebrate magnetic sense. *Nature*, 390, 371–376. Reprinted by permission of Nature. © 1997 Macmillan Magazines, Ltd.

Fig. 9.28. After: Walker, M. M., Diebel, C. E., Haugh, C. V., Pankhurst, P. M., Montgomery, J. C., and Green, C. R. (1997). Structure and function of the vertebrate magnetic sense. *Nature*, 390, 371–376. Reprinted by permission of Nature. © 1997 Macmillan Magazines, Ltd.

Fig. 9.29. After: Lohmann, K. J. (1992). How sea turtles navigate. *Scientific American*, 266(1), 82–88.

Fig. 9.30. After: Lohmann, K. J., and Lohmann, C. M. F. (1996). Orientation and open-sea navigation in sea turtles. *Journal of Experimental Biology*, 199, 73–81.

Fig. 9.31. After: Lohmann, K. J., Swartz, A. W., and Lohmann, C. M. F. (1995). Perception of ocean wave direction by sea turtles. *Journal of Experimental Biology*, 198, 1079–1085.

Fig. 9.32. After: Lohmann, K. J., and Lohmann, C. M. F. (1996). Orientation and open-sea navigation in sea turtles. *Journal of Experimental Biology*, 199, 73–81. Reprinted by permission of the Company of Biologists, Ltd.

Fig. Box 9.2. Courtesy: Fritz Albert.

Table 9.1 After: Hasler, A. D., and Scholz, A. T. (1983). *Olfactory imprinting and homing in salmon: investigations into the mechanism of the imprinting process.* Springer-Verlag, Berlin/Heidelberg/New York/Tokyo. Reprinted by permission of Springer-Verlag. © 1983 Springer-Verlag.

Chapter 10: Communication: the neuroethology of cricket song

Fig. 10.1. After: Loher, W., and Dambach, M. (1989). Reproductive behavior. In *Cricket behavior and neurobiology*, (eds. F. Huber, T. E. Moore, and W. Loher), pp. 43–82. Cornell University Press, Ithaca/London. Reprinted by permission of Cornell University Press. © 1989 Cornell University.

Fig. 10.2. After: Huber, F., and Thorson, J. (1985). Cricket auditory communication. *Scientific American*, 253 (6), 46–54; and: Huber, F., Moore, T. E., and Loher, W. (eds.) (1989). *Cricket behavior and neurobiology*. Cornell University Press, Ithaca. Reprinted by permission of Cornell University Press. © 1989 Cornell University.

Fig. 10.3. After: Dambach, M. (1988). Sozialverhalten und Lauterzeugung bei der Feldgrille. In *Praktische Verhaltensbiologie*, (ed. G. K. H. Zupanc), pp. 101–111. Paul Parey, Berlin/Hamburg.

Fig. 10.4. After: Bentley, D., and Hoy, R. R. (1974). The neurobiology of cricket song. *Scientific American*, 231 (8), 34–44.

Fig. 10.5. After: Huber, F., and Thorson, J. (1985). Cricket auditory communication. *Scientific American*, 253 (6), 46–54.

Fig. 10.6. After: Huber, F., and Thorson, J. (1985). Cricket auditory communication. *Scientific American*, 253 (6), 46–54.

Fig. 10.7. After: Huber, F. (1990). Cricket neuroethology: Neuronal basis of intraspecific acoustic communication. *Advances in the Study of Behavior*, 19, 299–356. Reprinted by permission of Elsevier. © 1990 Elsevier.

Fig. 10.8. After: Pires, A., and Hoy, R. R. (1992). Temperature coupling in cricket acoustic communication: I. Field and laboratory studies of temperature effects on calling song production and recognition in *Gryllus firmus*. *Journal of Comparative Physiology A*, 171, 69–78. Reprinted by permission of Springer-Verlag. © 1992 Springer-Verlag.

Fig. 10.9. After: Pires, A., and Hoy, R. R. (1992). Temperature coupling in cricket acoustic communication: I. Field and laboratory studies of temperature effects on calling song production and recognition in *Gryllus firmus*. *Journal of Comparative Physiology A*, 171, 69–78. Reprinted by permission of Springer-Verlag. © 1992 Springer-Verlag.

Fig. 10.10. After: Pires, A., and Hoy, R. R. (1992). Temperature coupling in cricket acoustic communication: I. Field and laboratory studies of temperature effects on calling song production and recognition in *Gryllus firmus*. *Journal of Comparative Physiology A*, 171, 69–78. Reprinted by permission of Springer-Verlag. © 1992 Springer-Verlag.

Fig. 10.11. After: Bentley, D. R., and Hoy, R. R. (1972). Genetic control of the neuronal network generating cricket (*Teleogryllus gryllus*) song patterns. *Animal Behaviour*, 20, 478–492. Reprinted by permission of Elsevier. © 1972 Elsevier.

Fig. 10.12. After: Hoy, R. R., Hahn, J., and Paul, R. C. (1977). Hybrid cricket auditory behavior: Evidence for genetic coupling in animal communication. *Science*, 195, 82–84.

Fig. Box 10.1. After: Schmitz, B., Scharstein, H., and Wendler, G. (1982). Phonotaxis in *Gryllus campestris* L. (Orthoptera, Gryllidae): I. Mechanisms of acoustic orientation in intact female crickets. *Journal of Comparative Physiology A*, 148, 431–444; and: Pires, A., and Hoy, R. R. (1992). Temperature coupling in cricket acoustic communication: I. Field and laboratory studies of temperature effects on calling song production and recognition in *Gryllus firmus*. *Journal of Comparative Physiology A*, 171, 69–78. Reprinted by permission of Springer-Verlag. © 1982 and 1992 Springer-Verlag.

Fig. Box 10.2. Courtesy: Günther K. H. Zupanc.

Chapter 11: Cellular mechanisms of learning and memory

Fig. 11.1. After: Milner, B., Squire, L. R., and Kandel, E. R. (1998). Cognitive neuroscience and the study of memory. *Neuron*, 20, 445–468. Reprinted by permission of Elsevier. © 1998 Elsevier.

Fig. 11.2. After: Kandel, E. R. (2001). The molecular biology of memory storage: a dialogue between genes and synapses. *Science*, 294, 1030–1038.

Fig. 11.3. After: Kandel, E. R. (2001). The molecular biology of memory storage: a dialogue between genes and synapses. *Science*, 294, 1030–1038.

Fig. 11.4. After: Kandel, E. R. (2001). The molecular biology of memory storage: a dialogue between genes and synapses. *Science*, 294, 1030–1038.

Fig. 11.5. After: Mayford, M., and Kandel, E. R. (1999). Genetic approaches to memory storage. *Trends in Genetics*, 15, 463–470. Reprinted by permission of Elsevier. © 1999 Elsevier.

Fig. 11.6. After: Bailey, C. H., and Chen, M. (1988). Long-term memory in *Aplysia* modulates the total number of varicosities of single identified sensory neurons. *Proceedings of the National Academy of Sciences U.S.A.*, 85, 2373–2377. Reprinted by permission of the National Academies Press. © 1988 National Academy of Sciences.

Fig. 11.7. After: Kandel, E. R. (2001). The molecular biology of memory storage: a dialogue between genes and synapses. *Science*, 294, 1030–1038.

Fig. 11.8. After: O'Keefe, J., Burgess, N., Donnett, J. G., Jeffery, K. J., and Maguire, E. A. (1998). Place cells, navigational accuracy, and the human hippocampus. *Philosophical Transactions of the Royal Society of London B*, 353, 1333–1340. Reprinted by permission of The Royal Society.

Fig. 11.9. After: Kandel, E. R. (2001). The molecular biology of memory storage: a dialogue between genes and synapses. *Science*, 294, 1030–1038.

Fig. 11.10. After: Kandel, E. R. (2001). The molecular biology of memory storage: a dialogue between genes and synapses. *Science*, 294, 1030–1038.

Fig. 11.11. After: Van Praag, H., Kempermann, G., and Gage, F. H. (2000). Neural consequences of environmental enrichment. *Nature Reviews Neuroscience*, 1, 191–198. Reprinted by permission of Nature Reviews. © 2000 Macmillan Magazines, Ltd.

Fig. 11.12. After: Van Praag, H., Kempermann, G., and Gage, F. H. (1999). Running increases cell proliferation and neurogenesis in the adult mouse dentate gyrus. *Nature Neuroscience*, 2, 266–270. Reprinted by permission of Nature Neuroscience. © 1999 Macmillan Magazines, Ltd.

Fig. 11.13. After: Shors, T. J., Miesegaes, G., Beylin, A., Zhao, M., Rydel, T., and Gould, E. (2001). Neurogenesis in the adult is involved in the formation of trace memories. *Nature*, 410, 372–376. Reprinted by permission of Nature. © 2001 Macmillan Magazines, Ltd.

Fig. 11.14. After: Barnea, A., and Nottebohm, F. (1994). Seasonal recruitment of hippocampal neurons in adult free-ranging black-capped chickadees. *Proceedings of the National Academy of Sciences U.S.A.*, 91, 11217–11221. Reprinted by permission of the National Academies Press. © 1994 National Academy of Sciences.

Fig. Box 11.1. Courtesy: McGill University Archives. Photograph Collection.

Fig. Box 11.2. Courtesy: The Nobel Foundation © The Nobel Foundation.

Index

Abel, Ted, 295
acetylcholine, 48
acoustic fovea, 104
action potential, 43–5
  all-or-nothing property, 43
  analysis by extracellular recording, 44, Fig. 3.5
  analysis by intracellular recording, 40, Fig. 3.3
  conduction, 44
  definition, 43
  generation, 43
  refractory period, 43
adrenaline *see* epinephrine
adult neurogenesis
  definition, 183
  involvement in behavioral plasticity, 184
  involvement in memory formation, 296–301
  involvement in spatial learning, 300–1
Aequidens pulcher (blue acara), Fig. 1.2
afferent neuron, definition of, 37
aggressive behavior
  cichlids, 3–6, Fig. 1.2
  crayfish, 195–6
  crickets, 251
aggressive song of crickets, 251
agonist, definition of, 48
alarm substance *see Schreckstoff*
Albertus Magnus, 13
alevin, 226
allostatin-related peptide, 194
American golden plover (*Pluviatis dominica*), 202, Fig. 9.2
American lobster (*Homarus americanus*), 194
amino acid transmitters *see also* GABA, glutamate, glycine, 48
γ-amino-butyric acid (GABA), 48, 176, 194
α-amino-3-hydroxy-5-methyloxazole-propionic acid (AMPA), 175
2-amino-5-phosphonovaleric acid (APV), 176
amnesia
  anterograde, 279
  definition, 279
  retrograde, 279
AMPA *see* α-amino-3-hydroxy-5-methyloxazole-propionic acid
AMPA-type of glutamate receptor, 116, 175

ampulla of Lorenzini, 222
ampullary receptor, Fig. 7.4, 170
AN-1 cell, 263
AND gate, 263
Andersen, Per, 293
anemotaxis, 81
angel fish (*Pterophyllum* sp.), 90, Figs. 4.7, 4.8
Anguilla anguilla (European eel), 230
angular nucleus, 143
antagonist, definition of, 48
anterior lateral lemniscal nucleus, 146
antibody *see also* immunohistochemistry, 54–6
  conjugation to fluorophore or enzyme, 55
  primary, Fig. 3.9, 55
  secondary, Fig. 3.9, 55
Apis mellifera (honey bee), 70–4
Aplysia californica (sea hare), 280–9
  facilitation, 285
  gill-withdrawal reflex, 282, Fig. 11.2
  long-term memory, 283, Fig. 11.3
  modulation by serotonin of gill-withdrawal reflex, Fig. 11.4, 285
  molecular biology of long-term sensitization, 286–9, Fig. 11.5
  molecular biology of short-term sensitization, 285–6, Fig. 11.5
  neural circuit of gill-withdrawal reflex, 283–5, Fig. 11.4
  sensitization, 282
  short-term memory, 283, Fig. 11.3
appetitive behavior, 16
APV *see* 2-amino-5-phosphonovaleric acid
Arctiidae (tiger moths), 106
Aristotle, 12, Fig. 2.1
Arnold, Arthur, 189
ascending neuron of type 1 *see* AN-1 cell
Asclepias (milkweed), 205
auditory interneurons in crickets, 262–5
auditory nerve
  crickets, 260
  owls, 142–3
auditory neuropil in crickets, 260

Australian field crickets (*Teleogryllus*), Fig. 10.1, 271
autism, 20
Autrum, Hansjochem, 26, 156
avidin, 58
axon, definition of, 37
axon hillock, 37

Bailey, Craig, 288–9
band-pass cell, 263
Bargmann, Cori, 31
barn owl (*Tyto alba*), 132–50
  accuracy of orientation response, 136–7, Fig. 6.12
  facial ruff, 139, Fig. 6.14
  neural circuit involved in sound localization, 142–50, Fig. 6.18
  physical parameters involved in sound localization, 137–42, Figs. 6.15, 6.16
  prey localization behavior, 132–4
barn owls (Tytonidae), 134
basilar pyramidal cell, 172
Bastian, Joseph, 157
bats (Chiroptera), 94–107
  big brown bat (*Eptesicus fuscus*), 96, 98
  echolocation, 59, 94–107
  greater horseshoe bat (*Rhinolophus ferrumequinum*), 97, 100, 103–5
  mustached bat (*Pteronotus parnellii*), 100
Baumgartner, Günter, 122
Beason, Robert, 221
bee language *see also* dance language, round dance, waggle dance, 27
behavior
  development, 14
  level of integration, 3, Fig. 1.3
  problems in quantifying, 5–6
behaviorism, 15, 22–4
Békésy, Georg von, 104
Benzer, Seymour, 286
Berthold, Peter, 205
big brown bat (*Eptesicus fuscus*), 96, 98
  sonogram of echolocation signals, Fig. 4.11(a)
biochemical switching of neural networks *see also* polymorphic network, 192–7
biogenic amines *see also* catecholamines, dopamine, epinephrine, histamine, 5-hydroxytryptamine, norepinephrine, 49
biotin, 58
black-capped chickadee (*Parus atricapillus*), 300
blackcap (*Sylvia atricapilla*), 205–6
  crossbreeding experiments, 206, Fig. 9.5
  routes taken during fall migration, 205, Figs. 9.4
Blakemore, Richard, 222
Bliss, Timothy, 293–4
blue acara (*Aequidens pulcher*), Fig. 1.2
bluntnose minnow (*Pimephales notatus*), 231
BNC-1 cell, 263
BNC-2 cell, 263

bobolink (*Dolichonyx oryzivorus*), 221, 223, Fig. 9.17
Bourtchouladze, Rusiko, 295
bouton *see* synapse
Bowen, Brian, 240
Bower, T.G.R., 65
Bradbury, Jack, 67
Braemer, Wolfgang, 237
brain neuron of class 1 *see* BNC-1 cell
brain neuron of class 2 *see* BNC-2 cell
brain stimulation experiments
  chicken, 26, 29
  importance for neuroethology, 29
  knifefish, 186
  toad, 131
Brainard, Michael, 150
BrdU *see* 5-bromo-2′-deoxyuridine
BrdU technique, 297
Breedlove, Marc, 188
5-bromo-2′-deoxyurdine (BrdU), 297
Brown, Graham, 115
Bufo bufo (common toad), 122–32
  brain recording experiments, 128–30, Figs. 6.8, 6.9
  brain stimulation experiments, 131
  connectivity of retina with other brain regions, 128–9, Fig. 6.7
  dummy experiments, 124–6, Figs. 6.2, 6.3, 6.4, 6.5, 6.6
  enemy recognition, 123–4, 126, Fig. 6.2
  lesioning experiments, 131–2, Fig. 6.10
  prey catching behavior, 123, Fig. 6.1
  prey recognition, 123–4, 126, Fig. 6.2
Bufonidae, 122
bug detector, 122
bullfrog (*Rana catesbeiana*), 93
Bullock, Theodore Holmes, 26, 28, Fig. Box 2.5, 112, 122, 155–7, 258
α-bungarotoxin, 115
Bungarus multicinctus (krait), 115
Burton's mouthbrooder (*Haplochromis burtoni*), 63, Fig. 3.18, 67, Figs. 3.21, 3.22

CA3 region of hippocampus, 299
Caenorhabditis elegans (nematode), 31
Cajal, Santiago Ramón y, 36, 46
calling song of crickets, 251, 253
  oscillogram, Fig. 10.2
cAMP-dependent protein kinase A (PKA), 285–6, 288, 295
cAMP response element (CRE), 286, 288
cAMP response element binding protein-1 (CREB-1), 295
cAMP response element binding protein-1a (CREB-1a), 286, 288
Cancer borealis tachykinin-related peptide, Fig. 8.7
Caretta caretta (loggerhead sea turtle), 240, Figs. 9.29, 9.32
Carlsson, Arvid, 281
Carr, Catherine, 144, 157
carrier frequency, 253, 259

catecholamines *see also* dopamine, epinephrine,
    norepinephrine, 49
  involvement in biochemical switching of neural networks,
    197
C/EBP, 288
cell theory, 35
cement gland, 113
central pattern generator, 26, 115
central posterior nucleus, 175
central posterior/prepacemaker nucleus *see also* CP/PPn-G,
    185–6, Fig. 8.3
CF area, 105
CF-FM signals, 96
Chelonia mydas (green turtle), Fig. 9.30
chemotaxis, 81
Chen, Mary, 288
Chiroptera (bats), 95
chirp
  crickets, 253, 266
  knifefish, 184, 186
chirp period, 266
cholecystokinin, 194
cichlid fish, 3
  aggressive behavior, 3, Fig. 1.2
  lateral display, 3, Fig. 1.2
Ciconia ciconia (European white stork), 202, Fig. 9.1
cilium, Fig. 4.2, 84
clawed-toad (*Xenopus laevis*), 111–19
  central pattern generator controlling swimming,
    115–18, Fig. 5.3
  coordination of oscillator activity involved in neural control
    of swimming, 118–19
  fictive swimming, 114
  physiological activity of spinal motoneurons, 113–15,
    Figs. 5.2, 5.4
    spinal circuitry controlling swimming *see* clawed-toad,
      central pattern generator controlling swimming
  swimming behavior, 112–13, Fig. 1.5
closer muscle, role in sound production of crickets, 255
CNQX *see* 6-cyano-7-nitroquinoxaline-2,3-dione
cochlea, 88
cochlear nucleus, 142–3
Coghill, George, 112
cognitive psychology, 278
coincidence detector, 144, Fig. 6.19
collateral, definition of, 37
color vision in honey bees, 27
Columba livia (Mediterranean rock dove), 208
common toad (*Bufo bufo*), 122–32
  brain recording experiments, 128–30, Figs. 6.8, 6.9
  brain stimulation experiments, 131
  connectivity of retina with other brain regions,
    128–9, Fig. 6.7
  dummy experiments, 124–6, Figs. 6.2, 6.3, 6.4,
    6.5, 6.6
  enemy recognition, 123–4, 126, Fig. 6.2
  lesioning experiments, 131–2, Fig. 6.10

prey catching behavior, 123, Fig. 6.1
prey recognition, 123–4, 126, Fig. 6.2
communication, 67–74, 251–73
  crickets, 251–273
  genetic coupling of sender and receiver, 269–73
  honey bees, 70–4
  principles, 67–8
  true, 68
comparative psychology, 15
complex recognition unit, 28
computational neuroscience, role in development of behavioral
    neurobiology, 30
conditioned reflex, 21
conditioned stimulus *see* conditioned reflex
conditioning
  bluntnose minnow, 231
  classical, 22
  coho salmon, 231
  delay, 299
  dog, Fig. 2.8
  instrumental *see* conditioning, operant
  operant, 23
  starling, 211
  trace, 299
  use in sensory physiology, 27
consummatory action, 16
convergence, principle of, 50
core, 146
Corey, David, 93
courtship song of crickets, 251
CP/PPn-G *see also* central posterior/prepacemaker nucleus, 175
Craig, Wallace, 16
crayfish, 195
CRE *see* cAMP response element
CREB-1 *see* cAMP response element binding protein-1
CREB-1a *see* cAMP response element binding protein-1a
CREB-2, 289
Creutzfeld, Otto, 122, 133
cricket song, 251–73
  behavioral analysis, 256–9
  biophysics, 253
  calling songs, 251, 253, Fig. 10.2
  carrier frequency, 253
  courtship songs, 251
  genetic coupling between sender and receiver, 269–73
  involvement in communication, 251
  mechanism of production, 253–5
  neural control of production, 255–6
  perception by sensory organs, 259–62
  recognition by auditory interneurons, 262–5
  syllable, 253
  temperature coupling, 265–9
crickets
  auditory discrimination experiment, Fig. 10.12
  auditory neuropil, 260
  Australian field crickets, Fig. 10.1
  central nervous system, 260, Fig. 10.5

crickets (*cont.*)
European field crickets, 254, 259
phonotaxis, 256
song *see* cricket song
crustacean cardioactive peptide, Fig. 8.7, 195
crustaceans, 195
curare, 162
Cuvier, Georges, 95–6
Cy3, 56
6-cyano-7-nitroquinoxaline-2,3-dione (CNQX), 175

Danaus plexippus (monarch butterfly), Fig. 9.3, 205
dance language *see also* round dance, waggle dance, 70–4
Darwin, Charles, 14, Fig. 2.2
de Bono, Mario, 31
decapod crustaceans, 192
Dekin, Michael, 195
delay line, 144
dendrite, structure, 37
dendritic plasticity, 184–91
role in chirping behavior of knifefish, 184–8
role in reproductive behavior of white-footed mice, 188–91
dendritic spine, 37
dentate gyrus, 290, 296
depolarization, 43
Dermochelys coriacea (leatherback sea turtle), 240
development of behavior, 14
digger wasp, 20
divergence, principle of, 49
Dolichonyx oryzivorus (bobolink), 221–3, Fig. 9.17
dopamine, 49, 194
Doppler, Christian Johann, 101
Doppler shift, 101, 261
Doppler shift compensation, 101
dorsal light reaction, 90
Dostrovsky, John, 292
Drees, Oskar, 66
drone in honey bees, 71
Drosophila (fruit fly), 286
dummy experiments *see also* supernormal stimulus
Burton's mouthbrooder, 63–4, Fig. 3.18, 67, Figs. 3.21, 3.22
common toad, 124–7
elephant nose, Fig. 1.5(f)
European robin, 60, Fig. 3.12
herring gull, 61–2, Fig. 3.15
honey bee, 74, Fig. 3.30
human infant, 65, Fig. 3.19
importance for behavioral analysis, 60
jumping spider, 66, Fig. 3.20
oystercatcher, 61, Figs. 3.13, 3.14
dunce mutant, 286
Dye, John, 157

earth, geomagnetic field, Fig. 9.14
Eccles, John, 25
echolocation, 94–107
adaptations of auditory system, 103–5

counter-adaptations of prey animals, 105–7
distance estimation, 98–100
Doppler shift analysis, 100–3
history of echolocation research, 95–6
ectothermic animals, 265
Edwards, Donald, 196
efferent neuron, definition of, 37
egg-retrieval response, 20
Eigenmannia sp. (knifefish), 155–77, 184–8
dendritic plasticity, 184–8
electric organ, 157, Fig. 7.2
electric organ discharge, 157, Figs. 7.3, 8.2
electrolocation, 159–61
electroreceptor organs, 158–9, Fig. 7.4, 169–70
jamming avoidance response, behavioral experiments, 161–4
jamming avoidance response, computational rules, 164–9
jamming avoidance response, evolution, 176–7
jamming avoidance response, neuronal implementation, 169–76
Eigenmannia lineata see also Eigenmannia sp., 157, Fig. 7.1
Eigenmannia virescens see also Eigenmannia sp., 157
eighth nerve *see* vestibular nerve
electric organ, 157, Fig. 7.2
electric organ discharge
elephant nose, Fig. 1.4
knifefish, 157, Fig. 7.3
electrical brain stimulation *see* brain stimulation experiments
electrocyte, 157
electroencephalogram, 231
electrolocation, 159–61
electromagnetic angle detector system, 135, Fig. 6.11
electroreceptor *see also* ampullary receptor, tuberous receptor, P-type receptor, T-type receptor, 158–9, 169–70
electrosensory lateral line lobe, 170–3, Fig. 7.10
elephant nose (*Gnathonemus petersii*), Fig. 1.5
elimination of single cells, 58
Emlen, John, 213
Emlen, Stephen, 213, 215
end foot *see* synapse
enriched environment, effect on adult neurogenesis, 296–9
entorhinal cortex, 290
epinephrine, 49
EPSP *see* excitatory postsynaptic potential
Eptesicus fuscus (big brown bat), 96, 98
sonogram of echolocation signals, Fig. 4.11(a)
equilibrium potential, 42
Erithacus rubecula (European robin), 60, Fig. 3.12, 216, Fig. 9.13
escape behavior, 113
esophageal ganglion, 192
ethogram, 3
ethology *see also* methods for studying animal behavior
concepts, 2, 59–74
history, 15–17, 19–20
importance for neuroethology, 2, 32
methods, 59–74
European eel (*Anguilla anguilla*), 230

European lobster (*Homarus gammarus*), 194
European robin (*Erithacus rubecula*), 60, Fig. 3.12, 216, Fig. 9.13
European white stork (*Ciconia ciconia*), 202, Fig. 9.1
evolutionary theory, impact on study of animal behavior, 14
Ewert, Jörg-Peter, 30, 123
excitatory postsynaptic potential, 46
external nucleus, 148
eye bar, effect on aggressive behavior of Burton's mouthbrooder, 63–4, Fig. 3.18

facial ruff, 139, Fig. 6.14
facilitation, 285
feature detector, 122–3, 127–8
fictive behavior, 114–15
fictive swimming, 114
fiddler crab, 69, Fig. 3.24
field studies
    importance for ethology, 20
    importance for neuroethology, 2
file, involvement in sound production of crickets, 254
file-and-scraper mechanism, 253
FITC *see* fluorescein isothiocyanate, 56
fluorescein isothiocyanate (FITC), 56
FM area, 105
FM-FM area, 100
FM signals, 96
focal brain stimulation *see* brain stimulation experiments
Forger, Nancy, 188
Fraenkel, Gottfried, 80
frequency coding, 51
frequency-modulated signals *see* FM signals
Frisch, Karl von, 19–20, 26–7, Fig. Box 2.4, 71, 74, 229, 258
fruit fly (*Drosophila*), 286
funnel technique, 213, Figs. 9.9, 9.10, 9.11

GABA *see* γ-amino-butyric acid
Gage, Fred, 296–7
galvanotaxis, 81, 83, 85–8
ganglion, 255
gap junction, 39
gastric mill rhythm, 194
generator potential *see* receptor potential
genetic coupling of sender and receiver in communication systems, 269–73
geotaxis, 81, 88
Gestalt principle, 64–5
Getting, Peter, 195
gill-withdrawal reflex, 282–3, Fig. 11.2
Gleich, Otto, 137
glia
    definition, 39
    involvement in multiple sclerosis, 39
    involvement in structural reorganization of neural networks, 191–2
glutamate, 48
glycine, 48

Gnathonemus petersii (elephant nose), Fig. 1.5
Golgi, Camillo, 36
Golgi method, 36
gonadotropic hormone, 234
Gould, Elizabeth, 298
graded potential, 43
gravireception, 88
greater horseshoe bat (*Rhinolophus ferrumequinum*), 97, 100, 103–5
    response of moth-echo-selective neuron in auditory cortex, 105, Fig. 4.14
    sonogram of echolocation signals, Fig. 4.11(b)
green turtle (*Chelonia mydas*), Fig. 9.30
Greengard, Paul, 281
Griffin, Donald, 96
Grüsser, Otto-Joachim, 28, 122
Grüsser-Cornehls, Ulla, 28, 122
Gryllus (field crickets)
    bimaculatus, 254, 266
    campestris, 254, 259, 266
    firmus, 266
Gunn, Donald, 80
Gymnarchus niloticus, 160
Gymnocorymbus sp. (tetra), 90
Gymnotiformes, 157
gyration, 82

H.M. (human patient used in memory research), 278–9
Haematopus ostralegus (oystercatcher), 61
Hahn, Janet, 271
hair cells
    effective physiological stimulus, 90–1
    involvement in echolocation of bats, 104–5
    physiological properties, 91–3, Fig. 4.9
    structure, 89–90, Fig. 4.6
half-center, 116
Haplochromis burtoni (Burton's mouthbrooder), 63, Fig. 3.18, 67, Figs. 3.21, 3.22
harp, involvement in sound production of crickets, 254
Hasler, Arthus Davis, 228–9, Fig. Box 9.2, 231
hearing
    bats, 94–107
    crickets, 251–73
    fish, 27
    owls, 132–50
    place theory, 104
Hebb, Donald Olding, 277–8, Fig. Box 11.1
Hebb synapse, 278
Heiligenberg, Walter, 30, 63, 67–8, 156–7, 185, 187, Fig. Box 7.1
Heinecke, Peter, 256
Heinroth, Oskar, 17, Fig. 2.5, 19
Helbig, Andreas, 205
Hemichromis fasciatus, 69, Fig. 3.23
herring gull (*Larus argentatus*), 61, Fig. 3.15
Hertwig, Richard von, 27
Hess, Walter Rudolf, 29, 258

heterogenous summation, law of, 62–4, 156
high-pass cell, 263
hippocampal formation, 290
hippocampus, 278, 281, 289–301
    involvement in memory formation of mammals, 289–99
    involvement in spatial learning of birds, 300–1
    long-term potentiation, 293–4
    molecular biology of LTP, 294–6
    place cells, 292–3
    structure, 291–2, Fig. 11.7
Hirsch, Peter, 229
histamine, 49, 194
Hochstetter, Ferdinand, 19
Hodgkin, Alan, 25
Hoffmann, Klaus, 211
Holst, Erich von, 25–6, Fig. Box 2.3, 29, 90–1, 258
Homarus
    americanus (American lobster), Fig. 8.7, 194
    gammarus (European lobster), 194
homing, 201–45
    pigeon, 208, 218–20
    Manx shearwater, 207–8, Figs. 9.15, 9.16
    salmon, 225–39
honey bee (*Apis mellifera*), 70–4
    color vision, 27
    dance language, 27, 70–4
    dummy experiments, 74, Fig. 3.30
    life cycle, 71
    round dance, 71–2, Fig. 3.26
    waggle dance, 72–4, Fig. 3.27
honey guide, 17, Fig. 2.6, 69
Hopkins, Carl, 157
Horridge, Adrian, 28
Hoy, Ron, 266, 271
Hubel, David, 26
Huber, Franz, 30, 258, Fig. Box 10.2, 263
Huber, Robert, 195
Hudspeth, James, 93
Huxley, Andrew, 25
5-hydroxytryptamine *see* serotonin
hyperpolarization, 43
hypothalamus, 29, 49
    anterior, 191–2

immunohistochemistry *see also* antibody, 54–6, Fig. 3.9
imprinting *see also* olfactory imprinting hypothesis, sequential
        imprinting theory
    birds, 15
    geese, 17, 231
    salmon, 228–37
impulse *see* action potential
in situ hybridization, 56–8, Fig. 3.11
inclination compass, 217
indigo bunting (*Passerina cyanea*), 213, 215, Figs. 9.10, 9.11
inferior colliculus, 99, Fig. 4.12
infrared receptor *see* pit organ
inhibitory postsynaptic potential, 46

innate behavior, 17
integration of sensory information and motor programs, 155–77
interaural intensity difference, importance for prey
        localization, 138
interaural time difference, importance for prey localization, 140
internal clock, 211
interneuron
    commissural, 117
    definition, 37
    descending, 116
    excitatory, 116
    inhibitory, 117
    local, 37
ionic channel *see* ionophore
ionophore, 40
IPSP *see* inhibitory postsynaptic potential

Jacobs, Werner, 258
jamming avoidance response, 155–77
    behavioral experiments, 161–4
    computational rules, 164–9
    definition, 161
    evolution, 176–7
    neuronal implementation, 169–76
Jeffress, Lloyd, 144
Jennings, Herbert Spencer, 81–2, Fig. Box 4.1
jumping spiders (Salticidae), 66
Jurine, Charles, 95

Kalmijn, Adrianus, 222
Kandel, Eric, 281, Fig. Box 11.2, 289, 295
Kant, Immanuel, 19
Katz, Bernard, 25
Kawasaki, Masashi, 157
Keeton, William, 218, 224
Keller, Clifford, 157
kinesis, 80
kinocilium, 89
knifefish *see also Eigenmannia* sp., 155–77, Fig. 7.1, 184–8
    dendritic plasticity, 184–8
    electric organ, 157, Fig. 7.2
    electric organ discharge, 157, Figs. 7.3, 8.2
    electrolocation, 159–61
    electroreceptor organs, 158–9, Fig. 7.4, 169–70
    jamming avoidance response, behavioral experiments, 161–4
    jamming avoidance response, computational rules, 164–9
    jamming avoidance response, evolution, 176–7
    jamming avoidance response, neuronal implementation,
        169–76
Knudsen, Eric, 133, 149–50
Konishi, Mark, 30, 133–4, Fig. Box 6.1, 144, 148
Köppl, Christine, 137, 148
krait (*Bungarus multicinctus*), 115
Kramer, Ernst, 256
Kramer, Gustav, 210–13
Kramer, Ursula, 67–8
Kramer locomotion compensator, 257–8, Fig. Box 10.1

Kramer treadmill *see* Kramer locomotion compensator
Kravitz, Edward, 195
Krebs, John, 300
Kühn, Alfred, 80
Kurz, Elizabeth, 189

labyrinth, 26, 89
Lack, David, 60
laminar nucleus, 144, Fig. 6.20
Langley, John, 46
Larus argentatus (herring gull), 61, Fig. 3.15
laser vibrometry measurement, 261
Lashley, Karl Spencer, 16, Fig. 2.4, 278
lateral giant interneuron, 196
lateral shell, 148
law of heterogenous summation, 62–5, 156
    behavior of cichlid fish, 63–4
    behavior of human infants, 65, Fig. 3.19
learning, cellular mechanisms of *see also* memory, 277–301
leatherback sea turtle (*Dermochelys coriacea*), 240
Leong, Daisy, 63, 67
Lettvin, J.Y., 122
ligand-ligand receptor mismatch, 196
Lindauer, Martin, 74
Lissmann, Hans, 112, 160–1
Locke, John, 24
locomotion vector, 269
Loeb, Jacques, 18
Loewi, Otto, 46
loggerhead sea turtle (*Caretta caretta*)
    life cycle, 240
    migration, 240, Fig. 9.29
    orientation on beach, 240–1
    orientation on land, 242
    waves as orienting cues, 242–5, Fig. 9.32
Lohmann, Catherine, 240, 242
Lohmann, Kenneth, 240, 242
Lømo, Terje, 293, 294
long-term potentiation (LTP), 293
Lorenz, Konrad, 17, 19–20, Fig. Box 2.1, 25, Fig. Box 2.3,
    26–7, 62, 156, 229, 231, 258
low-pass cell, 263
LTP *see* long-term potentiation
Lucifer Yellow, 58

Machemer, Hans, 83
macula, 89
magnetic compass, 215–23, 241
magnetic map, 215–23
magnetic sense, 222, 237
magnetite
    bobolink, 221
    chemical structure, 222
    magnetotactic bacteria, 222
    rainbow trout, 238
magnetoreception
    chemical magnetoreception mechanism, 222

electromagnetic induction mechanism, 222
    in birds, 220–3
    in salmon and trouts, 237–9
    in sea turtles, 241–2
    magnetite-based magnetoreception mechanism, 222
magnocellular hypothalamo-neurohypophysial system, 191
magnocellular nucleus, 143
Maler, Leonard, 157
Manley, Geoffrey, 137, 148
Manx shearwater (*Puffinus puffinus*), 207–8
map *see* neural map
MAPK *see* mitogen-activated protein kinase
Marder, Eve, 30, 195
Marler, Peter, 133
marsh tit (*Parus palustris*), 300
Marshall, Wade, 281
Matthews, Geoffrey, 207
Maturana, H.R., 122
McCulloch, W.S., 122
Mediterranean rock dove (*Columba livia*), 208
medulla oblongata, 157, 185
Megachiroptera, 95
membrane potential *see* resting potential
memory, 277–301
    declarative *see* memory, explicit
    episodic *see* memory, explicit
    explicit, 280
    implicit, 280
    long-term, 283
    nondeclarative *see* memory, implicit
    procedural *see* memory, implicit
    short-term, 283
memory suppressor gene, 289
methods for studying animal behavior, 59–74
    anecdotal approach, 12
    anthropocentric approach, 11
    anthropomorphic approach, 12
    black-box approach, 1
    comparative approach, 12, 14
    experimental approach, 14
    inductive approach, 12
    teleological approach, 12
    use of operational nomenclature, 23
    vitalistic approach, 12
Metzner, Walter, 157
Michelsen, Axel, 74, 261
Microchiroptera, 95
microelectrode, use in neurophysiology, 40
microtransmitter, use in navigation research, 209
migration, 201–45
    anadromous, 227
    approaches to studying animal migration, 208–9
    genetic control, 205–6
    in birds, 209–25
    in salmon, 225–39
    in sea turtles, 239–45
    modes, 202–5

migratory restlessness, 210
milkweed (*Asclepias*), 205
Miller, John, 58
Milner, Brenda, 278, 280
mind-body dualism, 13–14
mirror drawing task, 279–80, Fig. 11.1
mitogen-activated protein kinase (MAPK), 286, 295
Mittelstaedt, Horst, 26
model *see* dummy experiments
model systems in neuroethology, 6–8
modulation
    by neuromodulators, 194–7
    crayfish aggressive behavior, 195–6
    of amplitude and phase of electric organ discharge, 165–9
    of central pattern generator, 115
    of frequency and amplitude of electric organ discharge, 184
    of gill-withdrawal reflex, 285
    of modulators, 196
    of stomatogastric ganglion activity, 192–5
modulators, 39, 49
    dopamine as a neuromodulator, 49
    involvement in motivational changes in behavior, 197
    neuropeptides as neuromodulators, 196–7
    serotonin as a neuromodulator, 49
Moiseff, Andrew, 144
molecular genetics, role in development of behavioral
        neurobiology, 31
monarch butterfly (*Danaus plexippus*), Fig. 9.3, 205
Morgan, Conway Lloyd, 14–5, Fig. 2.3
Morgan's canon, 15
morpholine, 232–4, Fig. 9.22
mossy fiber pathway, 291
moths, 105–7
    Phytographa gamma, Fig. 4.13
    tiger moths, 106
motivation, 67, 181–97
    definition, 182
    mediated by biochemical switching of neural networks,
        192–7
    mediated by structural reorganization of neural networks,
        183–92
multiple sclerosis, 39
multipolar neuron, 185
Musolf, Barbara, 196
mustached bat (*Pteronotus parnellii*), 100
myelin, 39

Naesbye-Larsen, Ole, 261
nature *versus* nurture controversy, 24
nematode (*Caenorhabditis elegans*), 31
Nernst, Walter, 42
Nernst equation, 42–3
neural map
    auditory space, 149
    auditory-visual, 150
    motor, 150
    somatotopically ordered, 171

neuroendocrinology, role in development of behavioral
        neurobiology, 31
neuroethology
    aims, 1
    history, 24–30
    importance of field studies, 2
    model systems, 6–8
    whole animal approach, 2, 25, 32
neurogenesis *see also* adult neurogenesis, 183
neuromodulator *see* modulator
neuron, structure, 36
neuronal tract-tracing techniques *see* tracing
neuropeptide Y, 31
neuropeptides *see also* allostatin-related peptide, *Cancer
        borealis* tachykinin-related peptide, crustacean
        cardioactive peptide, cholecystokinin, neuropeptide Y,
        β-pigment-dispersing hormone-like peptide, proctolin,
        red-pigment-concentrating hormone, SDRNFLRFamide,
        somatostatin, tachykinin-related peptide,
        TNRNFLRFamide, 49
    involvement in biochemical switching of neural networks,
        196–7
Nguyen, Peter, 295
nitric oxide, 194
NMDA *see* N-methyl-D-aspartate
NMDA-type of glutamate receptor, 48, 116, 175
N-methyl-D-aspartate, 175
node of Ranvier, 39
non-basilar pyramidal cell, 172–3
non-NMDA-type of glutamate receptor, 48
non-synaptic release, 196
noradrenaline *see* norepinephrine
norepinephrine, 49
Nottebohm, Fernando, 300
npr-1 gene, 31
NPR-1 receptor, 31
nucleus of cells, 36
nucleus electrosensorius, 174

O'Keefe, John, 292
olfaction
    role in homing of pigeons, 224–5
    role in homing of salmon, 228–37
olfactory epithelium, Fig. 9.21
olfactory imprinting hypothesis, 228–37
oligodendrocyte, 39
omega-1 neuron *see* ON-1 cell
ON-1 cell, 262
Oncorhynchus
    kisutch (coho salmon), 225, Fig. 9.18, 231, Fig. 9.23,
        Table 9.1, Fig. 9.24
    masou (masu salmon), 227, 237
    mykiss (rainbow trout), 238, Fig. 9.27
    nerka (sockey salmon), 226, Fig. 9.19
open-loop condition, 137
opener muscle, role in sound production in crickets, 255
ophthalmic branch of the trigeminal nerve, 221

optic tectum
  owls, 149
  pattern of connectivity with retina, 37
  toads, 128
orientation
  accuracy of sound localization in barn owls, 136–7
  antidromic, 88
  homodromic, 87
  positional *see* orientation, primary
  primary, 80
  secondary, 80
oscillogram, 253
Ostwald, Joachim, 105
otolith organ, 88
overshooting potential, 43
ovipositor, Fig. 10.1
owls *see* barn owls
oystercatcher (*Haematopus ostralegus*), 61, Figs. 3.13, 3.14

P-type receptor, 170, Fig. 7.10
pacemaker, 115
pacemaker nucleus, 157–8, 185, Fig. 8.3
Papi, Floriano, 224
parallel processing, 171
Paramecium sp. (paramecian), 81–8
  chemotaxis, 81–3, Fig. 4.1
  cilium, Fig. 4.2
  galvanotaxis, 85–8
  phobotaxis, 84–5
parr, 225
Parus
  atricapillus (black-capped chickadee), 300
  palustris (marsh tit), 300
Passerina cyanea (indigo bunting), 213, 215, Figs. 9.10, 9.11
Paul, Robert, 271
Pavlov, Ian Petrovitch, 21, Fig. 2.7
Payne, Roger, 133–4, 137
pectoral spots, effect on aggressive behavior of Burton's
  mouthbrooder, 63–4
Penfield, Wilder, 278
Perdeck, Albert, 62
perforant pathway, 291
perikaryon, 36
Peromyscus leucopus (white-footed mouse), 188–91
pH, definition of, 81
phase ambiguity, 145, Fig. 6.21
phase locking, 142, Figs. 6.17, 6.21
β-phenylethyl-alcohol, 230, Fig. 9.22, 233
pheromone, 16, 69–70
phobotaxis, 83–5
phonotaxis, 81
  crickets, 256
photoinactivation, 58, 194
phototaxis, 81
Phytographa gamma (moth), Fig. 4.13
Pierce, George, 96
β-pigment-dispersing hormone-like peptide, 194

pigeons
  behavioral sensitivity to light of different wavelengths, 212,
    Fig. 9.8
  discrimination of rotated patterns, 15
  effect of magnetic fields on orientation behavior, 218–20,
    Figs. 9.15, 9.16
  evolutionary origin, 208
  homing, 208, 218–20
Pimephales notatus (bluntnose minnow), 231
Pires, Anthony, 266
pit organ, 28
Pitts, W.H., 122
PKA *see* cAMP-dependent protein kinase A
place theory of hearing, 104
place cell, 292
plasticity, neuronal, importance for motivational changes in
  behavior *see also* adult neurogenesis, biochemical
  switching of neural networks, dendritic plasticity,
  polymorphic networks, structural reorganization of
  neural networks, 183
Pluvialis dominica (American golden plover), 202, Fig. 9.2
polarity compass, 217
Pollak, George, 99
polymorphic network, 195
Porcelia scaber (woodlouse), 80
posterior lateral lemniscal nucleus, 148
postnatal neurogenesis *see also* adult neurogenesis, 237
postsynaptic potential *see also* excitatory postsynaptic
  potential, inhibitory postsynaptic potential, 46
prepacemaker nucleus, 175, 185
pressure gradient ear, 261
pressure receptor, 51
prey recognition in toads, 122–32
primer pheromone, 69
proctolin, Fig. 8.7, 194–5
projection of neurons
  definition, 37
  elucidation through neuronal tract-tracing techniques, 51
  retinotectal, 37
prothoracic leg, 259
Pteronotus parnelli (mustached bat), 100
Pterophyllum sp. (angel fish), 90, Figs. 4.7, 4.8
Puffinus puffinus (Manx shearwater), 207–8
pyloric rhythm, 194

QCF signals, 96
quality of perception, 50
quasi-constant-frequency signals *see* QCF signals
queen in honey bees, 71

radar, use in navigation research, 209
rainbow trout (*Oncorhynchus mykiss*), evidence for a magnetic
  sense, 238, Fig. 9.27
ramus ophthalmicus superficialis of the trigeminal nerve
  (ros V), 238
Rana catesbeiana (bullfrog), 93
reafference principle, 26

rebound firing
    definition, 118
    in spinal motoneurons of toad tadpoles, 117, Fig. 5.4
receptor cell, 49
receptor potential, 50
receptive field, 128, 149
recognition neuron, 263
recording
    extracellular, 44, 114
    intracellular, 40, 114
red-necked phalarope, 20
red-pigment-concentrating hormone, Fig. 8.7, 194–5
redd, 225
reflex, 111, 181
reflex chain theory, 25
reflex theory, 21
Regen, Johann, 256
reinforcement, 23
relative coordination, 26
releaser
    definition, 60
    effect on motivation of receiver, 67
releaser pheromone, 69–70
releasing mechanism, 59–60, 121
resonance, 254
resting potential, 39–40
reticular theory, 36
retinotectal projection, 37, 128
retinotopic map, 129
rheotaxis, 81
Rhinolophus ferrumequinum, 97, 100, 103–5
    sonogram of echolocation signals, Fig. 4.11(b)
rhythmogenesis, 117
Roberts, Alan, 111–12, Fig. Box 5.1
Roeder, Kenneth, 105, 258
ros V *see* ramus ophthalmicus superficialis of the
    trigeminal nerve
Rose, Gary, 157
round dance, 71–2, Fig. 3.26

Salmo salar (Atlantic salmon), 227, Fig. 9.27
salmon
    Atlantic (*Salmo salar*), 227
    coho (*Oncorhynchus kisutch*), 225, Fig. 9.18, 231, Fig. 9.23,
        Table 9.1, Fig. 9.24
    homing, 225–39
    life cycle, 225–8, Fig. 9.18
    magnetoreception, 237–9
    masu (*Oncorhynchus masou*), 227, 234
    olfactory imprinting, 228–37
    sockey (*Oncorhynchus nerka*), 226, Fig. 9.19
    sun-compass orientation, 237
    ultrasonic tracking, 233–4
Salticidae (jumping spiders), 66
satellite-based radiotelemetry, use in navigation research, 209
Sauer, Franz, 213
Schaffer collateral pathway, 292, 294

Scharrer, Berta, 25
Scharrer, Ernst, 25
Scheich, Henning, 156
Schildberger, Klaus, 263
Schmidt-Koenig, Klaus, 211
Schnitzler, Hans-Ulrich, 101, 105
Scholz, Allan, 234
Schreckstoff, 27, 229, Fig. 9.20
Schulten, Klaus, 222
Schwann cell, 39
Schwartzkopff, Johann, 133
Schwassmann, Horst, 237
scolopidium, 260
Scoville, William, 278
scraper, involvement in sound production of crickets, 254
SDRNFLRFamide, 194
sea hare *see Aplysia californica*
sea turtles, 239–45
    green turtle (*Chelonia mydas*), Fig. 9.30
    leatherback (*Dermochelys coriacea*), 240
    life cycle, 239–40
    loggerhead (*Caretta caretta*), 240, Figs. 9.29, 9.32
    mechanism of orientation, 240–5
    migration, 239–40, Fig. 9.29
seasonal changes in behavior
    knifefish, 184–8
    mediated by structural reorganization of neural
        networks, 192
    white-footed mouse, 188–91
Seitz, Alfred, 62
Selverston, Allen, 58
semicircular canal, 88
Sengelaub, Dale, 189
sensitization *see also Aplysia californica*, 282
sensory systems, 49–51
    amplification of energy, 50
    receptor cells, 49
    receptor potential, 50
    sensory modality, 49
    sensory organ, 49
    transduction, 49
sequential imprinting hypothesis, 236
serotonin
    chemical structure, 49
    effect on aggressive behavior of crayfish, 195–6
    effect on lateral giant interneuron, 196
    effect on stomatogastric ganglion, Fig. 8.7, 194
    modulation of gill-withdrawal reflex, Fig. 11.4, 285
Sherrington, Charles, 25, 46, 111, 115
Shors, Tracey, 298
sign stimulus *see also* releaser, releasing mechanism,
        supernormal stimulus
    effect on behavior of recipient, 59–60
    importance of condition of recipient, 65–6
signals
    acoustic, 251
    chemical *see also* pheromone, 69–70

electric, 6–8, 155–77, 184–8
  involvement in communication, 68
  visual, 69
Simmons, James, 98
Skinner, Burrhus Frederic, 22–3, Fig. 2.10
Skinner box, 23, Fig. 2.11
smolt, 226
smolt transformation, 226
snow bunting, 20
social strains in *Caenorhabditis elegans*, 31
sociobiology, 30
solitary strains in *Caenorhabditis elegans*, 31
soma, 36
somatostatin, Fig. 3.10
sound localization
  crickets, 256–65
  owls, 132–50
space-specific neuron, 148
Spalding, Douglas, 14–15
Spallanzani, Lazaro, 95–6
spatial map, 292
spermatophore, Fig. 10.1
Sperry, Roger, 25
spherical cell, 171
spike *see* action potential
spinal nucleus of the bulbocavernosus
  effect of sex steroid hormones, 189–91
  involvement in seasonal changes in reproductive
    behavior, 188–9
St. Paul, Ursula von, 210–11
star compass, 212–15
starling (*Sturnus vulgaris*)
  effect of clock shifts on orientation behavior, 211–12, Fig. 9.7
  spontaneous directional headings during migratory
    restlessness, 210, Fig. 9.6
stereocilium, 89, 93
Sternopygus, 177
stomatogastric ganglion, 30, 192
stomatogastric nervous system, 192, Fig. 8.7
stridulation, 253–4, Fig. 10.3
structural reorganization of neural networks, 183–92
Sturnus vulgaris (starling)
  effect of clock shifts on orientation behavior, 211–12, Fig. 9.7
  spontaneous directional headings during migratory
    restlessness, 210, Fig. 9.6
subicular complex, 290
sublemniscal prepacemaker nucleus, 175
Suga, Nobuo, 100
sun compass, 210–12, 237
superior colliculus, 37, 149
supernormal stimulus
  bill of herring gull, 61–2, Fig. 3.15
  eggs in oystercatcher, 61, Figs. 3.13, 3.14
  general features, 60
  human lips, 62, Fig. 3.16
  men's shoulders, 62, Fig. 3.17
surveillance radar, use in navigation research, 209

syllable
  definition, 253
  duration, 266
  period, 266
  rate, 259
Sylvia (warblers), 213
  atricapilla (blackcap), 205–6, Figs. 9.4, 9.5
synapse
  chemical, 37–9
  cholinergic, 48
  definition, 46
  electrical, 39, 171
  electrotonic *see* synapse, electrical
  muscarinic, 48
  nicotinic, 48
  structure of, 37
synaptic cleft *see* synaptic gap
synaptic gap, 38
synaptic vesicle, 38

T5(2) neurons in toads, 130–2
T-type receptor, 170, Fig. 7.10
tachykinin-related peptide, 194
tagging, use in navigation research, 208–9
Takahashi, Terry, 144
Takeda, Kimihisa, 155
Tauc, Ladislav, 281
taxis, 80
Teichmann, Harald, 230
Teleogryllus (Australian field crickets)
  commodus, Fig. 10.1, 271
  oceanicus, 271
temperature coupling, 265–9
terminal *see* synapse
tetra (*Gymnocorymbus* sp.), 90
thalamic-pretectal area, 128
thermotaxis, 81
thigmotaxis, 81
thyroid hormones, role in olfactory imprinting of salmon,
    234–7, Fig. 9.26
thyroid-stimulating hormone (TSH), 234
thyroxine, role in olfactory imprinting of salmon,
    234–7, Fig. 9.24
tiger moths (Arctiidae), 106
time-compensated sun compass *see also* sun compass, 211
Tinbergen, Elisabeth, 20
Tinbergen, Niko, 17, 19–20, Fig. Box 2.2, 24, 27,
    60, 62, 258
tip link, 90, Fig. 4.10
TNRNFLRFamide, Fig. 8.7, 194–5
toads, 111–19, 122–32
  brain recording experiments, 129–30, Figs. 6.8, 6.9
  brain stimulation experiments, 131
  clawed-toad (*Xenopus laevis*), 112
  common toad (*Bufo bufo*), 123
  connectivity of retina with other brain regions,
    128–9, Fig. 6.7

toads (*cont.*)
  dummy experiments, 124–7, Figs. 6.2, 6.3, 6.4, 6.5, 6.6
  enemy recognition, 122–32
  lesioning experiments, 131–2, Fig. 6.10
  prey catching behavior, 123, Fig. 6.1
  prey recognition, 122–32
torus semicircularis, 173
tracing, 51–3, 113, 185
  anterograde, 51
  combination with immunohistochemistry or *in situ*
    hybridization, 53
  fiber-of-passage problem, 52
  in fixed tissue, 53
  in vitro approach, 53
  in vivo approach, 53
  retrograde, 52, 187, 191
tracking radar, use in navigation research, 209
trail substance, 69, Fig. 3.25
transduction
  definition, 49
  in hair cells, 93–4
  in visual systems, 50
  tip-link model, 93–4, Fig. 4.10
transmitters *see also* amino acid transmitters, catecholamines,
    modulators, neuropeptides, 46
triiodothyronine, role in olfactory imprinting of salmon, 234
trill, 259
trophollaxis, 16
tropism, 18, 80
TSH *see* thyroid-stimulating hormone
tuberous receptor, Fig. 7.4, 170
tympanal organ, 260
tympanum, 259
Tyto alba (barn owl), 132–150
  accuracy of orientation response, 136–7, Fig. 6.12
  facial ruff, 139, Fig. 6.14
  neural circuit involved in sound localization,
    142–50, Fig. 6.18
  physical parameters involved in sound localization,
    137–42, Figs. 6.15, 6.16
  prey localization behavior, 132–4
Tytonidae (barn owls), 134

Uexküll, Jakob von, 17, 59, 121
ultrasonic tracking, use in navigation research, 233–4
ultrasound, 59, 94

Umwelt, 17, 59, 121
unconditioned stimulus *see* conditioned reflex

varicosity, structure, 37
vector, definition, 269
Vehrencamp, Sandra, 67
Verworn, Max, 82
vestibular nerve, 90
vestibular organ, 88
voltage-sensitive fluorescent dye, 39–40
VW effect, 220

waggle dance, 72–4, Figs. 3.27, 3.28, 3.29
Walcott, Charles, 218
Walker, Michael, 237
warblers (*Sylvia*), 213
Watanabe, Akira, 155
Watson, John B., 16, 22, Fig. 2.9, 82
weakly electric fish, 7, 28, 59, 155–77, 184–8
Wheeler, William Morton, 16
white-footed mouse (*Peromyscus leucopus*), 188–91
Whitman, Charles Otis, 16
Wiesel, Torsten, 26
Wilson, Edward O., 30
Wiltschko, Roswitha, 220
Wiltschko, Wolfgang, 216, 218, 220, 225
Wisby, Warren, 228, 231
woodlouse (*Porcelia scaber*), 80
worker in honey bees, 71
worm configuration in toads, 126, Fig. 6.5

Xenopus laevis (clawed-toad), 111–19
  central pattern generator controlling swimming,
    115–18, Fig. 5.3
  coordination of oscillator activity involved in neural
    control of swimming, 118–19
  fictive swimming, 114
  physiological activity of spinal motoneurons,
    113–15, Fig. 5.2
    spinal circuitry controlling swimming *see* clawed-toad,
    central pattern generator controlling swimming
  swimming behavior, 112–13, Fig. 5.1

Yeh, Shih-Rung, 196

Zupanc, Günther K.H., 187